THE OFFSHORE ISLANDERS

PAUL JOHNSON

The Offshore Islanders

Weidenfeld and Nicolson
5 Winsley Street London W1

ISBN 0 297 99466 2

Printed in Great Britain by
Cox & Wyman Ltd,
London, Fakenham and Reading

'Lords and Commons of England – Consider what nation it is whereof you are and of which you are the governors: a nation not slow and dull, but of quick, ingenious and piercing spirit; acute to invent, subtile and sinewy to discourse, not beneath the reach of any point that human capacity can soar to.'

JOHN MILTON: *Areopagitica*

Contents

Illustrations

Prologue

WHY should a journalist, early in the decade of the 1970s, sit down to write a history of the English people? Why should he renounce his proper function, which is to record and comment upon the present, and seek to explore the past, a task for which he is, perhaps, ill-qualified, even disqualified? My answer, in the first place, is that it is wrong to draw too sharp a distinction between the journalist and the historian. They are both in the same business: to communicate an understanding of events to the reader. Both are involved in the discovery and elucidation of truth – that is, the search for the facts which matter, and their arrangement in significant form. No one can possibly say where the historian's work ceases, and the journalist's begins. The present is continuously in process of becoming the past: the frontier of history ends only with yesterday's newspaper. A good journalist casts anxious and inquiring glances over his shoulder, and a good historian lifts his eyes from the page to look at the world around him. Sometimes the roles merge completely. Thucydides was writing not merely a history but an anguished record of contemporary events, in which he had acted and suffered. Bede, the first great English historian, living in a period of calm before the storm he sensed was coming, wrote not only, as he said, 'for the instruction of posterity', but also for the purposes of government; he told the King of Northumbria in his dedication: 'You are desirous that the said history should be more fully made familiar to yourself, and to those over whom the Divine Authority has made you governor, from your great regard for their general welfare.' Matthew Paris was a journalist as well as a historian. Walter Ralegh, in his *History of the World*, was directing a gigantic and angry editorial to the subjects of James I. Clarendon's history of the Great Rebellion was an essay in analytical and polemical journalism. Macaulay, recording the destruction of the Stuarts, was also subjecting his early-Victorian contemporaries to a subtle exercise in political education. Consciously or unconsciously, most great historians have influenced contemporary events, as all journalists seek to do.

In the second place, a journalist cannot divorce himself from history even if he wishes. He cannot prevent the past from intruding. The more he tries to understand the present, the further he is driven to probe into the past, in the search for explanations. In a sense, this is a reversal of F.W. Maitland's historical method, which he used in *Domesday Book and Beyond*, of advancing 'backwards from the known to the unknown, from the certain to the uncertain'. Seeking to peer through the mists of the present, the journalist uses as points of reference the established landmarks of the past. He sees people fighting in the streets of Belfast in the 1970s. Why? Because of certain events which took place in Londonderry in 1968? Partly. But partly also because of decisions reached in London in 1920, and of centuries of interrelated events before them, reaching back into the early Middle Ages and beyond, almost to the first recorded episodes in Anglo-Irish relations. Not all this material is important, or even relevant. But the journalist cannot be sure until he has examined it. He must continually turn aside from his typewriter and reach for his bookshelves. Of course Northern Ireland is a theatre of action where the past plays an unusually vivid role. But all events, however novel, have a history; every problem is a legacy. Why, in the 1970s, do local councils in England fight acrimonious battles over comprehensive schools? To understand, we must go back not merely to 1944, but to the roots of modern English education in the early nineteenth century, and to an examination of the systems which preceded it. Why is it so difficult to shape a wages policy for the Britain of the 1970s? It is pointless to ask the question unless we are prepared to travel backwards into the history of British trade unionism, and indeed examine the origins of the present industrial structure. Why are strikes so frequent in the British car industry? Part of the explanation lies in arrangements made in the two decades before 1914, themselves conditioned by attitudes shaped in the very earliest phase of the industrial revolution. Moreover, the journalist finds himself conjuring up the past not merely to provide answers to particular contemporary questions but to explain their relationship to each other. The historical structure of the British motor-car industry has a direct bearing on the struggle for a wages policy, and both are influenced by the evolution of the educational system. So the journalist plunges deeper and deeper into history, and on an ever-broadening front. Sooner or later he is tempted to write history himself, to satisfy his own legitimate and professional curiosity.

Therein lies the origin of this book. During the years 1965–70, as editor of a political journal, I had the duty, week by week, to comment upon – to try to understand myself and explain to others – the struggles and failures of one of the most tragically unsuccessful governments in English history. I was conscious all the time that the failures lay not merely in the limitations of the men and women who composed the government, but in the nation as a whole, in its institutions and the attitudes which shaped them. During the 1960s, this country underwent a profound and agonising experience. From year to year, almost from week to week, it shrank in its own estimation, and in that of the world. The Empire was gone almost before the decade commenced; but during it the loss was first felt, and the Commonwealth designed to replace it revealed as a paper sham. The decline of Britain as a world power, slow and almost imperceptible in the 1940s and 1950s, began to accelerate with unmistakable speed, and palpable results. This was accompanied by a growing awareness that the country was falling behind not merely in physical strength but in material prosperity. There was, too, no indication whatsoever that the declension could be arrested, let alone reversed: we faced a future not just of comparative weakness, but of relative poverty, and a future in which these characteristics would become more pronounced with every year that passed. Britain had entered the age of humiliations. The failure of a government simply epitomised and reflected the diminution of a people.

Was this process natural, indeed inevitable? Was it even desirable? What precisely did we mean by failure? The loss of imperial and world status might prove an advantage, a slow growth-rate a blessing. Power and wealth have never borne much relation to human happiness. On the threshold of the 1970s, the English could hardly be described as a suffering or an abject nation, nor even, by their own standards, a particularly discontented one. They enjoyed more freedom than ever before: not merely individual liberties, which had been greatly enlarged in the past decade, but the collective freedom from onerous responsibilities in the world. They enjoyed, too, a degree of civil peace and internal stability without precedent in their history, and without parallel abroad. They might take such things for granted: to most of the world these seemed enviable and elusive privileges. Was there not, perhaps, a certain logic in this national balance-sheet: the loss of power compensated by a real gain in security? If Britain were still running a world empire, operating as a great power, and throbbing with the rapid economic growth needed to sustain such efforts, could it possibly be an untroubled, law-abiding and stable country, let alone an agreeable one in which to live?

These questions naturally provoked others. What sort of people did the English wish to be, and what kind of country did they prefer to inhabit? Clearly, one could not begin to answer these without discovering how far the evolution of Britain, the type of country it was, and the position it had occupied in the world, was a matter of conscious choice by its predominant people, reflecting, with due allowance for the accident of events, their attitudes, aspirations and desires. In short, to make a worthwhile comment on the present predicament of the English, it seemed to me necessary to explore their history back to its very roots, to relate present to past, and on the basis of this connection to make some tentative projections into the future. I wanted to read a book which did this; but none such existed. So I decided to write it myself.

Such an audacious project is open to a number of powerful objections, of which I have been painfully aware. To begin with, the literature of English history is enormous and constantly increasing. Even by, say, the beginning of the Second World War, it was already difficult for a single writer to have read and absorbed the salient works of specialised history covering a period of more than 2,000 years. Since then there has been an explosion of English historical studies. One writer, summarising work on early English history since 1939, describes the production as 'gargantuan'; another, surveying the later Middle Ages, refers to recent research as 'a tidal river in full flood'; much the same could be said of later periods.* Moreover, English history since 1914, and even since 1945, now attracts a growing body of industrious and fertile scholars. A sizeable library could be formed from books dealing with aspects of English history published in the last 20 years, even discounting the enormous number of biographies which have poured from the presses; in addition there are thousands of monographs printed in scores of learned journals; and behind all these lie miles of archives and papers now open to inspection. One recent volume, covering less than a year of a single aspect of English history, involved the inspection of 60 hitherto unexplored collections of private papers. How can any one person – and a non-professional, too – hope to familiarise himself with such an enormous output, let alone master it?

Yet it would be a tragedy if writers of history were to allow themselves to become, like the physical scientists, the inhibited prisoners of available knowledge, and accept ant-like roles in a huge, impersonal industry, which no one mind felt capable of surveying as a whole. As one brilliant young historian has wisely observed, 'History does belong to everyman: that is a strength, not a weakness.'† The people have a right to be taught their history in a form they can grasp. If this is acknowledged to be impossible, then the labours of professional historians seem to me to be largely futile, self-indulgent, self-propagating

* I quote from *Changing Views on British History: Essays on Historical Writing since 1939*, edited by Elizabeth Chapin Furber (Harvard 1966). In the five years since this survey was published, vast and valuable additions have been made to English historical literature.

† Arthur Marwick in *The Nature of History* (London 1970), the latest and most comprehensive work on the theory and practice of history, and the evolution of historical studies.

exercises in mere antiquarianism. A certain ruthlessness is required, a willingness to accept the responsibility of making choices and forming judgments, a readiness to select, discount and discard.

Historical research tends to move in circles. A traditional view is inherited from the actual protagonists, and becomes orthodox, text-book history. In time, an enterprising historian comes along, subjects it to critical analysis, and produces a significantly new version. He breeds pupils, who form a revisionist school, and push his conclusions much further. With the advent of a new generation, there is a counter-revolution: the revisionist theory is itself assaulted. Sometimes a new synthesis is evolved. Sometimes the matter is now seen to be too complex to admit of any firm explanation, and the reader (who has followed the historians thus far) is left confused. More often, a modified version of the traditional view is re-established. Much academic blood is spilt, and little progress achieved. Moreover, professional historians are human, indeed all too human; often the smoke of controversy, of theory and counter-theory, conceals personal antagonisms rooted in ancient common-room brawls, or in disputes which have nothing to do with history. J.H. Round's ferocious assaults on Professor Freeman, for instance, were motivated, at least in part, by Round's hatred of Mr Gladstone, Liberalism in general, Little Englandism in particular and, not least, the anti-blood-sports lobby. One could quote modern examples, of which there are many.

More seriously, much research tends to obscure, rather than reveal, the truth; or, most depressing of all, to suggest that truth cannot be finally established, often on matters of outstanding importance. Just as astronomers seem unable to agree on the salient point of whether the universe is expanding, contracting or standing still, so historians constantly reveal new areas of doubt, or violent disagreement, on points which had once seemed clear. Thus: the Roman city was a failure in Britain; it was a substantial success. The Anglo-Saxon Church (and Anglo-Saxon society as a whole) was backward; its cultural and artistic achievements were immense. There was no 'feudalism' in England before the Conquest; there was 'feudalism'. The English population rose in the early fourteenth century; it fell dramatically. The fifteenth century was a period of economic decline; it was a period of exceptional dynamism. Similar black and white contrasting versions, held with angry tenacity and backed by massive documentation, envelop the nature of the Tudor monarchy, the origins of the Civil War, the loss of the American colonies, the politics of George III's England, and the origins and chronology of the industrial revolution, to mention only a few vital aspects of English history. Sometimes historians meet in seminar to debate their disagreements, not, as a rule, to much purpose. The layman can only survey the battlefield from a quoin of vantage, and make up his own mind about the honours of victory. Pierre Mendès-France used to say, to his divided cabinet, *'Gouverner, c'est choisir'*. To write general history it is necessary to make choices, almost on every page. This I have done, without bravado but also without fear; and if I am often wrong, I have the comforting words of the present Regius Professor of Modern History at Oxford, who has observed that there are times 'when a new error is more life-giving than an old truth, a fertile error than a sterile accuracy'.

There is a further objection to such a book as this: that it rests on the assumption that what happened in the past has some constructive relevance to our own times. This view would be wholly repudiated by many historians. Some have gone further. The great historian of the seventeenth century, S.R. Gardiner, for instance, held that the avowed or unavowed comparison with the present is 'altogether destructive of historical knowledge'. 'He who studies the society of the past,' he wrote, 'will be of the greater service to the society of the present in proportion as he leaves it out of account.' I do not agree; indeed, it is an impossible aim. Every historian has his contemporary bias; better to acknowledge it explicitly than to assume, wrongly, that it does not exist. It is no accident that Bishop Stubbs, writing in the golden age of the parliamentary statute, should have seen English history as primarily the development of constitutional forms, above all of Parliament; or that Professor Tout, whose own lifetime saw the birth and growth of 'big government', should have sought the key to English history in administration. Every age rewrites the history of the past in its own terms. We each have only one pair of eyes to see, and they are modern ones. In *History as the Story of Liberty*, Benedetto Croce pointed out that:

> The practical requirements which underlie every historical judgment give to all history the character of 'contemporary history', because, however remote in time events thus recounted may seem to be, the history in reality refers to present needs and present situations wherein those events vibrate.

This seems to me almost beyond argument, because it is impossible to still those vibrations. The writing of history, as Professor E.H. Carr puts it, is a 'dialogue between the present and the past'. Each age makes a different analysis of what has gone before, and extracts from it significant pointers, lessons and warnings. It is in the nature of man to pray to his ancestors for guidance. He may, of course, receive nothing but riddles. Lord Acton, in one of his lectures, overstated the case when he claimed: 'The knowledge of the past, the record of truth revealed by experience, is eminently practical, as an instrument of action and a power that goes to the making of the future.' The truth is often unclear, and statesmanship (not least in our own lifetime) frequently founders on false analogies. But most sensible men, in all ages, have been closer to Acton's view than Gardiner's. History has always, and properly, been regarded as 'the school of princes'. We should not hesitate – we should be eager – to make it the school of peoples.

A third objection to this book is that, in its exclusive preoccupation with the English, or rather with the peoples who have occupied the land we call England, it presupposes that history is Anglocentric, and is therefore irrelevant in an age when the centres of world power have shifted elsewhere. Many modern historians, notably Professor Geoffrey Barraclough in his admirable book, *History in a Changing World*, have urged that we should abandon the habit of writing history based on the assumption that a particular race is the sole active agent. Such advice has been widely followed. One American scholar notes sadly the decline of English historical studies in the United States in a period when

> ... the subject has to fight hard for a toe-hold in curricula in which students are invited to study such topics as the dynamics of Soviet power, under-development among the African peoples, the renascence of Moslem culture, or parliamentary institutions in Asian countries, and when English history has been dropped altogether from the curriculum of most schools.

Now I object strongly to this drift away from English history, which is part of a wider movement away from European and North Atlantic history. Virtually all the ideas, knowledge, techniques and institutions around which the world revolves came from the European theatre and its ocean offshoots; many of them came quite explicitly from England, which was the principal matrix of modern society. Moreover, the West is still the chief repository of free institutions; and these alone, in the long run, guarantee further progress in ideas and inventions. Powerful societies are rising elsewhere not by virtue of their rejection of western world habits but by their success in imitating them. What ideas has Soviet Russia produced? Or Communist China? Or post-war Japan? Where is the surge of discovery from the Arab world? Or liberated Africa? Or, for that matter, from Latin America, independent now for more than 150 years? It is a thin harvest indeed, distinguished chiefly by infinite variations on the ancient themes of violence, cruelty, suppression of freedom and the destruction of the individual spirit. The sober and unpopular truth is that whatever hope there is for mankind – at least for the foreseeable future – lies in the ingenuity and the civilised standards of the West, above all in those western elements permeated by English ideas and traditions. To deny this is to surrender to fashionable cant and humbug. When we are taught by the Russians and the Chinese how to improve the human condition, when the Japanese give us science, and the Africans a great literature, when the Arabs show us the road to prosperity and the Latin Americans to freedom, then will be the time to change the axis of our history.

Meanwhile, the story of the English is an instructive one, for others as well as themselves. It has strong elements of continuity, so that one can detect attitudes and characteristics, shaped by geography, among the islanders long before they acquired their mature racial composition. It has the true and graceful symmetry of art: a backward island gently washed by the tides of Continental cultures; its separate development rudely forced out of true by colonisation; independence seized, repeatedly lost, at last firmly established within a complex racial mould; the intellectual divorce from the Continent; the expansion overseas; the crystallisation, within the island, of an entirely new material culture, which spreads over the earth; the moment of power and arrogance, dissolving into ruinous wars; the survival, and the quest for new roles. This is not the stuff from which gigantic and delusive theories of history can be built. There is nothing in it which is inevitable; but nothing purely accidental either. English history is the study of recurrent and changing themes, and the evolution of national paradoxes. It is a story well worth telling, and one which each generation of us will wish to tell afresh.

Iver, Buckinghamshire

January 1972

PART ONE

The Pelagian Island

[100 BC–AD 600]

IN the year AD 410 Britain ceased to be a Roman colony and became an independent state. The inhabitants of the offshore island – or rather the settled lowland parts of it which we now call England – shook off the shackles of a vast European system, which tied it politically, economically and militarily to the Continental land-mass, and took charge of their own destinies. This event is usually presented in English history as a catastrophe, in which the protective umbrella of Rome was removed, and the defenceless inhabitants of the island exposed to the fury of the barbarians: civilisation in Britain was extinguished for centuries and the island vanished into the long night of the Dark Ages.

But the truth is more complex, more interesting and, in the light of the island's later history, more significant. The difficulty is that we have only scraps of information from which to compile an account of what happened; and any such account must be based to a large extent on interpretation, and even guesswork. But it is worth our while to make a reconstruction, because it can tell us something important about the history of the offshore islanders, and show how geography, as well as racial composition, shapes English history.

During the last decades of the fourth century, the British provinces of the Roman Empire had been progressively denuded of regular imperial troops. Already the authority of Rome did not run beyond York in the north, and Chester and Gloucester in the west; and even this authority was maintained rather by imperial expeditions, sent from the Continent under specially assigned generals, than by a standing garrison. By the turn of the century, the Roman military organisation in Britain had virtually ceased to exist, though a few units remained, and the civil administration was still carrying out its functions. But about this time we begin to detect faint traces of the emergence of British public opinion. Until now the Britons had played no perceptible role in imperial politics. Since the revolt of Boudicca, nearly three and a half centuries before, they had appeared to be model, or at least docile, colonial subjects. But the decline in Roman authority – the growing evidence that the Empire was incapable of discharging its military responsibilities – produced two distinct currents of political thought among the native inhabitants of the colony. On the one hand there were those who believed that Rome was still capable of re-establishing its powers; that only Rome was able to maintain internal order and external security; that without Rome civilisation would disappear, and the lives and property of all be at risk, and that therefore the only hope for the islanders was to re-forge and strengthen the imperial links, and place their trust wholly in the resources of civilised Europe. Untethered from the Continent, Britain would drift into anarchy, and life would become brutal, nasty and short.

On the other hand there was the independence party, the nationalists. They could argue that the forces which were tearing the empire apart were irreversible: that it was foolish and dangerous to place any confidence in a revival of Roman military power; that in any event Britain had a low place in Rome's scheme of priorities, and that her interests would be sacrificed without compunction to the needs of the imperial heartland. In recent years such Roman military bosses as had set up station in Britain had been more anxious to carve out sub-empires for themselves on the Continent than to protect British lives and property. They had become tyrannical adventurers, and had taken the precious regular units in Britain across the Channel on personal expeditions, leaving the Britons unprotected. Consider the events of four years before, in 406. The remnants of the Roman force in Britain, under pressure from local public opinion, had chosen a native, Gratian, as their local emperor. The act was plainly illegal. The historian Orosius called him *municeps tyrannus;* he presumably came from London, where he held office in the local administration; and his appointment was an unwarranted act on the part of the army and the British civic communities. It made sense, however, from the point of view of British interests, if Gratian could keep the forces together, and use them solely to defend the island from external attack. But this they declined to accept. When, four months after his appointment, Gratian made it plain they had to stay in Britain, they murdered him; instead, they gave the command to a new and foreign usurper, who called himself Constantine III; and he took all the regular units across the Channel to create a Gallic empire. In 410 these events were remembered with bitterness by the Britons. They provided arguments for the independence party which were difficult to refute. What use was Rome to Britain? Britain had been for centuries exploited as an economic colony. She had been accorded only the barest measure of local self-government. If Roman authority was fully re-established, the process of exploitation would merely be resumed. But in the meantime Rome was impotent, and the time had come to assert British independence.

These practical arguments on both sides were overshadowed by an intellectual debate which was both religious and political. For a century, since the conversion of the Emperor Constantine, the official religion of the Empire had been Christianity. Though the administrative centre of the Empire had been transferred to Byzantium, the state religion was still centrally conducted from Rome. Already indeed its chain of command, and its contacts with outlying regions such as Britain, were maintained in a more regular fashion than the political and military functions of the Empire. Christianity still had a working international infrastructure. This religion, by its very nature, was centralised, universalist, authoritarian and anti-regional. It was run by a disciplined priestly caste, commanded by bishops based on the imperial urban centres, under the ultimate authority of the Bishop of Rome himself, the spiritual voice of the western Empire. Its doctrines were absolutist, preaching unthinking submission to divine authority: the Emperor and his high priest, the Bishop of Rome, in this world, and a unitary god, who appointed the Emperor, in the next. Man was born in sin, and must accept tribulation as inevitable; he could indeed be redeemed, but only by an authority external to him – God in the next world, the Emperor in this. Salvation, now and for ever, lay solely with the Christian Empire. These attitudes and doctrines underlay the political posture of the pro-imperial party in Britain.

They had, however, come under increasing challenge from a theologian who took an altogether less pessimistic view of the human condition, and of the divine dispensation for man. Significantly, this theologian was British. Pelagius was born in Britain, of native stock, about AD 350, and was about thirty when he first travelled to Rome. He had had a good education, in the legal traditions of the Empire, but his outlook had been shaped by the local environment – physical, political and economic – of a distant province, which had never been more than semi-Romanised, and which was a very peripheral factor in imperial policy. Pelagius attacked the prevailing orthodoxy of Roman Christianity. When Adam sinned, he argued, he injured himself only: it was nonsense to pretend his fault was transmitted to every human being, to be effaced only by divine grace; a child was baptised to be united with Christ, not to be purged of original sin. Man was a rational, perfectible creature: he could live without sin if he chose; grace was desirable, but not essential. Man was a free being, with the power to choose between good and evil. He could become the master of his destiny: the most important thing about him was his freedom of will. If he fell, that was his own fault; but by his actions he could rise too.

Pelagianism was the spiritual formula for nationalism, for the independence movements breaking out from a crumbling empire. In the year 410 Pelagius was still in Rome, leaving it just before the city was sacked by the Goths. His work was by no means complete, and had not yet been anathematised by a Church which saw it as a threat to its universalist authority. But his views were already widely known and arousing fierce controversy. They were hotly repudiated by the orthodox political and religious element who saw the re-establishment of the Empire, in all its plenitude, as the only hope of salvation from the barbarian. But they were eagerly accepted by those who thought that the Empire was already dead, and that individual communities must look to their own defences. Man could save himself by his exertions, and others by his example: in this world as well as in the next. The Empire could not, by a miraculous infusion of grace, turn back the savages from the gates: only organised local resistance could do that. Possibly even the barbarians themselves could be brought within the pale of civilisation, and unite with local citizens in building viable societies to their mutual profit. Pelagius had pointed out that free will existed even among the barbarians; they too were perfectible, could choose freedom and profit from it.

These arguments had a particular appeal in Britain, which had always felt itself a neglected, despised and expendable outpost of the Continental imperial system. There is no evidence Pelagius ever returned to Britain. But he was not the only British member of his school; one of the most energetic and vehement of his companions was also a Briton, and there may have been others. At any rate his beliefs were widely held in Britain by 410: there was a strong Pelagian party among the British propertied class. There, orthodox Christianity was no more than a powerful, officially endorsed sect; perhaps not even the predominant one. Not all the leading Britons were convinced that Christianity was the only religion. In the late fourth century there had been a pagan revival in Britain, which has left traces in the splendid shrine of Nodens, in the west country, built possibly as late as AD 400. Among the British Pelagians, at least, there was an ambivalent attitude to other religions, a refusal to recognise Christianity as the exclusive route to salvation, a willingness to do business with the unconverted. This could be expressed in political and military, as well as religious, terms. Tolerance may have been dictated by common sense. Nearly 150 years later, the monk Gildas, writing from the standpoint of orthodox Christianity, blames the destruction of an independent Britain by barbarous invaders on the moral failings of the British, their lack of resolution in their faith. Echoing him, Bede says that the British were submerged because they made no attempt to convert the heathen to Christianity. But Gildas's account is avowedly didactic, not historical; he was a partisan, among other things an anti-Pelagian. His reconstruction of events after 410 distorts what actually happened, for he made himself the mouthpiece of the pro-imperial party. To negotiate with the barbarians, on the basis of a mutual tolerance of race and religion, was an obvious course for the British nationalists, who were also Pelagians. Saxons had been established, as military settlers federated to the provincial authorities, on parts of the East Coast for many decades. They were part of Britain's defensive system, such as it was. It was sensible to encourage others, of Jutish and Frisian and Frankish origin, moving across the narrow seas, to settle themselves in Kent in organised, law-abiding communities, working in co-operation with the British authorities for the defence of all the island's peoples. These settlers had been touched by civilisation; they were not outer barbarians but military tribes who could be used against them. The story of the British Vortigern, or High King, and Hengist and Horsa, reflects an arrangement which made good political and military sense at the time. It ended in tragedy, according to the subsequent gloss of both British and English Dark Age historians. But it may, in fact, have successfully ensured a limited period of peace in which newly independent Britain could organise itself. And the collapse of the British State, which endured in some form for nearly 150 years, seems to have been brought about by civil war rather than external attack; moreover, our only account of what happened comes from Gildas, who was a leading member of one of the British factions.

At any rate, in 410 the Pelagian nationalist party in Britain took control, though its authority, and policy, were qualified. We know roughly what happened from the historian Zosimus. He says that in 410 an enormous army of barbarians crossed the Rhine, without effective resistance from the imperial authorities. The British revolted from Roman rule, and established a national state. They took up arms, freed their cities from the barbarian invaders, expelled the remaining members of the imperial administration and set up their own system of government.

This was, in one sense, an anti-colonial revolution, the execution of the political programme of the Pelagian party. But it was significantly more than this. The pro-imperial, or pro-European party, was sufficiently influential to impose its own limitations on this course of action. Possibly it was felt that Roman power might eventually re-establish itself, and steps must therefore be taken to cover such an arbitrary act with some show of constitutional legality. There may have been a compromise between the two parties. Under Roman imperial law, the British were permitted one form of organised political activity. The settled part of the country – which is all that concerns us here – was divided into regions, originally on a tribal basis, administered from city-capitals. They were, in effect, cantons, with elected magistrates, who lived mostly on their country estates, but who spent some months of each year in the cantonal cities on legal and financial business. Periodically the senior magistrates were allowed to attend councils of the whole province, to organise the administration of the State religion; originally they had elected the imperial high priests at such assemblies. Hence the only form of representative national government took place in a religious context; and this is one reason why the Pelagian issue was of such importance to the events of 410. In that year members of the council met in emergency session, to coordinate resistance to the invasion and determine political policy. As we have seen, they opted for independence, and took vigorous measures to secure it. Roman Britain was in many respects a multi-racial society. Though predominantly Celtic, many Britons were descended from settlers and soldiers from a great variety of races, chiefly German. The magistrates, assembled in London, the administrative and commercial centre, spoke Latin, the language of government, in a pure and uncorrupted form, which was already foreign to Rome itself. Their native tongue was a mixture of Celtic dialects. Some of those present, representing the eastern settlers, may have spoken only Germanic dialects, with a smattering of Latin. It must have been a heterogeneous collection of notables united only by their common predicament.

Nevertheless, what they did was a unique act of statesmanship. Having seized power for themselves, they wrote to the Emperor Honorius asking formal and legal authority for what they had done. They had got independence *de facto*; they now wanted it *de jure*, a written acknowledgment from the imperial power that Britain had been decolonised with the permission of the authorities. More specifically, they wanted exemption from the famous *lex Julia de vi publica*, the bedrock statute of the Roman Empire, which forbade civilians to bear arms except when hunting or travelling. In due course they got it. Honorius sent his rescript, or reply, accepting the *fait accompli*, and instructing the *civitates* of Britain to look to their own defences. Thus the ancient world ended, and the independent history of Britain was resumed, in a thoroughly legal and constitutional manner. There was no provision in Roman law for a territory to leave the Empire. But by an ingenious use of the *lex Julia*, the British got round the difficulty, and severed their links with the Continent by a process of negotiation which legitimatised their use of force. It was a unique event in the history of the Roman Empire; it was based on no precedent, and had no parallels elsewhere. For the first time a colony had regained its independence by law; and it was to remain the last occasion until, in the twentieth century, the offshore islanders began the constitutional dismantlement of their own empire.

What in fact the British were doing was resuming their pattern of insular development, dictated by climate, ecology and geography. The lowland parts of Britain are unique in our hemisphere. The climate is temperate, there is just enough sun, and just enough rainfall, to permit settled cultivation; too few mountainous areas to impede it. The soil is fertile; rivers are conveniently small and abundant, communication is possible. The terrible excesses of nature are absent: floods, droughts and tempests operate within a tolerable range of magnitude. It is possible to create a prosperous and self-generating economy here, as it is not in Scotland, Ireland or Wales; the Channel is wide enough to permit a degree of social independence, but narrow enough to serve as an access to Continental cultures. In prehistory lowland Britain was always a receptacle of population movements from the east and south, settling in numbers limited by the hazards of the Channel crossing, cutting their social, but not cultural, bonds with the Continent. In the highland areas it was far more difficult for settled, farming societies to establish themselves. But in lowland Britain there is a continuous process of cultural and economic progress, with marked characteristics not to be found on mainland Europe. In Palaeolithic and Mesolithic times there were perhaps no more than 3,000 people in this area, living exclusively by hunting. In the Neolithic age, from 3,000 BC, a form of primitive farming began to emerge: scrub and woodland was cleared and burnt, a corn crop sown and harvested, then the process repeated elsewhere; small herds of cattle and sheep were kept. Even so, the population rose very slowly: it was perhaps only 20,000 at the beginning of the Bronze Age. Seen in the long perspective of history, Britain was a very late developer. When the Greek colonists began to build their great city of Syracuse in Sicily, Britain was still wholly locked in the restrictive culture of the late Bronze Age, with a population of less than 100,000. The use of iron was unknown here when Solon ruled Athens, when Croesus was King of Lydia, when Cyrus took Babylon. The iron culture reached Britain only at the end of the sixth century BC, and it spread far more slowly than on Continental Europe. The British do not seem to have constituted an innovatory society in any way. But some of their creations were remarkable. Stonehenge was a kind of state cathedral, of great size and complexity, altered and re-built several times during the period 1900–1400 BC; the rings of

Avebury were still larger. The British hill-forts, too, were larger and more numerous than anything produced by similar cultures on the Continent. Nevertheless, Britain remained in every respect a cultural and economic backwater until the last wave of settlers, the Belgic peoples of northern Gaul, reached the island just before the beginning of the first century BC.

The Belgae were Celts, but they incorporated certain characteristics of the German forest-dwellers, and they had also been touched by the outermost ripples of advancing Roman civilisation. Settling in Kent, Sussex, Hampshire, Essex and the Thames Valley, they introduced agricultural methods which allowed, for the first time, the systematic cultivation of the heavy and productive lowland soils. They probably did not possess ploughs armed with a coulter, capable of turning the sod. But they used iron in much greater quantity than any previous society in Britain: they had many more ploughs, and other implements, and above all thousands of axes. They cleared the forests on a considerable scale, and settled in the valleys on sites which have been occupied ever since. For the first time the topographical axis of agriculture, and thus of society, began to shift from the uplands to the lowlands, and the new areas thus brought under cultivation made possible an increasingly rapid growth of population. The business of clearing the forest was to last for 1,000 years, and was the first decisive economic event in the history of the offshore island.

Hence, in the first century BC, lowland Britain was a territory in the course of rapid economic, social and indeed political development. In the terms of the Ancient World, it had reached what can be called a take-off stage in its history. Between the beginning of this century and about 50 BC, the population probably doubled, from a quarter to half a million. Much larger tribal units, and later tribal confederations, began to emerge. Their kings were identifiable personages, exercising authority over large areas. They traded extensively, replacing the iron bars originally used for exchange by regular coins, first brought from Gaul, later minted locally. Here was a living, expanding, progressive society, whose members were conscious of radical, even revolutionary, changes taking place in their own lifetimes. But it was at precisely this moment that Britain came in contact with Rome. This has produced a fundamental distortion of history: not only is British development henceforth seen entirely in terms of the growth and decline of the Roman Empire, but it is seen exclusively through Roman eyes. Britain was incorporated into Continental Europe, and its history became a mere peripheral function of the history of a great land-civilisation.

To get a truer perspective, we must switch the angle of vision from the Roman to the British and try to examine events as they would have been seen through intelligent British eyes. They are the eyes of a pre-colonial, a colonial, and a post-colonial people. During the period of Caesar's conquest of Gaul, the British kings and their advisers watched with growing anxiety the rapid approach of a great Continental military power. For the first time a political society existed in Britain capable of opposing a cross-Channel invasion, and therefore able to formulate a conscious policy towards the Continent. But it was also aware of the definite material advantages of Roman civilisation, and realised that its growing prosperity depended in great part on cross-Channel trade and contacts. How could it get the best of both worlds – that is, exploit the opportunities offered by an expanding European culture and market, without risking incorporation, and thus exploitation, in the political and military system of the land-mass? This is the fatal question which has always confronted the inhabitants of lowland Britain. It has never received a final answer, and perhaps no final answer is possible.

At the time, the British reacted in a manner characteristic of pre-colonial peoples. They prevaricated; they were indecisive; they were ambivalent. They gave some assistance to the Gallic tribes which were fighting Caesar: not enough to stem his advance, but enough to give him a pretext to invade. They were willing to make treaties, but not to keep them if they involved real sacrifices of economic and political sovereignty. They were always anxious to play for time, hoping, no doubt, for some *deus ex machina* in the shape of a change of policy in Rome. But they were also divided. Some British chieftains were active supporters of Caesar. One or two even worked with his invading forces. At every British court there was a pro- and an anti-Roman faction. In some cases the pro-Roman faction triumphed: the Trinovantes of Essex, for instance, feared the aggressive expansion of the tribal confederation north of the Thames, and adopted a pro-Roman posture: their alliance with Caesar made possible the limited success of his second invasion. And at other tribal courts, if the anti-Roman faction triumphed, ousted politicians often sought refuge with the Roman Authorities.

On the whole, Caesar's two invasions must have persuaded a majority of the British political *élite* that, by one means or another, Rome could be held at bay. Caesar, in his commentaries, puts the best possible gloss on his expeditions; but both came near to disaster, and were marked by recklessness, lack of preparation and a confusion in Caesar's own mind as to what his objectives really were. They did not impress the British, nor, in the end, did they impress Rome. After the first one, the Senate, relying on Caesar's dispatches, accorded him an unprecedented triumph. But the second, much more costly, was accompanied by many independent observers, who wrote letters home; and this time Caesar's withdrawal was greeted by a resounding silence in Rome. Moreover, his ineffective manoeuvrings across the Channel clearly helped to inspire revolt on the mainland. The British could reasonably assume that the Romans would not return, and for the next hundred years it looked as if they were right.

During this period, the evidence reveals an unprecedented growth of prosperity in Britain. The British were indeed getting the best of both worlds. They imported huge quantities of pottery from the Continent, but also began to make their own sophisticated models. They exported a wide range of products, and developed their own mines. They had their own coinage, and not just in the areas of Belgic settlements. Strabo, the Roman court geographer, claimed that Britain's rulers had made 'the whole island almost a Roman colony'; but the stress should be placed on the 'almost'. The British were deriving all the benefits of economic contacts with a great Continental market, with none of the disadvantages of economic and political subjection. Living standards were rising fast, probably much faster than on the Continent. Equally important, Britain was making rapid progress towards political unity. By the time of the Claudian invasion in AD 43, a single paramount power was emerging in the south-east. Given a few more decades, it is possible that the whole of lowland Britain would have been absorbed into a single military state, making an invasion and occupation of the island beyond Rome's resources. If so, the history of north-west Europe for a thousand years would have been radically different, for a unitary kingdom in lowland Britain has always constituted a formidable power.

Fear of an emergent British kingdom was undoubtedly one factor in persuading the Romans to annex Britain, though another was clearly the growing prosperity of the British lowlands. There was always a fierce argument in Rome as to whether the Empire should expand or not. The prospect of acquiring wealth from new territories had to be balanced against the enormous cost of fixed garrisons, and especially the legions, each of which, in terms of finance and skilled manpower, was the equivalent of a nuclear aircraft-carrier today. Rome lacked a modern economy. It had no developing technology and no industrial base, because it did not know how to create demands for new goods and services, or even how to create mass-markets for what it already produced. It could not, or at any rate did not, raise the purchasing power of the overwhelming majority of its subjects. It simply spread a thin and static level of economic culture wherever it went, exporting craftsmen and techniques rather than goods, and failing wholly to develop the specialisations which are the key to self-sustaining economic growth. The Empire had to expand to survive at all; once it ceased to expand, its currency collapsed in inflation, and there was no way to pay for the armies to defend the imperial frontiers.* These problems, though not understood, were already making themselves felt at the time of the Claudian conquest. Britain had been left alone because Caesar's experiences had given the island the reputation of being difficult to deal with, and not worth the trouble. But evidence of rising prosperity and developing unity in Britain tipped the balance of argument at the Roman court in favour of conquest. But it was a near thing: a few decades later Rome might have decided otherwise.

For the mass of the British, the Roman occupation was a disaster. It is true that some tribes welcomed the Romans, or at any rate found it prudent to sign treaties with them rather than fight. Caratacus, a man of great resources and pertinacity, was never able to create anything approaching a national confederation against the invader. Many chieftains found it worth their while to accept the role, titles and dignities of puppet sovereigns. Some allowed their followers to be disarmed. The propertied class found access to the Rome credit market a new adventure, and quickly borrowed huge sums which they used to buy the new range of sophisticated trinkets touted in the wake of the legions. But the experience of the first generation of colonial rule was decisive in turning the British against their conquerors. What is significant about Boudicca's rebellion was that it was a mass-uprising among both a tribe which had been conquered by Rome and one which had freely submitted. Evidently all sections of opinion in Britain came to resent the occupation, which was marked by blatant racism and the systematic exploitation of all classes. The rising was savage enough to bring about a change in Roman policy: even the Romans came to recognise that they must govern with some element of consent. All the same, the rapid rise in living standards, which had been such a striking feature of the last pre-colonial century, was halted and then reversed. The mineral wealth of the country passed wholly into Roman hands, exploited directly by the imperial government, or under licence by Roman firms. Tin-mining was halted so as not to interfere with the

* In the first century AD, a pound of gold was the equivalent of 1,000 denarii, the basic silver coin. The silver coinage declined steadily in relation to gold, and in the mid-third century the monetary system disintegrated. In 301 Diocletian attempted to stabilise the currency on the basis of 50,000 denarii to a pound of gold; but a decade later the figure was 120,000. By 324 it had risen to 300,000 and by 337 it was 20 million; in the 350s it was 330 million. The Roman Empire was destroyed by inflation, though this itself was the result of deeper causes. See Sture Bolin: *State and Currency in the Roman Empire to AD 300* (Stockholm, 1958).

tin-profits of imperial Spain. Many forms of economic activity were banned. Huge tracts of the best land became imperial estates, worked by slave-colonies. A small British propertied class was allowed to survive, to ape Roman customs and even to discharge minor functions; but it did not get citizenship as of right for 150 years, and by then the privilege had lost much of its value. Most of the British were pushed down the scale, both socially and economically; they received nothing from Rome, though some of them picked up a smattering of its language. It is an astonishing thing that, in 350 years of Roman occupation, only a tiny handful of British-born subjects achieved even the most junior prominence in the Empire. And we cannot be sure that these, whose names we know, were British by race.

The predicament of the British was not improved by the uncertainties and abrupt reversals of Roman policy. Indeed the British must have been puzzled and angered by the evident inability of the Romans to decide what they wanted to do with the island. The Roman occupation always had an air of improvisation. It was a badly planned experiment, which successive generations of Roman statesmen tinkered with, and then abandoned without finishing. At one time or another, most of the best brains in Rome took a hand in British affairs: Caesar, Claudius, Vespasian, Hadrian, Septimus Severus, Constantine. But to all of them it was a marginal problem: it never focused itself at the centre of Rome's preoccupations. Rome treated Britain as, later, the English were to treat Ireland: as a tiresome and unresolved problem, to be dealt with only when it reached crisis-point, and then to be forgotten. Only Agricola, who devoted a large part of his life to Britain, seems to have had a deliberate and consistent policy: he wanted to conquer the whole of the British Isles, but was recalled when his projects were seen to be ruinously expensive.

It was money which damned the Roman experiment in Britain. It was impossible to create a profit-making colony which was also defensible. If vigorous measures were taken to guarantee the security of the lowland zone, the colony immediately went into deficit. The Romans originally intended to hold the Trent–Severn line, which incorporated all the more profitable agricultural areas. Then they discovered that this excluded most of the mineral wealth. For the next 20 years they pushed into the north and west, to find that this raised still more difficult frontier problems. Where was the frontier to lie? The Romans never found an answer. For 300 years over 10 per cent of all the Empire's land-forces were held down in Britain, perhaps the least significant of the colonies. This enormous expenditure could not be justified in economic terms. But how could it be reduced without imperilling the colony? Hadrian thought he could solve the dilemma by building fixed defences from Tyne to Solway, and thus economise on manpower. His wall involved shifting 2 million cubic yards of soil and subsoil, and absorbed over a million man-days: it was the greatest single artifact in the history of the Empire, and probably the most costly. But in the end it did not even save manpower. Moreover, the Romans could never decide whether it was in the right place. A generation later they built another wall on the Clyde–Forth line, and then abandoned it. Some of these northern fortifications absorbed a significant proportion of the entire resources of the Empire. The legionary fortress at Inchtuthil in north-east Scotland required seven miles of timber walls. When it was evacuated, unfinished, 11 tons of unused iron nails were buried there. All these materials had to be brought hundreds of miles up north. The Romans were constantly building bases in Britain which were soon abandoned, often before they were finished. (This was also a striking feature of the late British Empire.)

In theory at least, Britain was supposed to pay for this huge military expenditure, and to support an army and administrative establishment which was up to 5 per cent of the total population.* But this cannot have been possible, even allowing for the fact that taxation kept British living standards at a permanently depressed level. The Romans lacked the technology to exploit Britain's mineral resources effectively. Lead was mined in considerable quantities for cupellation into silver, but it was of notoriously poor quality. Mining for tin, Britain's leading export in pre-Roman times, was held down until the Spanish mines ran out in the late Empire. Some corn was exported, under compulsion. But most British exports were luxuries: fine-quality woollen goods, two items of which figure on Diocletian's price-control list, semi-precious stones, and hunting dogs – Irish wolfhounds, bulldogs, spaniels and greyhounds. If we add all these together, they could not balance the flood of pottery, metalwork, manufactured goods, wines and luxury foods which poured in from the Continent to satisfy the needs of the Roman establishment and the British upper class. Roman Britain must have had an adverse balance of trade with the rest of the Empire throughout most of its existence, and trade was balanced by the one great 'invisible', the spending-power of the occupying army.

With such a distorted economy, it is not surprising that the effort

* Professor Sheppard S. Frere, in his *Britannia, A History of Roman Britain* (Oxford, 1967), calculates that the population at the end of the second century AD was about 2 million. But this is based on many arguable assumptions. It may have been as low as 1 million. From Domesday Book we can calculate that the figure in 1086 was about 1,100,000 (excluding Wales), with much more land in cultivation. My view is that Roman Britain could not have supported a population much above the million mark.

to Romanise Britain failed. The British, indeed, rejected Roman civilisation because they rejected its instrument: the city. To the Romans, the city was not just the centre of government and the economy but a living theatre in which all the rites of civilisation were enacted; planted in the wake of the legions, it underpinned their rule and acted as the conduit of their civilising mission. Through the cities they built, all Italy, Spain and France were Romanised, with a thoroughness which enabled the Romanic element to survive through centuries of political and economic confusion, and vast movements of population, as the dominant cultural pattern. But the Roman city was an expensive luxury: it was essentially parasitic. It was not so much an administrative and service centre for the rural economy as an artificial and exotic creation, an end in itself, which the rural economy had to support. It provided a range of amenities out of all proportion to its size: a city hall big enough to hold all free men and women for the transaction of public business; theatres and arenas where all could be entertained; baths which the entire public could use daily; temples for universal congregations. These cities were immensely costly to build, and they needed a fortune to maintain.

The British economy could not support such a system, any more than it could support the occupying forces. Though the area of cultivation was being extended by the introduction of heavy ploughs and drainage, it is by no means sure that agricultural productivity was rising; it may even have fallen in Roman times. Roman farming technology under the Empire was stagnant, in some respects decadent. Roman estates in Britain may well have been less efficient than the small farms and holdings which they often displaced. The economic basis for a flourishing urban civilisation did not exist in the British colony. In any case the British did not want it. The building of Colchester was one of the main factors in producing the mass-revolt which Boudicca led, and the principal animus of the insurgents was directed against civic buildings there, and in London. Only the most vigorous pressure from the authorities got cities built at all. Thirty years after the first landing, Agricola was dismayed by the slow progress and launched a massive programme of construction. Half a century later Hadrian found it necessary to do the same. Cities were indeed built, but they did not flourish. Silchester, the only one to have been fully excavated, had some of the apparatus of Roman civilisation in the third and fourth centuries: administrative buildings, a market place, four pagan temples, a Christian church, baths, an imperial post office, an amphitheatre. But it was small: there were only about 25 large houses, for tribal magnates; 25 smaller ones for administrators and merchants; the rest of the inhabitants, about 2,500, lived without dignity. The conventional picture of gracious living in the Roman city does not apply to Britain.*

Only in Bath were the highest levels of Roman sophistication reached. But Bath was little more than a resort, a rest and recreation centre for soldiers and expatriates. The wealthiest Britons no doubt patronised it, but it must have seemed to most of the natives, if they ever heard of it, as incongruous as the Indian hill-stations of the British Raj, or the leave-centres which the American forces have built in Asia. Some of the Roman–British cities were never finished. Only Lincoln had a sewerage system built to Roman standards. Leicester never got a regular water-supply. It is significant that very few Roman civilians could be tempted to settle in Britain; if anything, cultured Britons emigrated south. It was fashionable for Romans to sneer at the British for their savage ways; the Romans maintained they still wore woad, though they had long ceased to do so even in Caesar's day. These feelings were doubtless reciprocated. The British may have come to welcome the security Rome provided, but as a race they never accepted its civilisation. Most of them never learnt to speak Latin, except a few phrases for functional purposes; the wealthy few who did so spoke it as a cultural supplement to their natural tongue, as the Tsarist aristocracy spoke French. When pressure from the authorities relaxed, the city-sites in Britain tended to degenerate into purely economic instruments. In the late Empire, Roman civilisation in Britain withered, and the cities acquired a pragmatic British flavour. There is ample evidence not so much of a discontinuity in city life but of a change in its function, from an artificial cultural creation to a viable, albeit austere, trading centre. City mansions were taken over by craftsmen; at St Albans the amphitheatre was turned into a market-place. City populations may even have increased, but the cities tended to serve the countryside, not vice versa.

The only Roman institution the British welcomed was the country villa, though they invested it with characteristics of their own. The Roman upper class, and its Continental imitators, saw the villa as a place for rest from the cultural ardours of city life, especially in the summer. The British upper class reversed the system. They spent most of the year on their estates, living in villas which were working manor houses.† They went to the cities only for essential business; many of them did not even possess town houses. They formed a rural gentry,

* The most recent estimates give maximum city populations as follows: London 30,000, Colchester and St Albans 15,000 each, Lincoln and Gloucester 5,000. Most other towns were 3,000 or less.

† We know of about 600 villas in Britain; there may have been about 800 in all. For their distribution see the Ordnance Survey *Map of Roman Britain* (1956).

and it is not absurd to project backwards into their attitudes a love of country life, especially hunting, an intimate connection with their tenant-farmers, a close attention to estate management, and a condescending view of the city – all of which became salient characteristics of the more affluent offshore islanders in later ages. They took little interest in the Empire. They did not seek, at any rate they failed to obtain high positions in the imperial service. They were an upper class, but in no sense a ruling class. The later Empire was a centralised tyranny. Under the pressure of uncontrolled inflation it had changed from a constitutional republic into an oriental despotism, with the state directly controlling vast sectors of the economy. In such a desperate and unstable structure there was no place for colonial self-government. In their last decades as a colonial people, the British lived under military rule, such as it was. When the soldiers left, and they themselves expelled the administrators, there was no one trained to work the machinery of government. The British were a colonial people, abruptly deprived of the protection, the guidance, the political skills and the markets of an Empire; and they were surrounded by enemies.

Yet an independent British society survived in the lowlands, or large parts of them, for a century and a half – a history longer than the Tudor dynasty, much longer than united Italy or Germany, almost as long as the United States. This phase in British history goes almost unregarded, because it is virtually unrecorded, but it was a considerable achievement. The removal of the dead hand of the Roman Empire unleashed the dormant energies of the British people. The Empire had been economically stagnant by AD 250, 150 years before its military and political collapse: during this period it held itself together at the price of creating serfdom on the land, State capitalism in industry, and a theocratic totalitarianism in religion and politics. The removal of this festering incubus gave the British the chance to think and act for themselves; it is not surprising that they embraced the free-will doctrines of Pelagianism, their native brand of Christianity – which eventually, by a process of insular transmutation, became Anglicanism.

The tragedy of the post-colonial British was that they failed to achieve, or at any rate to maintain, their unity. Disunity has always been fatal to the offshore islanders, or whatever race. The reason why they were divided was that one remaining link to the Continent held, at least for a time—Christianity, or rather the centralised Roman version of it. Roman Christianity did not exactly flourish in the ruins of the Empire, but it managed to hold most of its ground and even to devote a considerable portion of its energies to the extirpation of what it regarded as heresy. For some decades it kept watch on its outlying provinces. As we have seen, the British in 410 were divided into a nationalist–Pelagian party and an imperial–papal one; the nationalists won, but on the basis of a compromise which observed the legal niceties of a world system. But the anti-nationalist faction remained active. In the decades after 410, as we know from Gildas's account, they twice appealed to the imperial authorities to restore the links with Rome. They got no response from the secular arm, but the Church rallied to the defence of its lost province. On at least two occasions before 450, clerico-military expeditions were sent from France under the leadership of fighting bishops, notably St Germanus of Auxerre. Germanus, who had been a senior military commander, led the British (we are told) to a victory over the heathen invaders, with 'Alleluia' as a war-cry. But his principal purpose in coming to Britain seems to have been to combat Pelagianism. Politics and religion were inseparable: he was in fact intervening in a civil war, on behalf of the Continental party.

This internal conflict seems to have continued throughout the history of the independent British state, and was indeed the chief cause of its extinction. Direct contact with Rome was lost some time after 455, but both orthodox Christianity and Pelagianism continued to fight for supremacy over the British people. The sources are fragmentary, contradictory, always suspect for one reason or another; some are lost entirely, though we can detect distant echoes of them in the works of twelfth-century writers such as William of Malmesbury and Geoffrey of Monmouth. Piecing together these scattered clues, it is possible to reconstruct a history of the period which makes sense. By the mid-fifth century, when the Saxon raids began to turn into a mass migration, lowland Britain had become a confederation of regional kingdoms, with a tendency to acknowledge a single powerful king as overlord (a practice later transmuted into the English institution of the paramount ruler, or bretwalda). This leader was called the Vortigern, and in the latter part of the century his name was Ambrosius Aurelianus. Ambrosius was probably an orthodox Christian, with marked Continental leanings, 'the last of the Romans' as he is called. One of his army commanders, from a West Country landed family, was called Artorius, or Arthur, a Roman name given as a token of the family's imperial allegiance. Arthur was born about 475 and, shortly after the turn of the century, when Ambrosius died, succeeded him as the senior military commander. He fought the invaders as the general, and later as the overlord, of the British kings. Using disciplined units of armoured troops, he was highly successful: he won 12 engagements in various parts of lowland Britain, culminating in the battle, or siege, of Badon in 516, in which a Saxon army of about 900 men was annihilated. This victory was followed by a reverse migration of many of the Germanic settlers.

Some 20 years later, probably in 537, Arthur's kingdom collapsed, and he himself was killed in the course of a civil war. This was the prelude to the final triumph of the Germanic settlers in lowland Britain. What was the civil war about? It certainly had a religious flavour. Significantly, Gildas, though he refers to these internal disputes, does not mention Arthur; and this looks like a deliberate omission, indeed a suppression. Gildas was not writing history, but a politico-religious diatribe, a work of propaganda and exhortation. His life and Arthur's overlapped, for Gildas died about 570. According to his biographer, Gildas had a dispute with Arthur, whom he hated. Gildas was, of course, a vigorous exponent of main-line Christianity, as he conceived it (for contact with Rome had been lost). It seems probable that Arthur, in the course of his career as the paramount British leader, had become a convert to the insular nationalism of which the Pelagian doctrines formed the theoretical basis. The fact that Arthur carried into battle the emblem of the Virgin Mary would have little weight with Gildas, to whom Arthur was not only a heretic but a renegade. It may be that Arthur's wife, or queen, remained orthodox; there is a tradition that Arthur went to seek her at Glastonbury, where she found refuge, or possibly imprisonment with the monks. Perhaps he was killed there, or near by; he may even be buried there. But his career seems to suggest that the two factions in lowland Britain were still evenly balanced; too evenly balanced, indeed, for either to subdue the other, and together they brought the state to ruin.*

The Arthurian traditions survived and proliferated in the Celtic fringes of Britain. They were of no interest to the English, but they quickly captured the imagination of the Normans – who felt, indeed, some affinity with the Celts in their common hostility to the English. The Normans took the Arthurian legends to the Continent. Thanks to Geoffrey of Monmouth, a literary propagandist of genius, and by a delicious series of ironies, Arthur became England's first great cultural export. Carried forward on a wave of anti-French sentiment, Arthur, as the King of Romance, displaced the far more solid and authenticated Charlemagne until the end of the Middle Ages. He and his knights made their Continental début on the north doorway of Modena Cathedral, certainly not later than 1120. He appeared in every kind of work

* Arthur may have been a violent and brutal man; it is possible his name signifies 'bearish'. Recent studies of Arthur, written for the general reader, include Geoffrey Ashe: *From Caesar to Arthur* (London, 1960), and Christopher Hibbert: *The Search for King Arthur* (London, 1970).

of art in France, Germany, Ireland, Greece, Scandinavia and Switzerland, in Spain, Portugal, Cyprus and Sicily. The crusaders brought him to Beirut and western Asia. Every generation seemed to have something new to say about him. He provided the inspiration for Edward III's charmed circle of the garter, a form of male fellowship widely imitated even by the English middle classes in the later Middle Ages, which evolved into the characteristic English institution of the club. His knights found a place in Dante, and he himself, superbly cast in bronze by Dürer, helps to guard the tomb of the Emperor Maximilian in Innsbruck. Arthur had proved even more vigorous in death than in life. The Roman Church strongly and repeatedly condemned Round Tables; perhaps it had a long memory; perhaps it instinctively knew that Arthur was a heretic. At any rate, it was to the Arthurian legends, and in particular to the belief that Arthur had ruled a British empire, casting off allegiance to Rome, that Henry VIII turned in search of historical ammunition to fire at the Pope.

But Arthur's real achievement was that he delayed, indeed for a time reversed, the progress of Germanic settlement. This had important consequences, for it prevented the British from being exterminated in, or wholly expelled from, the lowland area. It is true that British culture disappeared almost completely. As a colonial people they had rejected the civilisation of Rome, but in the centuries of subjection they had lost much of their indigenous culture, for their upper class had been unable to patronise it, and they had been forced to accept an alien religion; their post-colonial history had been too brief, and troubled, to permit the development, or re-emergence, of a powerful life-style of their own. The culture the Germanic settlers brought with them was rustic and humble but immensely pervasive. Hence the native population accepted the manners of its conquerors, their laws and customs, habits and predilections, political organisation and methods of warfare, religion, arts, crafts and attitudes, most of all their economic ways and structures. In Gaul names based on Gallo-Roman estates remain even today one of the commonest elements in the village names of France. In Britain, even in Kent where there were other elements of continuity, estate names and boundaries disappeared completely. There was in time a complete break with the agricultural past. The manors of late-Saxon England have no demonstrable connection with the Roman–British past. All the same, large numbers of the British survived, though generally at the lowest levels of society. They lived on in the uplands, forests and marshes. Their existence even leaves some faint tracings. In the Humber area and Wessex, for instance, some of their personal names are found. A score or more can be detected in Domesday

Book seven centuries later. One Saxon royal house seems to have intermarried with them, on more than one occasion, an example which humbler Saxons would have followed. The laws of Ine and Alfred gave recognition to a distinctive 'Welsh' that is, British class in the social system: not only Welsh slaves but Welshmen holding up to five hides, with wergilds of 600 shillings, and three other categories of Welsh freemen. Most of the British became rural slaves, and lost all sense of cultural and racial identity. But they nevertheless contributed to the composition of the English people; they help to explain why the English became what they are; they served as a human bridge between the remote past and the future of England.

PART TWO

Unity, Stability, Continuity

[600–1154]

IN the autumn of the year 663 a remarkable group of men and women assembled at Whitby Abbey in Yorkshire to take a decision of momentous importance for the future of England. English society was still in its early stages of development; the only available and systematic machine through which literacy could be spread and civilisation advanced was the Church; it was the supreme instrument of de-tribalisation. But the question was: which Church? For there were, in effect, two. The Celtic Church of Scotland, Ireland and Wales had pursued a course of separate development since it had lost contact with Rome after 455. It observed a different date for Easter; it had its own form of tonsure, and many other practices. More important, it had a wholly different system of organisation, based on rural monasteries rather than urban bishoprics. Its outlook was ascetic, other-worldly, anti-hierarchical, contemptuous of the temporalities of religion. It preferred stone cabins to great basilicas, and self-denial to triumphant ritual. It was still permeated by the insularity of Pelagianism, and took its colouring from the lands and peoples which nourished it.

On the other hand there was the Church of Rome, representing the universalist order of the late Empire, its bishoprics based on the old city and provincial administration, radiating from the ultimate authority of the eternal city itself, its ceremonies and buildings and vestments echoing imperial grandeur, its hierarchy and discipline upholding the principles of a world theocracy, with power finally resting in the hands of one man, the vicar of the Christ-Emperor. The Roman Church still spoke for the Empire. Britain had cut itself off from the Empire 250 years before, but on the Continent the mainland rump had absorbed the Germanic invasions, and imposed its civilisation and languages upon the settlers it had received; only in the last decade or two had its soft underbelly, in the Mediterranean, been ripped asunder by the new Oriental power-religion of Mohammedanism. In Merovingian Gaul, the urban civilisation of Rome, dominated by romance-speaking peoples, was still the basis of society. It was not then unthinkable that the Christian Western Empire could be restored in all its plenitude, and Britain, its lost province, rejoined to it.

Lowland Britain, now settled by Angles, Saxons and Jutes, and increasingly called England by its inhabitants, thus became an ideological battlefield. If Celtic Christianity triumphed, the Channel must inevitably become a religious and cultural barrier, as it was already a political and military one. The whole of Britain would, in effect, cease to be part of Europe. On the other hand, if Roman Christianity became established there, the Celtic world could not survive alone, but would be increasingly pulled into the European pattern. The actual issue to be decided at Whitby was the date of Easter; but all else flowed from the verdict. Rome had put out a tentacle to England 70 years before: an expedition under Augustine to set up two Christian provinces, at London and York, on the old imperial model. It had met with moderate success and many setbacks; its headquarters had, in fact, been established at Canterbury, the only city-site in Britain where continuity had been maintained into the English age. The Roman attempt to convert Northumbria had ended in disaster; and Christianity had finally been established there from Iona, the headquarters of Celtic Christianity. In many parts of England, Celtic and Roman missionaries were now coming up against each other, as Englishmen and Frenchmen were to meet in the heart of Africa, in the last decades of the nineteenth century.

The situation was confused, and the forces at Whitby evenly balanced. The Abbess Hilda herself was a Celtic Christian; so were her cousin Oswui, the King, and Colman, the chief bishop of the area; indeed the last had been a disciple of the original Celtic missionary, St Aidan, and he acted as spokesman for the Celtic case. But the internationalist party was also strong. It included the deacon James, a direct link with the original Augustinian mission, Bishop Agilbert from Wessex, who was to end his life as Bishop of Paris, and the King's own son and heir. Its spokesman was Wilfred, a young, ruthless and enthusiastic Romanist, who had spent five years in Italy and Gaul, and who was now in charge of a cadre of Romanists at Ripon. The King presided at the debate, and eventually gave his decision for Rome. The arguments have come down to us only through Romanist sources, and they are dressed up in the technical language of theological controversy. But what seems to have convinced the King was Wilfred's passionate contention that England, an obscure and remote island, could not cut herself off from the very sources of European civilisation and progress; she would thereby condemn herself to stagnation and impotence. The King took what was, in essence, a secular decision: the links to the Continent must be maintained.

Yet the decision was a very English one: it was not clear cut, it was not carried through to its logical conclusions, it was heavily qualified, and left its interpretation and enforcement to be shaped by local conditions. It was a constructive compromise; some would say a muddle, but a muddle of the type the English are adept at contriving for their own purposes. Bishop Colman returned to Iona defeated. But Celtic Christianity remained in the north, and was quietly absorbed into a new English pattern. Colman's abbey of Lindisfarne flourished. St Cuthbert, himself taught by one of Aidan's original pupils, became the most influential figure in the English Church. The Roman pattern was formally adopted, eventually throughout England – and the Celtic Church in time conformed to it. Rome sent important international figures to England to reinforce what it thought a victory: Theodore from Tarsus, and Adrian from North Africa. But the Church which emerged was essentially *sui generis*, a Church of England which took from the Celtic world and from Rome certain elements which it blended into a national composition with a new flavour of its own. The plans of Rome for the structure of the English provinces were never carried out in full; a much more haphazard organisation grew up.

Wilfred's attempt to recreate a Roman Christian State in England was thus thwarted. His contemporary, Cuthbert, was universally venerated; but Wilfred made himself thoroughly disliked, and was twice expelled from the Northumbrian court. He was too deeply imbued with the Continental tradition for the English taste. His emphasis on the temporalities was too marked. He became a notorious pluralist, amassed great wealth, was attended by a huge retinue, and sought to play a dominant role in secular as well as Church affairs. When defied, he introduced the dangerous practice of appealing to Rome: he was the first of the great clerical litigants, and his activities kept the English Church in forensic uproar for most of his life. True, his energy was enormous; he converted the heathen of the Frisian coast, Sussex and the Isle of Wight; he established Christianity on a permanent basis in Mercia; he used his money to build fine churches and introduce a splendid ceremonial. But his alien enthusiasms were distasteful to the English, clergy and laity alike; even the old Greek Archbishop Theodore realised that Wilfred's extremism was unsuited to an English context. Despite his force and ability, Wilfred never became master of the English Church, and his influence was negligible. Thanks to the vigorous propaganda of his disciple and biographer, Eddius Stephanus, he was canonised; but, unlike Cuthbert, he never became an object of popular veneration. He remains an outstanding example of the minority tradition of Continentalism which flows through English history.

Seventy years, almost exactly, after the Synod of Whitby, the historian Bede sat down to write a long and thoughtful letter to his pupil Egbert, now Archbishop of York. In the interval, the constructive compromise of Whitby, blending Celtic and Roman elements into the main-stream of English development, had produced an extraordinary flowering of culture in north-east England; Bede, in his person and in his work, epitomised its achievement. He was a new phenomenon – a civilised Englishman – and he is worth examining at some length. He had the salient qualities of the new English Church: tolerance, moderation, exactitude in scholarship, a high regard for truth, an appropriate degree of unworldliness leavened by common sense. His life was fortunate. The rise of a strong Northumbrian Kingdom, coinciding with the settlement of the Church's internal disputes, produced a rapid growth there of monasticism on the Roman model, existing side by side with older Celtic houses such as Lindisfarne. In 674 Benedict Biscop, Abbot of St Peter's in Canterbury, founded a Roman monastery at Wearmouth, and in the next decade a sister establishment at Jarrow. He brought the nucleus of a library from Rome, and his successor added to it; in Bede's day it was one of the finest collections in north-west Europe. At its height, the twin foundation housed over 600 monks, many of them distinguished scholars, artists and craftsmen. Bede came there at the age of seven, an orphan of good family, and spent his entire life at Jarrow. It was a very insular existence; Bede left the monastery only twice, once to go to York and once to pay a fraternal visit to Lindisfarne, where his name was written in the *Liber Vitae*. But the culture of the house was cosmopolitan. Relations with Lindisfarne were friendly; Bede himself wrote a remarkable life of its honoured son, Cuthbert. There were frequent contacts with Gaul and Rome, and a constant stream of visitors. Bede was thus the beneficiary of both Latin and Celtic cultures, and he had by inheritance a third, English. He wrote a pure and simple Latin, understood Greek and even a little Hebrew. Through books he absorbed virtually all the knowledge then available in western Europe; and his own writings cover a vast field.

Most of Bede's time was devoted to annotating the scriptures and translating sacred texts into English; but he also wrote biographies, history, hymns, epigrams, homilies, and grammatical and scientific treatises. Bede was fascinated by chronology, and wrote two surveys of the subject, the second and more important being his *De Temporum Ratione*, finished in 725. This adopted the method, first developed by Dionysius Exiguus in the sixth century, of calculating year dates from the Incarnation. The practice was virtually unknown in western Europe; it was first heard of in England when Wilfred explained it at Whitby. But not until Bede's book was circulated did the English accept the new system. His manuscript was soon taken to the Continent, where it was copied and recopied in scores of religious houses; thus it was Bede who popularised the modern method of dating in the West. Indeed all his works travelled abroad: he was the first scholar from Britain since Pelagius to have an impact on the world outside.*

Yet Bede's real importance was in helping to create a specific English consciousness. There was in him a deep, if gentle and unassertive, strain of patriotism and racial pride; he venerated the royal house of Northumbria and its achievements; he loved the English people and their language; and he had an overwhelming affection for their Church, now a century old. All these found expression in his *Ecclesiastical History of the English People,* which he completed about 731. It is perhaps the most remarkable work of the entire Dark Ages; in some ways it is a finer piece of scientific history than anything produced in the Ancient World. Bede not only possessed the critical faculties of a professional historian, he took great pains to exercise them. He understood the nature of evidence, the evaluation of sources, and the crucial importance of original documents. He sent to Rome for copies of letters in the papal archives, and reproduced them. He searched the library for relevant material and used it in a selective and judicious spirit. He got information from all over England, and interviewed old men who had taken part in the events he described; he tapped local and family traditions. His account is thus lit by flashes of colour and detail which only eyewitnesses could have supplied. Bede had the true humility of the scholar whose only object is the truth. He submitted his drafts to his informants, such as King Ceolwulf, and incorporated their factual corrections. He makes it clear to the reader that there are important *lacunae* in his materials, and indicates plainly when his statements rest on dubious authority. Bede takes the reader into his confidence, and inspires confidence in return. Succeeding generations of Englishmen, especially after Alfred had the text translated into English, felt strongly that here was the authentic record of their past, a true and fair

* But Bede has had a poor deal from the English. Owing to lack of English pressure, Rome did not recognise him as a saint until 1899; and even today he is honoured with the grotesque title of 'Venerable', which is also used indiscriminately of Anglican archdeacons while they are still alive. Bede met many posthumous misfortunes. Early in the eleventh century his bones were removed from Jarrow by a professional relic-thief called Father Alfred, and taken to Durham. They were rediscovered in 1104 in the coffin of St Cuthbert, and placed in a casket of gold and silver; but in 1541 they were scattered by an ignorant and rapacious reformer. After the fall of the Old English State, his works were more honoured on the Continent than in his own country; he was first printed in Strasbourg and Milan (1473). Though Bede was the epitome of moderate Anglicanism, he was foolishly classified by the English Reformers as an obscure Romanist. Collected editions of his works were printed in Paris (1544–5), Basle (1563) and Cologne (1612); England had to wait until 1843–4. But he has recently been made the object of a fine study, Peter Hunter-Blair: *The World of Bede* (London, 1970).

account of the events that had made England and the English what they were.*

Moreover, Bede invests his narrative and analysis with his own peculiar virtues. The English, he felt, were capable of great endeavour, but also liable to folly. His history is the story of what had been, but also of what might be, if the English learned to conquer their weaknesses and develop their strengths. Bede was a good as well as a great man, and a recognisable human character. We are touched when, at the end of his masterpiece, he diffidently inserts some scanty facts about his own life, and lists the books he has written. He was capable of anger: when brazenly accused of heretical opinions in one of his chronological treatises, he refers to his critics, in a letter intended to be read by his archbishop, as 'drunken rustics'. But he was essentially mild and unassuming. He held no high office, we may be sure, entirely by his own choice. He was kind and tolerant by nature; his instincts and training as a historian led him always to see both sides of the case. His beloved Northumbrians, he said, were wrong to attack the Irish: it was a deplorable act of aggression. His message was one of peace and compromise. Both the Celtic and the Roman Churches had merits. A great organisation like the Church must be wide enough to contain a Cuthbert as well as a Wilfred. Even a heretic was capable of virtue, and possessed rights. Though a monk himself, Bede was remarkably free from the aggressive self-esteem of his order: he not only recognised but advanced the proper claims of the secular clergy and the hierarchy. Strong opinions must be reconciled; argument must replace force, and itself be resolved by compromise. Only thus would civilisation be advanced. We can well believe Bede was loved by those who knew him. The famous account, by his pupil Cuthbert, of his last hours propped up in bed, dictating the final lines of his translation of St John's Gospel, may be an edifying invention, but it carries conviction all the same. It was in Bede's character to urge his scribe repeatedly to 'write faster', thereby rebuking those modern historians who fear to commit themselves to paper; and it was in character, too, to distribute his little stock of personal possessions – incense, writing paper – to the young monks clustered in his cell.

But Bede was not just a mild old scholar. He was a shrewd observer of contemporary events; he kept himself exceptionally well-informed; and he did not hesitate to express decided views on what was going on.

* Bede was the first person to think of himself as an Englishman. His writings transcend the tribal divisions and introduce the concept of 'England' as a political society, and 'the English' as a race. See the lecture by J. D. A. Ogilvy: *The Place of Wearmouth and Jarrow in Western Cultural History* (Jarrow, 1968).

When he sat down to write his long letter to Archbishop Egbert he felt he was nearing the end of his life, and what he wrote is an ecclesiastical and political testament of an old and wise historian, surveying the society he loved – and feared for. After giving the archbishop much sound, detailed and practical advice about the management of the northern province, Bede went on to express grave concern about certain developments in Northumbria. The Church had flourished mightily in its first century in England; and secular patronage of monasticism had enabled civilisation to flower. But there were in this process seeds of decay. Too many pseudo-monasteries, he said, were being created by the leading families and royal officials, with the object of exempting their lands from taxes and services to the State. This was not only tax-evasion but socially destructive. It was bad for the Church, for such monasteries brought it into disrepute, but even worse for the State, for young men needed to form the cadres of the army were unable to get land and raise families, and were going elsewhere in search of it. Bede's letter gives us a valuable insight into the reasons for the decline of the Northumbrian kingdom. It testifies to his belief that the interests of Church and State, properly conceived, were the same – to press one at the expense of the other would be fatal to both. Bede understood that the strength of Old English society rested on the ability of Church and State to work in the closest harmony. The State upheld the doctrines, and ensured the material prosperity, of the Church; equally the Church must reinforce the authority and efficiency of the State: that way lay progress for society as a whole.

And of course Bede was right. England was the first society to create a strong and civilised central authority on a permanent basis; it lies at the root of such felicity as this country has enjoyed throughout its history. It was what the post-colonial state of the British so conspicuously lacked; it was what the English were eventually able to create – so that William I inherited the oldest and strongest monarchical state in Europe; and it was made possible because a national Church, identifying itself with the public interest, underwrote the institution of popular monarchy. But Bede, with his historian's long perspective, was also aware that the process was far from complete. He recognised the fragility of his own country; and, perhaps imperfectly, he saw that the systematic exploitation of landed resources was the key to irreversible progress.

We come here to a little-understood point about the origins of English society. Looking back on English history from the last decades of the twentieth century, we lay too much stress on the development of sea-power, and maritime commerce, as the dynamic of English progress.

England's use of the sea to acquire wealth, power and influence has indeed been unique. But the strength of England, on which this expansion was based, lay in the land, and in the creation of a political and social system geared to agricultural advance. As a pre-colonial society, lowland Britain had made spectacular agricultural progress in the century before the Roman occupation. Under colonisation, that progress had been slowed down, halted, perhaps even reversed, because Britain was attached to an empire which was city-orientated; an empire unable to develop the technological and economic advantages of a city culture, and which financed its security and its civilisation from a stagnant and wasteful use of the land. Freed from the incubus of empire, the peoples of lowland Britain had a fresh opportunity to pursue the natural development of its resources. The Celtic British had rejected the Roman city-concept. They did not like cities, but used them for functional purposes. The new settlers from Germany and Denmark positively hated them. Except in Kent, where the settlers included elements from Frisia and Frankish territories touched by Roman influence, there was no continuity in city life. The English came from a race of forest-dwellers; their technology was the axe, the ox and the heavy plough. Resuming the work of the Belgic settlers – to whom they were akin – they settled in the river valleys and took up the task of exploiting the rich lowland soil.

But England in Bede's day was still caught in the offshore current of an empire which took centuries to die. Until the Arabs demolished its southern structure, and closed the Mediterranean to Roman Christian commerce, England was still to a significant degree part of a Continental, maritime economic system. The English settlers arrived here in open, oared ships, without masts and sails, or the keels which made these possible. These voyages took anything up to two months. The view that they then settled down exclusively to agricultural development is false. Wealth and state power could be created much more swiftly by the use of the sea, and by the trade carried on it. Almost exactly ten years before the Synod of Whitby met, the mourners of a pagan king of East Anglia, using an elaborate system of log rollers, dragged up from a creek near Sutton Hoo a great ship which they transformed into a cenotaph for their chief. His body was lost, at sea or in battle, but they placed in the ship before they covered it with sand and earth, a selection of his possessions. These included a monarchical standard (of a type Bede saw carried before the Northumbrian kings), a carved stone sceptre, and an abundance of gold and silver artifacts. One was a great silver dish from Byzantium, with an emperor's date-stamp; there were others of foreign manufacture, from a variety of places; foreign coins; and some beautiful English pieces, of great weight and elaboration. When the hoard was uncovered in 1939, an entirely new light was shed on English kingship and society in the mid-seventh century. The progress of the English had been much more rapid and spectacular than anyone had hitherto thought possible. But what has only recently been appreciated is that such regional societies were the products essentially of sea-power and a sea-culture.* They owed their wealth and culture essentially to maritime trade; they were still, economically, part of a decaying Continental system, the sub-Roman empire of the West, with its offshoots into the Mediterranean. What was true of East Anglia was true also of Kent, of the kingdoms of the south coast and the Thames valley, still more perhaps of Northumbria. The civilisation in which Bede flourished owed its dynamic – because it owed its communications, its contacts, its wealth – to the sea. Only when the Roman-style economy finally dissolved, in the aftermath of the Arab conquests, did the English shift their economic axis inland, and find their true basis for development in the exploitation of the land.

In 795 King Offa of Mercia received an unpleasant letter from the Emperor Charlemagne, as he called himself. These two men were each supreme in their own regions, paramount kings, and they corresponded on a level of solemn equality, though Offa had to insist rather more sharply on his dignity than Charles. They had engaged in intermittent trade-war, shutting up their ports to each other's ships. Now Charles summarised the matter: if the English complained about the stones sent from France (probably from Tournai, for use in church fonts), then he had an equal right to complain of the shortness of the cloaks sent from England. Charles, as we know from another source, hated these mini-cloaks. They were too short, he said, to cover him in bed, or to protect him from the rain when riding, and they were a nuisance when he went to the lavatory. What is so striking about this dispute is the evidence it provides of the poverty of the economic contacts between these two major states. Maritime trade had ceased to play a significant role in either. The old Roman economic system had broken up, and nothing had replaced it. Societies were turning inward, to the land; the instrument of economic progress – indeed of all progress – had changed from the sea- and river-port to the manorial estate. This

* The best recent survey, for the general reader, of the Sutton Hoo finds and their significance is Charles Green: *Sutton Hoo, the Excavation of a Royal Ship Burial* (London, 1963).

process was far more significant in England, cut off from the rest of the Continent by water, than anywhere else. It intensified the isolation, and made the development of distinctive national characteristics much more profound, and rapid. Power shifted from the littoral societies of the east and south and became balanced in the midlands. It is to Mercia, in the heart of England, that we owe the true origins of the nation, its institutions, its language and its attitude to public life.

The change was clearly marked by a reconstruction of the monetary system. England moved from a gold to a silver standard. This is evidence not so much of declining wealth, but of growing common sense. Gold was the exchange-medium of the international merchant, silver of the progressive farmer. Gold coins had been minted in England in pre-colonial times; local minting had been resumed in the late seventh century; but this had been a function of a littoral economy, the last stigma of post-colonial status, what we would now call neo-colonialism. The economic independence of the new English State was symbolised about 780, when Offa issued a regular silver penny, at a standardised weight of 22 grammes. The name was ancient, of unknown origin. But the new currency was, from the start, essentially modern in concept and execution. Offa's penny was to hold its quality for 500 years. It was the basis of all later improvements. In the tenth century, Edwin of Wessex adopted the device of calling in and re-issuing all coins at regular intervals; foreign coin was melted down and re-struck; mints were farmed to professional moneyers, but they were obliged to put their names on the reverse, and penalties for debasement were heavy and ruthlessly enforced. Thus England developed a currency which the most powerful of Roman emperors would have envied. It became a recognised medium of exchange from Scandinavia to the Balkans, and a ubiquitous, ocular testimony to the stability and wealth of the English state. A puritanical devotion to their currency has always been a salient feature of the English public consciousness.

Yet in one sense this strong currency was merely the consequence of two even more deep-rooted English characteristics: the use of the land as the ultimate index of wealth and status, and a marked preference for strong, efficient and honest government. By creating a state which gave them expression, Offa laid the foundations of English public life. It is a tragedy we know so little about him as a man; modern England probably owes more to him than to any other individual. But of course he built on earlier foundations. England was born of a fortunate marriage between geography and race, between fertile lowland soil, and hard-working Germanic immigrants. England created the English; it was the land which shaped the people. Though well aware of their

Germanic history and traditions, the English settlers were bound much more firmly to the soil they acquired. Their arrival was framed in heroic legend, but this was background and entertainment; the reality of their lives was dominated by farming. Theirs was essentially an agricultural, not a military, conquest. When the Normans came, half a millennium later, they remarked that 'the English thought of nothing so much as the cultivation of their lands'. Their forebears had been industrious and energetic farmers in Germany. Now, in England, the opportunities which the countryside gave them were eagerly seized. The farming patterns of the Britons were largely rejected. They marked out their own fields and villages, established their own methods of communal production. The plough-team of up to eight oxen was the biggest single factor in the shaping of Old English society, for it was crucial to their methods. The salient feature was the exploitation of huge open fields, often of a hundred acres, on a social basis. The low-land soils demanded a heavy plough and a powerful team to pull it. Yet few men possessed a whole team; equally, a team, to reach its maximum efficiency, required vast fields to be ploughed in strips. Thus the operational needs of the team became the units of measurement, of ownership, of wealth. An acre was what one family could plough in a day, a hide the area that gave them work, and livelihood, for a year. Ploughing dictated the need for huge fields; this in turn meant a measure of communal effort, for the phases of the agricultural cycle had to be coordinated and jointly determined. But within this communal structure, individual ownership, rights and wealth were fiercely upheld and narrowly calculated. A man's land, and his share of the crops, depended on his contribution in working capital and labour; he might supply a plough, or a team, or both, or a share of either; and he drew his rewards accordingly. His obligations embraced many tasks besides ploughing, and there were strict penalties for failure to discharge them, as the earliest laws make plain.

It was a mixed economy: but the element of private ownership was there by choice, that of communal effort only by necessity: a typically English approach. There were many and increasing gradations of wealth. At one end of the social scale there was a large slave class, composed initially of Britons, but augmented by convicts, captives, human purchase and degradation. Even peasant farmers, with their single hide of land, owned slaves; like oxen they were part of a man's working capital. A man was free by virtue of his ownership of land; nothing else mattered. A landowner was entitled a thegn if he owned five hides or more; no matter how rich he was (except in the case of sea-going merchants) he could not claim rank 'unless he hath the land'. The lord of the village, or the manor, which were often coterminous, was by virtue of his estate the symbol of ownership, the guarantor of protection, the chief arbiter of opinion (and so of justice), and the agent of authority. Often the village bore his name, and does so to this day. Ownership of land was the key to the legal system, for it determined both the nature of crime and the methods of law-enforcement.

Old English society was preoccupied with two categories of offences, both with agricultural roots: murders and blood-feuds arising over tenure and boundary-disputes, and the theft of cattle. The first was settled by the payment of wergilds which varied with ownership: 100 shillings for a free farmer, equivalent to the 100 oxen he was supposed to be worth; 600 shillings for the landed gentry. There were intermediate categories and local variations, but the principle of compensation was always related to notional concepts of landed worth. The determination of guilt, in both categories, rested on a man's oath, whose value, again, was related to notional concepts of his agricultural status and property. If a man's oath were not equal to the magnitude of the charge, others had to swear for him, and his accusers would do the same; elaborate computations were made of the value of conflicting testimony to determine the verdict: we come across phrases like 'an oath a pound in value' and 'let him deny it with an oath of three twelves'. The system was thus squarely based on popular concepts of natural justice, for what could be more obviously fair than that a man should rest his case on the sworn evidence of people who knew him, were in a position to watch his daily movements, and whose desire to uphold local stability and order was *ipso facto* guaranteed by their ownership of property in the district? Such a realistic concept, moreover, ensured that the social system remained flexible and never acquired the rigidity of caste: judicial worth, and therefore status, had to be determined not by birth but by current landed possessions. There was at the heart of Old English society a tremendous dynamic to get on by exploiting the land, and all institutions were geared to keep this dynamic alive.

Yet the real genius of the English consisted in harnessing this dynamic to the functions of the State. How could this self-regulating structure of village life, with its built-in economic impetus, be reproduced at national level? It was at this point that the English Church, with its close identification with the secular authorities, made a decisive contribution. The Church had inherited from the Roman Empire instruments which Germanic society conspicuously lacked: the ability to construct an ordered and regular hierarchy of command, and operate within its limits; not only literacy, but written law and documented transactions, particularly the land-deed; the impulse to delve below habitual custom to first principles; and a cosmopolitanism which accelerated the flow of ideas. The Celtic Church lacked these gifts; thus the Whitby decision gave England something unobtainable from native intellectual resources. It is no accident that the administrative developments in Mercia followed the missionary activities there of Wilfred, the ablest of the Romanists. The Church became the principal instrument of civil government; the bishops were the King's chief advisers, his chapel the centre of administration and record-office, his chaplains civil servants as well as spiritual ministers. The Church codified the law, and put it in writing. Even before the Church came, English society was developing a definite structure: but the Church supplied the literate manpower and expertise to build a State machine.

We see the process at work in eighth-century Mercia. The key to all State authority is finance, the means to purchase the power of compulsion. This, in turn, depends on the regular collection of adequate taxes. And taxes require a currency realistically related to the working of the economy. Offa's establishment of a silver-standard coinage as the regular medium of exchange between farmers was thus the first in a long chain of events which built up a mighty state. But taxes must not merely be imposed; they must be seen to be justly imposed. Their efficient, comprehensive and equitable collection is the foundation of healthy and stable government. Such a system demands knowledge and documentation. The English grasped this very early in their history, and it has remained for them a central preoccupation. The foundation was laid in eighth-century Mercia in a document known as the Tribal Hideage; it sets down the number of hides subject to tax in every province of the state.* Without an accurate basis of assessment, any tax is a selective tyranny, and its collection incompatible with the growth of free institutions and of government by consent. Such records are difficult to compile; it was here that the help of the Church was vital. The Hideage was the first giant step towards modern government. It contained, in embryo, the concept of a territorial pyramid: from the village to a district of a hundred hides, from hundreds to shires, from

* It was very likely based on earlier models, since the *bretwalda* was an overlord who exacted tribute from vassal kingdoms, assessing their liability according to an overall system. Presumably, then, the earlier *bretwaldas* had hideages; and it is possible that Bede used, for reference, a Tribal Hideage compiled under the bretwaldaship of one of the Northumbrian kings. But the Mercian hideage was unique. 'As a precise cadastre and as a detailed record of historical topography, it has no parallel for its period in the whole of western Europe. . . . The whole work bears the stamp of authority, combined with great administrative ability': see Cyril Hart: 'The Tribal Hideage', *Transactions of the Royal Historical Society*, November 1971.

shires to nation. The concept was refined later to produce a Burghal Hideage for towns, and a County Hideage for shires; and this last, in turn, made possible Domesday Book. Domesday Book adumbrated the growth of the Exchequer, and its characteristic instrument, the pipe-roll, which survived as the record of central finance until 1832, when England was already a great industrial nation, and the heart of a world-empire. This crude summary, of course, ignores an infinite multitude of complexities; but it is still true to say that the rural society of eighth-century Mercia developed the matrix of modern England.

The strength of Old English society was thus based on a well-informed central authority, which used its knowledge to pay its way. But the England of the English was still highly vulnerable: a million people sitting on some of the best land in the world, developing it steadily to make it a still more tempting target for the violent and predatory forces of north-west Europe. When England turned inwards in the eighth century, it became essentially a civil society of farmers. The English manor never became a military institution; not even the Normans, who were geared to little else but warfare, could make it one. The Channel and the North Sea provided powerful natural barriers to aggression; but both could be crossed, and they constituted a standing temptation to ignore the unpleasant and expensive realities of a world ruled by force. The English never developed a professional army. Except in brief moments of extreme crisis they could not even produce an amateur one able to keep the field. Their efforts to create a navy nearly always ended in lamentable failure. It was not that the English lacked aggression; they have always been among the most aggressive peoples on earth. But they seemed incapable of any sustained attempt to harness their aggression to a national purpose. They accepted the concept of a national defence force. They had the administrative machine to produce it on an equitable basis – an armed man for every two hides, making the fyrd equal to about 1 per cent of the total population. The conscripts, with much reluctance, would assemble; they would even fight fiercely, if battle was not delayed; then their only thought was to get back to their farms, and their blood-feuds. For most of the time the English State was, for all practical purposes, disarmed. The wonder is that the English contrived to survive at all. They might so easily have become another lost people of history. There was absolutely nothing inevitable about their durability. In the ninth, tenth and eleventh centuries they were the victims of overwhelming aggression. Why were they not extinguished? There is no simple answer. History is not propelled by single causes. The English were saved once by a great man, once by the cunning and resourcefulness of their ruling class, and once by the resilience of their institutions and their language. Each episode is worth examining.

In 865 a Scandinavian army of unprecedented size moved into England with the object of setting up a permanent system of exploitation. In the next 13 years it destroyed all the English kingdoms except Wessex, and in most districts began the partition of the land for settlement. In 878 the odds were overwhelming that English civilisation would be destroyed; that its forms of government, speech and culture would disappear; that an alien ruling class would be established and a mass-migration take place under its aegis; and that the English would survive, like the Britons before them, only as a servile class, gradually adopting the dominant culture. When Alfred took refuge with his personal followers in the marshes west of Selwood, this was the imminent prospect facing his country and people.

But a civil society based on a degree of consent has enormous reserves. It is one of the most comforting lessons which history teaches us. The resources of civilisation are not easily exhausted. A society banded together for aggressive purposes, whose ethics, criteria and hierarchy are exclusively military, led by men whose status rests solely on force, possesses great initial advantages. But its strength is more apparent than real; it has no self-sustaining moral authority, no internal discipline other than violence; it can satisfy only a limited spectrum of human desires; it is inherently corrupt; it possesses no collective wisdom, except in the narrow field of military expediency; it can tolerate no freedom of discussion, and therefore has no capacity to respond to changed conditions; its victories generate anarchy, and its defeats despair, for it has nothing worth-while to defend. By contrast, a civil society can more easily survive setbacks and learn from them; it has a sense of righteousness which breeds determination and, if necessary, unparalleled ferocity: it confronts instinct with reason, formulates long-term policies and new forms of discipline and organisation. Once grant it a breathing-space, after the initial shock, and it will quickly develop a strategy of survival and forge the instruments of victory. In the long run it holds all the moral and intellectual cards, and these are decisive in combination.

But the breathing-space is vital; and it is usually left to an individual to make it possible. There is always a role for a great man in the clash of collective forces; no one who studies English history can be in any doubt on this point. The opportunity exists; the moment is ripe; the resources are there; but unless the man to set them in motion is available, the occasion will pass, and perhaps never recur. One solitary person, with clarity, single-mindedness, energy and will can thrust his shoulder against the hinge of history, shift the equipoise, and thus accomplish the work of multitudes. In retrospect it looks inevitable, but without him it would not have taken place. Such a man was Alfred. The legends which surround him cannot obscure the extraordinary facts of his life. As we study them, we feel at times that he was taking upon himself the responsibilities of an entire nation: saving the state, rescuing civilisation from ruins, building a fleet, organising a system of urban defence, creating a militia, setting up a diplomatic service, educating a ruling class, importing scholars, transforming his court into a centre of learning, administrative innovation, and systematic justice – doing all this, as it were, with his bare hands. Whenever the documents allow us to glimpse him at close quarters, we see an essentially solitary figure: harassed by a multitude of worries, overburdened by conflicting demands on his time. In one letter his bedroom is shown invaded by a pack of arguing litigants; he looks up – he is washing his hands – and gives a cool and sensible judgment. He is always thoughtful, with the originality of a man who has come to education late, has received no packaged opinions, and has worked things out for himself; an ingenious man, forced by events to devise solutions to entirely new problems. A naïve man, in some ways, and an eccentric – the first English eccentric – designing curious mechanical gadgets. Not a man ever allowed to relax for long from the most crushing cares of State, but one whose thoughts were none the less haunted by the deepest mysteries of existence. What is life? Why are we here? What, then, must we do?

Alfred was not a lucky man. Most of his life he was sick.* His family background was first strained, then tragic. His father, then in succession his three elder brothers, died at brief intervals. He had virtually no education as a child. He inherited no advisers of any ability, and always had difficulty in finding trustworthy subordinates. The machinery of the State was running down, and he had to rebuild it. He won no easy victories. He devoted his whole adult life, with many setbacks, to securing the minimum of national security. What is remarkable about his achievement is not its magnitude but the means he employed. We see him on the one hand as a successful soldier and administrator; on the other as a man of wide tastes who brought about a renaissance of English civilisation. From the standpoint of our age, the two roles seem

* His disease, said Asser, baffled all the doctors. It may have been epilepsy, or a violent skin infection (very common among the Old English) or piles. See Wilfred Bonser: *The Medical Background of Anglo-Saxon England* (London, 1951), pp. 109–10.

incompatible, almost in conflict. They would not seem so to him. A king would, he felt, be a better soldier and administrator by acquiring the civilised disciplines and applying them to his public functions. It was his theory and his practice of government. A realm was not worth defending, unless it itself defended worth-while things; standards and honourable conduct mattered as much as life and property. An enemy must not merely be defeated but reformed, and induced to come within the lighted circle of civilisation. He never appears to have felt any racial hatred for the Danes, or contemplated a war of extermination. Perhaps he realised that a Scandinavian element in English society was now inevitable; it must be absorbed on a basis of peace, in which the alien presence would be made acceptable by acquiring first the veneer, then the substance, of English culture. And the conductor in this process was Christianity. Attempts to tame the Danes by baptism were common enough; the Danes complied, when forced to, and sneered afterwards. But Alfred saw that the method, if pursued with patience and persistency, would work in the end. Treaties sealed by baptism might be broken; but it was sound policy, as well as Christian duty, to use diplomacy as well as war; every respite could be put to use, not least for military ends. He had often to revert to war; but each time his sense of purpose was clearer, his means more adequate, his strategy more decisive.

What, in effect, Alfred did was to apply the Mercian concepts of civil administration to the business of winning a war, and thus impel the State to take a giant leap forward in sophistication. He operated a regular budget, for the first time, and placed public responsibilities – for the army, for the fleet, for the construction and defence of fortified towns – on a systematic basis of shared responsibility. These measures created the infrastructure of a united kingdom, as much by the process of putting them into effect as by the security they provided. In 886 his forces entered London and, says *The Anglo-Saxon Chronicle*, 'all the English people submitted to Alfred except those who were under the power of the Danes'. As for the settled Danes, Alfred grasped the point that the very process of acquiring real estate, a stake in the country, was a solution to the problem of perpetual warfare between them and the English – they now, like the English, had a great deal to lose. His final treaty with the Danes not only demarcated the frontier but interlocked the legal systems by establishing an agreed scale of wergilds: the effect was to produce a degree of inter-racial harmony at a personal as well as at a State level, and so expose the Danes to English cultural penetration. Alfred never seems to have doubted that, under the rule of law, English civilisation could absorb the Danes without resort to force.

His treaty was an early example of English confidence in the power of diplomatic effort.

Alfred indeed seems to have reposed unlimited faith in the civic virtues. He believed in the moral authority of a civilised people. In law, that is the moral framework of government, the King was necessarily the final arbiter, but his decisions were not arbitrary; he merely judged whether or not the law had been observed. The law itself evolved from the collective wisdom of many men; the King codified it, and in that sense it became his law; he might even create new laws, but this was done in consultation with his council or *witan*; and it was a prerogative exercised sparingly, and not necessarily binding on his successors. That the King felt himself to be subject to the law is made touchingly clear by Alfred's will, in which he is at pains to show executors and posterity that his dispositions are fair, just and legal; and in setting out his own law-code, he is admirably succinct on how he thinks the legislative process ought to work:

> [Holy bishops and other distinguished wise men] in many synods fixed the compensation for many human misdeeds, and they wrote them in many synod-books, here one law, there another. Then I, King Alfred, collected these together and ordered to be written many of them which our forefathers observed, those which I liked; and many of those which I did not like I rejected with the advice of my counsellors, and ordered them to be differently observed. For I dared not presume to set in writing at all many of my own, for it was unknown to me what would please those who should come after us.

Thus to the Mercian creation of an equitable system of direct taxation – the basis of effective and acceptable government – the kingdom of Wessex added a clear doctrine and practice of legislation. The law was ultimately based on collective custom; inspired by rational concepts of evidence, proof and fair play; tidied up by men trained in methods derived from imperial Rome; added to by King and council as and when required; but never ossified – always left open to revision and repeal and augmentation. Here, at the end of the ninth century, we already have an organic structure of public affairs; the political and legal pattern is established, infinitely capable of growth and development, but conditioning and controlling the process with great tenacity. The structure could and did absorb alien elements, but it could no longer be fundamentally changed. The primary cause of the continuity and stability of our offshore island's history was the strength of Old English society. The strength was based on an accurate balance between the needs of the State and the rights of the individual; and the balance, in turn, was were maintained by a law to which all, the State included, subject – a law founded on custom and modified by consent.

With this system Alfred gave the English people, and the British lowlands they inhabit, a unity which they had never before possessed, and which they have never since lost. His successors extended this new unity to the entire English territory. But it was a unity based on, and strengthened by, diversity. Alfred won the allegiance of Mercia and Northumbria by diplomacy, conciliation and by the devolution of authority; his laws incorporated elements from Mercia and Kent as well as Wessex, and respect for local custom was a salient principle of his rule. The Danish settlers, too, as they entered into the kingdom, kept their own organic elements of law and tenure.

Thus English unity was created not by force but because men were persuaded, by a political genius who was also a transparently good man, that they needed it. To Alfred unity was, I think, more a cultural matter than something which revolved around race and politics. Coming to literacy and learning late in life, amid the terrible pressures of an active and anxious career, it seemed to him a miraculous gift, a window into a better and purer world, which it was his duty to share with all. Of course it had its practical purposes: it was essential to the administration of just law, which Alfred rightly recognised is the foundation of human happiness in this world. It was fear of the King's rebukes, says his biographer Bishop Asser, that made

> ... the ealdormen and reeves hasten to turn themselves with all their might to the task of learning justice ... so that in a marvellous fashion almost all the ealdermen, thegns and reeves, who had been untaught from their childhood, gave themselves to the study of letters, preferring thus toilsomely to pursue this unaccustomed study than resign the exercise of their authority.

But Alfred certainly did not regard learning as merely utilitarian, nor the exclusive right of the ruling and administrative class. He 'with great care collected many nobles of his own nation and boys of humbler birth and formed them into a school'. The learning was cosmopolitan. Few of Alfred's cultural advisers came from Wessex; four, including Plegmund, whom he made Archbishop of Canterbury, were Mercian; one imported scholar was French, another German; and Bishop Asser himself was Welsh. Alfred corresponded with a wide range of foreign scholars, including the Patriarch of Jerusalem, a professional beggar on behalf of his see, who got money from the King in exchange for medical recipes. The culture Alfred tried so desperately to embrace was the universal culture of the ancients and the fathers, expressed in Latin; it was a dramatic moment for him when, in 887, he first began to make sense of the language. But he had been literate in English since the age of 12, and he recognised that, except for the minority, this must be the language of cultural progress. So he began an elaborate programme of translating key texts into English, taking a leading part in the task himself. As he says in a letter to the Bishop of Worcester, translations were not made earlier because scholars could not believe learning would fall into decay. They could not have foreseen the terrible events of his own lifetime. An English literature was a necessary guarantee against future catastrophe, and the only means by which large numbers of people could get at the truth:

> Therefore it seems better to me . . . for us also to translate some books which are most needful for all men to know into the language which we can all understand, so that we can very easily bring it about, if we have tranquillity enough, that all the youth now in England, if free men who are rich enough to devote themselves to it, be set to learn as long as they are not ready for any other occupation, until they are able to read English writing well; and let these who are to continue in learning, and be promoted to a higher rank, be afterwards taught more in the Latin language.

Alfred could not foresee that the humble tongue he thus encouraged would in time wholly supersede Latin as the international language of culture and scholarship, would conquer the world as the chief vehicle for political, economic, scientific and technological advance, and poise itself for its ultimate role as the first universal language, spoken and written by countless millions in countries he did not even know existed. He would have rejoiced at this astonishing prospect. He believed in the English. He had Bede's famous *History* translated, and he sponsored a systematic record of English events which, in its various texts, we know as *The Anglo-Saxon Chronicle*. Such works were made in multiple copies, as were the laws and other public documents, and sent all over the kingdom to be preserved for use and reference in the libraries of cathedrals and monasteries.

What a strange man Alfred was; an archetype of all that was best and yet most mysterious in the curious race whose destinies he helped so decisively to shape. We can trace the development of his own thoughts from his early, literal translations, to the much freer ones towards the end of his life, in which he interpolates fragments of knowledge from visitors to his court, whom he subjected to relentless questioning, and his own private reflections. It is an extraordinary privilege thus to be allowed to peer into the mind of this great king and man of action, who has been dead more than a thousand years. In some ways it was a clumsy mind, grappling awkwardly with abstract concepts which were beyond it, and taking refuge in laboured metaphors. Alfred came from a society which had advanced its economy by the conquest of the forest, and his images revolve around wood: wooden ships, wheels, buildings. Here is the practical, English mind, trying to come to terms with Latin abstraction. Behind it lies a fierce energy, an implacable refusal to accept defeat, the same obstinacy and resolution which marked his public life. There must be answers to the deepest questions of existence, as there were practical answers to military, naval, judicial and administrative problems. 'I would know,' he wrote, 'whether after the parting of the body and the soul, I shall ever know more than I now know of all that which I have long wished to know; for I cannot find anything better in man than that he know, and nothing worse than that he be ignorant.' Or again: 'Man must increase his intelligence while he is in this world, and also wish and desire that he may come to the eternal life, where nothing is hid from us.'

We can detect, in this endeavour to enlighten the earthly world, a strain in Alfred which transcended the limitations of the medieval Christian mind, which imprisoned man by insisting that material progress was futile, and release would come only through eternity. Alfred was to all appearances an orthodox Christian, but he had in him the instinct of his native Pelagianism: a belief in free will and in the ability of humanity not indeed to perfect itself but to raise its status and improve its condition. One day the Renaissance and the Reformation would break down the prison walls; in the meantime Alfred already sounded a new English note of earnest moral conviction that the prevailing darkness could be pushed back. If his characteristic tone is pessimistic, it is relieved by the hope which comes from struggle and achievement, however incomplete. 'I can understand little of Him, or nothing at all, and yet at times, when I think carefully of Him, inspiration comes to me about the eternal life.' The battle for knowledge in Alfred's life mirrored the battle to preserve a kingdom and rescue a civilisation. He might have said, 'Wessex has saved herself by her exertions, and England by her example'; as, centuries later, England would repeatedly save Europe and the world. Most of Alfred's work achieved fruition only after his death; he certainly found no tranquillity in this life; we get hints of a certain weariness, of impatience and anger with officials who failed to carry out his orders, whether in building fortresses or administering the law. Alfred never seems to have possessed lieutenants (except his splendid daughter Æthelflæd, and her husband Ealdorman Æthelred of Mercia) who measured up to his own standards of responsibility. This is not surprising; he would have been outstanding in any age or society. But greatness makes for loneliness. Alfred yearned for men of stature to share his burdens, and thought anxiously of the future after he was gone. By the time he died, in his mid-fifties, he was – like all great English monarchs – a very tired man. Using his familiar metaphor, and with a final, sad phrase which catches at the heart, he left his gospel of work and aspirations for English posterity:

> Then I gathered for myself staves and props and bars, and handles for all the tools I knew how to use, and crossbars and beams for all the structures I knew how to build, the fairest piece of timber I knew how to carry. I neither came home with a single load, nor did it suit me to bring home all the wood, even if I could have carried it. In each tree I saw something that I required at home. For I advise each of those who are strong and have many waggons, to plan to go to the same wood where I have cut these props, and fetch for himself more there, and load his waggons with fair rods, so that he can plait many a fair wall, and put up many a peerless building, and build a fair enclosure with them, and may dwell therein pleasantly and at his ease, winter and summer, as I have not yet done.

In the year 1014, Archbishop Wulfstan of York preached a remarkable sermon in his cathedral. It must have made an immense impression on those who heard it; it was repeated on several occasions, and, under the title of 'Sermon of the Wolf to the English', was written down and copied in many manuscripts. It is the first recorded instance we have of a dramatic and sombre appeal to the English to save themselves from destruction, and thus part of a long tradition, running through the speeches of Henry V, Elizabeth and Pitt, and culminating in the great Churchillian broadcasts. Wulfstan was a formidable personage: the leading churchman of his age, an experienced legalist, a secular statesman, and the unofficial head of that powerful and mysterious body which we can, for the first time, dimly perceive: the English establishment. A century and a half before, England had been saved by a great king. Now it was to be saved by a class, and the man who spoke for it. In the interval, the unitary kingdom established by Alfred had acquired all the accretions of stable and ancient authority: sonorous titles for its monarch, an elaborate coronation service at which he was invested with them; a proliferating hierarchy of honour, office and wealth; traditions and ceremonials which already inspired foreigners with awe. But it had not acquired lasting military security: this was something beyond the capacity of the Old English State to achieve. Wulfstan was profoundly aware that England's ultimate defence lay in the integrity of its civilisation – the system of laws and government, of public and private standards, built up on the work of Offa and

Alfred. This could be fatally diluted, not by an infusion of race, for the English could always cope with that, but by the heedless acceptance of alien modes of conduct. It was not the swords of the heathen he feared so much as their lack of probity. Wulfstan inspired a sinister entry in *The Anglo-Saxon Chronicle* for the year 959, in which he accused King Edgar of having too much truck with the alien world: 'He loved evil foreign customs . . . and attracted hither foreigners and enticed harmful people to this country.' In 975 occurred a national catastrophe. Edgar died unexpectedly, leaving sons by different mothers, and the elder, Edward, only in his teens. He quickly became involved in a violent conflict with a group of landowners, almost certainly on the issue of tax-evasion through the endowment of family pseudo-monasteries – the very evil against which the aged Bede had warned 250 years before. Three years later he was murdered by the servants of his young half-brother Æthelred, who thus became the beneficiary (if not the author) of the worst State crime in Old English history.

This terrible event led to a steady and in the end dramatic decline in English public standards. Æthelred took no steps to find and punish the murderers. The episode cast a lengthening shadow over his reign, and the presence on the English throne of a compromised king inevitably attracted hostile foreign attention. In 981 the Danish raids were resumed. Loyalty was the salient principle of the Old English State. Undermined by the throne itself, it collapsed. Some of Æthelred's own appointees changed sides several times. He himself married, as his second wife, a Norman princess, Emma, of barbarous Norse forebears. Great private landowners made their own arrangements with the invaders. The Danes themselves were often disloyal. One of Æthelred's few successful commanders was a Danish deserter, and his son and heir, Edmund Ironside, drew more effective support from the men of the Danelaw than from Wessex itself. It became increasingly difficult for men to know where their true interests and allegiance lay. Money replaced patriotism as the instrument of national survival. In 991 Æthelred bought a peace treaty with the Danes for £10,000. Until then to pay Danegeld was not necessarily dishonourable or imprudent; Alfred himself had sanctioned the practice as a useful expedient to gain time. But Æthelred made it into a principle of government: the sums demanded rose to £16,000, to £24,000, to £36,000 – in 1012 to the colossal figure of £48,000. It was pointless to blame the Danes for breaking these treaties; the English King himself was equally unscrupulous. In 1002 he ordered his Danish hostages to be slaughtered. Such a policy would have seemed inconceivable to Alfred, with his policy of combining firmness with reconciliation. Moreover, it was not merely a crime but a blunder, for among those killed was the sister of Swein, King of Denmark. Her murder persuaded him to turn large-scale piracy into a national invasion and seize the throne itself – a project triumphantly completed by his son, Cnut.

The *Sermon of the Wolf* accurately reflects the prevailing atmosphere of broken morale and national self-abasement. Wulfstan paints a devastating picture, in considerable detail, of the collapse of the social system. He speaks of 'wavering loyalties among men everywhere'. He says that 'too often a kinsman does not protect a kinsman any more than a stranger', that men sell their relatives into slavery, that women are openly purchased, girls and widows forced into marriage for money, thegns reduced to slaves, and slaves, by desertion, become lords. Self-respect has been lost:

> The English have been for a long time now completely defeated and too greatly disheartened through God's anger; and the pirates so strong with God's consent that often in battle one puts to flight ten . . . and often ten or a dozen, one after another, insult disgracefully the thegn's wife, and sometimes his daughter or near kinswoman, whilst he looks on, who considered himself brave and mighty and stout enough before that happened. . . . But all the insults which we often suffer we repay with honouring those who insult us; we pay them continually and they humiliate us daily.

Wulfstan had been brought up in the English tradition which relied on a strong central government to secure the safety of the realm and the health of society; and that government was embodied in the royal line of Wessex, already the oldest and most distinguished in Europe, occupying a throne with a longer continuous existence than any other Christian institution, except the papacy itself. He had thought deeply about kingship, and the qualities required of the men who discharged its duties. He had written a book on the subject, his *Institutes of Polity*, the first original English work of political theory. When writing Æthelred's laws, Wulfstan had placed tremendous emphasis on the dignity and power of the office. The King was Christ's earthly vicar in the realm of England. He was, to use an expression later employed on Edward the Confessor's behalf, the judge set up by God to rule Church and State and arbitrate between them. What, then, was to be done when the King was inadequate or betrayed his office? The only answer was that those around the King, the wise clerics and substantial lords of his kingdom, should by one means or another, act in concert. In 1012–13 occurred the first significant constitutional crisis in English history. The territorial aristocracy refused to lead the levies into battle unless Æthelred attended in person. He left the country and fled to his wife's relatives in Normandy. He was eventually permitted to return, but only on condition that he signed a document explicitly promising wholesale reforms in his methods of government. There can be no doubt that Wulfstan was the controlling agent behind this solution. It adumbrates Magna Carta almost exactly by two hundred years. This bargain between King and subjects introduced a new principle into the system of English monarchy, which henceforth was never allowed to lapse entirely. It illustrates the axiom, by no means confined to England, that military disaster is the father of constitutional change – by consent in England, by revolution elsewhere. For the first time government had become contractual, and the concept of a commonwealth was born.

Under Wulfstan's guidance, England survived not only this crisis, but the death of Æthelred in 1016 and his son a year later. There were now only two sources of authority in the country: the English establishment, and the impending military tyranny of Cnut. But Wulfstan was an audacious man. He decided to marry the first to the second and invest a gifted savage with the apparatus of constitutional English regality. Once again, the resources of English civilisation were not exhausted. The powers of the English monarchy were there to be exercised; the administration still existed; the Church was still the repository of learning and the link with the international civilised community. Wulfstan was a smooth exponent of the wiles of the establishment, adept at compromise, able to flatter a powerful outsider out of his senses, willing to take on the job of taming a barbarous Danish warlord, as his kind would later tame socialist cabinet ministers. Cnut was an apt, indeed eager, pupil. His Christian background was uncertain; but he recognised that enthusiasm for the Church was the mark of civilised statesmanship, and he adjusted his religious ideas accordingly. He was only too willing to submit to the guidance which Wulfstan gracefully proffered. Thus the old English prelate and the young Danish general went into partnership together, and one of the most successful experiments in English history commenced.

The truth is Cnut was as anxious to come in out of the cold as the English were to receive him in their warm places. They wanted peace; he wanted to become respectable. They thought they could do a good civilising job on him; and they were right. Cnut felt the time had come to wipe the blood off his hands, and learn a new trade as a civilised ruler. No upstart adventurer has ever settled down more complacently with a rich heiress of ancient lineage. Cnut wanted power, but he also wanted to go up in the world, to be recognised as a great Christian gentleman as well as a warrior. He yearned for the flattery of the warm south, as have so many of his race since. To sit on the English throne, as its recognised, conformist and legitimate tenant, was the key which unlocked all these doors. The instincts of a ruffian remained: he quickly disposed of, without trial, several inconvenient relics of the old reign. But he then proceeded to become an enthusiastic English monarch. He dutifully married Æthelred's widow, no easy assignment. He cut military expenditure and reduced taxation, always a high road to English hearts. He allowed Wulfstan to codify the laws in such a thorough and comprehensive manner that the text was still regarded as an authority in twelfth-century England. When in Denmark, in 1019–20, he delighted the English by sending them an open letter, to be read at the shire courts, in which he reported progress and gave instructions for the laws to be justly enforced. His impeccable behaviour survived Wulfstan's death. In 1027 he went to Rome for the coronation of the Emperor. He was given a splendid reception, not only by the Pope but by the assembled European dignitaries, which he rightly guessed was due more to his status as an English king than to his reputation as a northern warrior. His letter from Rome to the English people naïvely records his pleasure at this honour, and also lists certain important commercial advantages which he was able to negotiate with the Emperor, the Duke of Burgundy and the Pope, releasing English merchants from irksome tolls. He was a great credit to the old archbishop. He kept a modest but effective fleet of 16 warships. He promoted English trade. He told the people what he was doing. He advanced Englishmen, rather than Danes, to positions of authority, so that by the end of his reign his government personnel was almost exclusively English. He had all the qualities of a popular English monarch. If the national game had existed, he would doubtless have played cricket too. He thus became a revered, semi-mythical figure for the offshore islanders, who were capable, then as now, of rewriting history while it is still happening. In fact he was a creature of the English establishment. A barbarous king, who was not surrounded and protected by able ecclesiastics, speaking his tongue and wholly creatures of his making, had no chance in confrontation with the Old English State. When Cnut's line died out, the English quietly put the House of Wessex back on to the throne. It was as though the Danish monarchy had never been. The Danish settlers became Englishmen, making their own distinctive contribution to our language and our free institutions.

But could England and the English survive if the establishment itself committed suicide, as it did in 1066 and the years that followed? Here was the real test of English resilience. In the year 1085 William I spent Christmas in the abbey at Gloucester, and after the feast and the traditional crown-wearing

> ... the king had much thought and very deep discussion with his council about this country – how it was occupied or with what sort of people. Then he sent his men all over England into every shire and had them find out how many hundred hides there were in the shire.

This survey was exceptionally thorough and detailed, so thorough indeed that it aroused the disgust of the Anglo-Saxon chronicler, who thought it shameful that a king should be avaricious enough to want to know how many pigs a man possessed. But the chronicler, in his hatred of the Normans, was being disingenuous. Though the survey was directed and supervised by Archbishop Lanfranc and the able group of ecclesiastical and lay barons of William's council, it was actually carried out by the Anglo-Saxon civil service along familiar English lines which had their origins in the eighth century. Indeed England was the only state in Europe which had established the concept of a direct tax on land, and therefore possessed the method and machinery to conduct such an inquiry. To suit William's purposes, the findings were to some extent rearranged on a personal basis (tenancies-in-chief) rather than the strictly geographical basis of hundreds which the English preferred. But the concept of the shire was maintained, and in all other respects the Domesday survey was a characteristically English administrative operation. Carried out 20 years after the Conquest, it testifies to the durability of the English infrastructure.

Yet the facts and figures in Domesday also testify to the complete destruction of the English ruling class. Lands belonging to 4–5,000 English earls and thegns had been redistributed among 180 barons of Continental origin. Only two Englishmen held lands directly from the King as tenants-in-chief; both came from families of minor importance at the time of Hastings, and had clearly prospered by working with the regime and acquiring the confiscated lands of other Englishmen. One fifth of the land was controlled directly by the King himself; a quarter by a powerful ring of senior vassals, bound to the King by marriage, official status and long friendship; a quarter by the Church; and the rest by other barons, almost all of them from France. A few important old English families survived as sub-tenants in a small way. Some of the lesser families kept their lands, adopted the Continental culture, and were to re-emerge in the thirteenth century as magnates – the Berkeleys, Cromwells, Nevilles, Lumleys, Greystokes, Audleys and Fitzwilliams were of Anglo-Saxon origin, despite their names. But on the whole

William, over 20 years, made a clean sweep of the English establishment. At his death political power was confined almost exclusively to Continentals, as his charters testify. No more than six Englishmen had any say in government. They held only two bishoprics and two major abbeys, and all six were old men who had been appointed before 1066, for William gave senior Church posts only to Continentals. His household and chancery were controlled by Frenchmen, and nearly all the sheriffs were French.

But this Continental takeover was simply a matter of personnel, and personnel chiefly in the higher reaches. Some monks were imported from the Continent; small colonies of Frenchmen were set up in certain key towns, for defensive purposes but no doubt for trade also; groups of Jews came to England, for the first time, in the Conqueror's wake, and other middle-class cosmopolitan elements found a home here. Great nobles brought retinues, though from the earliest times they evidently picked many of their servants from the English. The institution of the *murdrum*, in which local hundreds were held collectively responsible for a heavy fine unless they could prove that a murdered man found in their area was not a Frenchman, indicates not merely a certain amount of racial tension but the presence in England of Continentals of comparatively humble status, whose disappearance would not immediately be noticed. But there is no evidence that Frenchmen in large numbers came, or even wished to come, to England. When William dismissed his mercenaries in 1070, nearly all returned to France. Even the prospect of vast possessions over here was not always tempting. In 1080 William made Aubrey de Courcy Earl of Northumberland, but he soon resigned and returned to France, even though this meant he forfeited his other English estates. The probability is that the Continental settlement did not involve more than 10,000 people – and perhaps as few as 5,000 – out of a population of well over a million. England simply acquired a new ruling class.

What, then, happened to the old one, which so successfully absorbed the Danes and turned their mighty monarch into a satisfactory English gentleman? The answer is that it destroyed itself. It lost its self-confidence and unity as the custodian of English culture. Already in Æthelred's day, as we have seen, there had been a confusion of identity among the English ruling class. With the development of fast and reliable sea-transport, eleventh-century England was increasingly exposed to the geopolitics of north-west Europe; it was the greatest prize in that part of the world, and vulnerable to the aggressive and active races of Scandinavia, and their settlements in north-west France. There was a mingling of cultures at the courts and, more important, intermarriage among the great. Wulfstan had perceived the threat, and his political genius had enabled England to surmount it. But he had been dealing with Danes. Scandinavian culture, despite its military superiority, was no match for the ancient, Christianised civilisation of England; the Danish presence here was formidable, and felt well beyond the Danelaw itself; but it soon became subordinate. The English legal system, springing from common roots, soon adjusted itself to Danish customs and categories, which were allowed to prevail where the Danes were predominant; the English have always been prepared to tolerate foreign importations, except of course in essentials. In cultural matters the Danes were humble and easily suborned.*

The Norman-Scandinavians were a different matter. They had adopted French speech and culture, and thus inherited the extraordinary aggressiveness and self-confidence of that civilisation. Through Æthelred's unfortunate marriage to Emma, they acquired a toehold at the heart of the English establishment. Emma, the queen-spider figure in a tangle of relationships, was half-barbarian; but in so far as she was civilised she was Norman-French. Her son Edward the Confessor was likewise Norman by culture, association, inclination and perhaps by speech too. During his reign there was a manifest and violent conflict of cultures at his court and in his administration. Its outcome was all the more uncertain in that none of the principal actors was wholly sure of his or her cultural and racial identity. In eleventh-century England, the confusion produced by intermarriage across the narrow seas was almost absolute. Take Emma herself. She was Norman, the great-aunt of William, who by the 1040s was the effective ruling Duke. She had married Æthelred the Englishman, then Cnut the Dane, and had had sons by both. Where did her loyalties lie? With the English line, represented by Edward? With Cnut's children? With her own Norman house? Royal intermarriage, in theory designed to promote international amity, is far more likely in practice to provoke disputed claims across frontiers, and so racial tension; it is the same today with international sport. The royal houses of both Norway and Denmark had blood-claims to the English throne, and both were also related to men who stood at the centre of English politics. Earl Godwine, the most powerful English politician and landowner at the mid-century, was a self-made man of lesser gentry stock. He probably had Scandinavian blood anyway. To advance himself with Cnut, he had married Cytha, the sister of Cnut's brother-in-law; she was a savage lady who, among other things, bought beautiful girls in England and shipped them as slave-prostitutes to Denmark. Of her many sons, one, Tosti,

* See Appendix I.

eventually identified himself with the Scandinavians; most of the rest, led by Harold, thought of themselves as Englishmen, at least for political purposes. The problem might have been solved if Edward the Confessor had produced an heir. Godwine persuaded him to marry his daughter Edith (who was of course half-Scandinavian); but no child was born. When the Confessor died, all the chief claimants were related to each other. Even Harold had a marriage-relationship with his mortal enemy, the Conqueror. It was a small world in the eleventh century; but a violent world, in which blood-links raised more problems than they solved.

When Harold seized the throne immediately after the Confessor's death, the English ruling class was not only racially confused, it was also politically divided. There had been a major internal crisis in the early 1050s, and another in 1065, when Harold's brother Tosti had been ousted from Northumbria, possibly with Harold's connivance, and had then thrown in his lot with a Scandinavian claimant. England was coming apart at the seams. The growth of huge territorial earldoms threatened the unity of the kingdom; they might eventually have developed into semi-autonomous territories, as in Germany and France. The Confessor's properties brought him in only about £2,500 a year, not much more than that of several of his subjects; to run the State he relied on the *geld* which produced a further £6,000; he was becoming merely a *primus inter pares*.* Thus William did not so much conquer England as save it from disintegration. The haste with which Harold acted after the Confessor died indicates the weakness of his position. The other earls, apart from his own brothers, did not attend his coronation. He quickly married the sister of Edwin and Morcar, the two most important, but even this gesture could not persuade them to fight with him in battle. The men he led at Hastings were almost all mercenaries. The Church, which had in the past so successfully underpinned the unity of the State and ruling class, was in great difficulty. The Archbishop of Canterbury, Stigand, could not play his role as unofficial head of the establishment, as Wulfstan had done. He was not only a pluralist but had got Canterbury as a result of the political crisis of the 1050s, in an uncanonical manner; he had been declared deposed by successive popes. Harold would not allow Stigand to crown him, and got the Archbishop of York to do the job instead. This was another element of weakness in Harold's position. It is evident that many important people did not want him as King, and felt he was motivated purely by personal and family interests. He was thus forced to rely on the sole

* The Confessor's finances are examined in Frank Barlow: *Edward the Confessor* (London, 1970), Chapter 7.

arbitration of battle, and this in the end failed him. The truth is that the bulk of the English establishment contracted out of the 1066 crisis.

But at least Harold was a purposeful and decisive man. After his death, the behaviour of the English ruling class was both foolish and contemptible. It deserved to disappear, for in effect it abdicated. This is what made Hastings one of the truly conclusive battles of history. It did not appear so immediately: William's first action, when the field was won, was to send to the Continent for reinforcements. The citizens of London were anxious to resist, and make Edgar, grandson of Edmund Ironside, King; the Earls of Mercia and Northumberland promised support. But no one actually did anything. As the *Chronicle* puts it: 'But always the more it ought to have been forward the more it got behind, and the worse it grew from day to day, exactly as everything came to be at the end.' In the event, the establishment, such as it was, decided to submit. They met William at Berkhamstead and went with him to Westminster, where he was crowned on Christmas Day. He swore to uphold the laws and customs of thc Confessor, and the mixed congregation of English and Continental notables was asked by the Archbishop of York (in English) and the Bishop of Coutance (in French) if they accepted him as King. They assented.

Thus William's occupation of the throne took place within a framework of law – English law. It signified nothing more than the transfer of supreme authority, within the existing structure, to an alien family group. That this group became a class was entirely the fault of the English. William's original intention was to run the kingdom through a mixed Anglo-Norman aristocracy, in which the native element would swiftly have become predominant. True, he confiscated the property of those who 'stood against me in battle and were slain there'. Some of his officers were sent to key points throughout the country and ordered to construct and garrison castles. But there were no mass confiscations on racial grounds. William's charters of 1068–9 were signed by leading English landowners, churchmen and royal officials inherited from Harold. Until 1069 most of the sheriffs were English, and indeed Englishmen received important fresh appointments. Unlike some of his followers, William was not a racist. He had no animus against his new subjects; he even tried to learn their language, a formidable task for a professional soldier and politician of his age: it is not surprising he gave up baffled. He seems to have liked many Englishmen. In 1070/1 he judged a dispute between the English Bishop of Worcester and the Norman Archbishop of York; although all his court and expert advisers, with the sole exception of Lanfranc, favoured York, William settled for Worcester – a very difficult and courageous decision. Domesday Book is a mysterious document, but (*pace* the *Chronicle*) it was aimed much more at the Norman element than the English.

But William's policy was frustrated by the blind irresponsibility of the English ruling class. While incapable of organising concerted resistance, they repeatedly engaged in piecemeal or regional revolts, and in a manner calculated to rouse William's fury, allying themselves with anyone – Irish, Scots, Welsh, Norwegians, Danes, disaffected Normans and Frenchmen – willing to challenge the existing order. In some cases they did not even enjoy popular support. The *Peterborough Chronicle* makes it plain that the Fenland revolt associated with Hereward the Wake was bitterly resented by at least a section of local public opinion. After the northern rebellion of 1069 William scrapped his general policy of associating the English aristocracy with his government, but he still tried to be generous to individuals. In 1072 he gave the earldom of Northumbria to Waltheof, head of the leading English family in the north; but three years later Waltheof let him down and conspired with two of William's own barons. The forces William raised to break this plot included, significantly, units provided by two senior English ecclesiastics. William was so angry and disappointed that he had the earl executed, a painful decision for the King, who was opposed on principle to capital punishment, and hardly ever permitted it. But William was a very serious-minded and responsible head of government, who believed his primary duty of maintaining order overrode all other considerations; he found that in practice he had to work through his confederates, and this meant that the English upper class had to be stripped of their lands.

What is most interesting about William's handling of the conquest and its consequences is the way in which his mind changed its focus after he became King. He invaded England at the head of what was a European crusade, whose object was to drag the offshore island back into the Continental system. Rome had always viewed the English Church and monarchy with intense suspicion. This is curious, for in the eighth century the English St Boniface and many hundreds of courageous English men and women had carried out the conversion of Germany; the English were enthusiastic proselitisers – one might say that the first English empire was a spiritual one. But Rome was profoundly ignorant of what went on in England, and tended to regard the English as barbarous and heterodox. Even Wilfred, the arch-papalist, felt himself an outsider at the Lateran court. On the rare occasions when papal legates came to England they seem to have seized eagerly on the wrong ends of any number of sticks. In 786 legates reported to Pope Hadrian that the English settled legal disputes by casting lots, and were accustomed to mutilate and eat their horses – a likely tale! Some of the papal letters which have survived betray a bewildering ignorance of English conditions. Thus about 877 Pope John VIII boasted that he had 'admonished' King Alfred, the best friend the Church in England ever had, for infringing the rights of Canterbury. No doubt Alfred had been insisting on the legal military service from Church lands – natural enough at a time when the kingdom, and indeed the Church itself, was in mortal peril from the Danes. About 891 Pope Formosus wrote a querulous letter to the English bishops complaining that he had heard 'the abominable rites of the pagans have sprouted again in your parts', and adding that he had considered placing the country under an interdict; it would be difficult to conceive of a more complete misapprehension of the true situation in England, then being ravaged by what the *Chronicle* calls 'the great heathen army'. Even more irritating must have been the letter Alfred received from Fulk, Archbishop of Reims, in response to a royal request (accompanied by the present of a fine pack of wolfhounds) that Fulk should lend him the services of the scholar Grimbald. With unctuous condescension, Fulk emphasises his hesitation (happily overcome by the gift) at committing a civilised and pious Frenchman to the care of such a 'rude and barbarous race' as the English.*

These misunderstandings continued, despite the fact that England was one of the few countries which regularly provided Rome with funds in solid sterling silver. At any rate, in 1066 there was a general impression in western Europe that the English Church was in a disgraceful condition, and that it would be an act of piety to invade the country and bring the English up to Continental standards. William may or may not have shared this view, but he played on it skilfully to improve his own chances. In Normandy itself he had used the Church to consolidate his position as duke, and he had a high reputation in Rome as a reforming sovereign. Rome, under the influence of Archdeacon Hildebrand, was then stirring with a new movement to assert universalist papal claims, and was anxious to play politics everywhere. When the Confessor died, William had the audacity to take his case to Rome. He insisted that Harold had broken his oath; and that his coronation was illegal because (so William said) he had been crowned by Stigand, who held his

* Whether Formosus or Fulk had the right to rebuke anyone is doubtful. After Formosus died, his body was dug up by a rival papal faction, put on trial, condemned, mutilated and thrown into the Tiber. The insufferable Fulk fared no better. In 900 he quarrelled with Baldwin of Flanders, Alfred's son-in-law, and was murdered by the count's men. In the next generation his province was treated as a plaything by the depraved women who then controlled the papacy, and a five-year-old child was appointed to his archbishopric. Such things did not occur among the 'barbarous' English.

see uncanonically; that the English Church was virtually in schism, and that its restoration to orthodoxy and godliness would be a natural consequence of the successful establishment of his claim. None of this was true, but it was what Rome wanted to hear. So William's claims were pressed by Hildebrand himself before an eager pontiff. Harold was not invited to be represented, may not even have heard of the suit until after the decision went against him. William thus went to England not only with the emotional support of Rome but with its explicit and formal authority; he fought under a papal banner, and carried into battle a string of papal relics round his neck. The Emperor, the King of France and most of the other potentates of Europe endorsed his claim, and he fought Hastings as the champion of Continental Catholicism. Philip II of Spain could not have asked for more.

Whether William took himself seriously as a crusader is doubtful, but if he did his views underwent a mighty and marvellous conversion once he found himself safely established on the English throne. The rapidity with which he acquired an English perspective testifies to his political realism, and must have come as an unpleasant surprise to Hildebrand, who was now Pope himself. When he instructed his legate to demand from William both the resumption of payments to Rome and a formal act of homage, the King sent him a letter which was brief and very much to the point. The Pope could have the money but nothing else: 'I have not consented to pay fealty, nor will I now, because I never promised it, nor do I find that my predecessors ever paid it to your predecessors.' The English kings had always governed the Church of England, and William laid down a string of regulations making it plain he intended to follow, and indeed reinforce, custom. He made indeed one serious mistake by allowing Church courts to be established for the first time; his motive was purely practical, for he felt that Church business was cluttering up the work of the county courts, but it was an error of judgment which caused immense trouble to his successors. In all other respects, however, he resisted the Hildebrandine aggression.

So, to their credit, did his sons William and Henry. William II has had his reputation blackened by monkish scribes, who then possessed a monopoly of the writing of history; he suffered from the further disadvantage of having red hair, always a handicap to politicians.*

* Three red-haired prime ministers, Peel, Baldwin and Churchill, all had trouble from their own parties, and it is difficult to explain on purely rational grounds the intense suspicion they aroused. Barbara Castle suffered from the same misfortune when trying to get her trade union bill before parliament in 1969; I have heard one trade union leader refer to her as 'that red-headed —'. His language would have been more guarded had she been Jewish or coloured.

He was accused of effeminacy, though he was a first-class and very active soldier, rarely out of the saddle; and of homosexuality, though he fathered two bastards. He was in fact a king cast in the Conqueror's mould, shrewd, industrious, energetic and highly professional. It is significant that the Conqueror preferred him to his brothers, and gave him the kingdom. His early death in a hunting accident robbed England of a great king. But he lacked his father's polish, and his celebrity, and the respect the old man aroused through close association with the reform movement. He swore in public, and his oaths – 'by the Holy Cross of Lucca', 'by St Luke's face' – were notorious. He viewed the Church with a certain cynicism, contrasting its extravagant claims with the manifest failings of many of its senior office holders. He had a Jewish doctor and was friendly with the Jewish community; he challenged them to convert him – a joke felt to be in very bad taste. So far as the bishops were concerned, William treated them like other tenants-in-chief: they must make their full contribution to the services of government, military and financial. As they were immune to the accidents of wardship and marriage, an important source of royal revenue, he kept their sees vacant for long periods, and took the proceeds himself. It was a rough and ready method, which aroused the fury of the clericalists. William was unscrupulous; but so were they. They were not only sharp lawyers but skilful forgers. Monks tampered with their charters, even fabricated entirely new ones; rings of professional forgers operated on both sides of the Channel; the clerks of the Lateran Palace were the greatest forgers of all. The more extreme papal lawyers had the audacity to argue that no document of secular origins should be quoted as authoritative. It is no accident that, about this time, the English State began to keep its records systematically, as religious establishments had long done; the Exchequer probably came into existence in William's reign. The State was defending itself, a little belatedly, against a movement not unlike twentieth-century Communism, combining militant idealism with systematic mendacity.

While recovering from an illness, perhaps while still delirious, William had appointed an Italian scholar, Anselm, to succeed Lanfranc at Canterbury. Anselm was a pious intellectual, unsuited to a role of great worldly responsibility. His outlook was wholly Continental; he never seems to have grasped the peculiar structure of the English State, and the traditional place of the Church within it. Once in office, he revealed himself as an ardent papalist, fixing his narrow, philosopher's mind on delicate and abstruse points of principle to the exclusion of every other consideration, including common sense. William wanted straightforward dealing from his prelates; what he got from Anselm was a babble of canon law; as William saw it, it was as though the regimental chaplain was trying to teach the Commanding Officer military law. So Anselm went into voluntary exile, declaring with the serenity of the fanatic: 'I would not dare to appear before the judgment seat of God with the rights of my see diminished.' Using the terse language of his father, William wrote to Pope Urban II:

> I am astonished you should take it upon yourself to intercede for Anselm's restoration. Before he left my kingdom I warned him I would seize all the revenues of his see if he departed. I have done what I threatened, and what I have a right to do, and you are wrong to blame me.

It is clear that William had majority opinion in England on his side, including nearly all the bishops: and it is significant that his successor and brother, Henry I, who went to considerable lengths to treat the Church with courtesy and respect, was soon driven into exactly the same disputes with Anselm, and reacted in exactly the same manner.* The truth is that no English king who sought to uphold the rights of the State, as established for centuries, could afford to compromise with Continental papalism. Only when, under Stephen, the State was weakened by internal disputes, did the Continentalists make progress.

By resisting the encroachments of an international organisation based on Rome, the Norman conquerors thus maintained the continuum of English historical development. They did so in a number of other important respects. Generations of historians have analysed the merging of English and Normans in terms of moral, racial and cultural supremacy, and taken sides accordingly. They thus tend to cover themselves in ridicule. Take, for instance, Carlyle, a confirmed 'Norman':

> Without the Normans what had England ever been? A gluttonous race of Jutes and Angles capable of no great combination; lumbering about in pot-bellied equanimity; not dreaming of heroic toil and silence and endurance, such as lead to the high places of the Universe, and the golden mountain-tops where dwell the spirits of the Dawn.

Freeman, an 'Anglo-Saxon', went to the opposite extreme:

> We must recognize the spirit which dictated the Petition of Right as the same which gathered all England round the banners of Godwin, and remember that the 'good old cause' was truly that for which Harold died on the field and Waltheof on the scaffold.

The argument persists even among sophisticated and erudite historians

* The monkish scribes gave Henry a better press than his brother William. Yet his morals were certainly worse; he fathered at least 19 bastards, more than any other English monarch.

today. But a more realistic approach is to see England in terms of the institutions and manner of life shaped by her geographical predicament. These are more powerful factors than racial habits. What the Normans found in England was a unitary society, underpinned by a sophisticated legal system and a strong popular monarchy. They embraced this valuable inheritance, they identified themselves with it, they developed it; they did not fundamentally change it. They certainly did not impose an abstract conception called 'the feudal system'.* In terms of land tenures, English and north-west European society had been developing on roughly similar lines since the eighth century. The difference lay in the fact that English State administration was organised on a civil basis, made possible by the barriers of the Channel and the North Sea; Continental states, with fluid and insecure land frontiers, were forced to organise themselves on a military basis. William I and his successors could not entirely free themselves from their Continental background; and thus to some extent they changed the viewpoint of English administration. Old English Society was essentially agricultural in outlook; the society William represented was military. The English had flourished by taming the land, the Normans by taming men. Thus they approached administration, and above all fiscal obligations, from different angles. The acre, the hide, the hundred, the shire: these were the English units of computation. But of course they were fiscal, not actual units. The Norman unit was military: the armoured knight. The number of these who could be provided or paid for was their measure of wealth and therefore fiscal obligation. And since the knights, in practice, served under the banners of great lords, the Normans saw the countryside, for administrative purposes, as a collection of great estates, owned by responsible individuals, rather than as territorial units, where responsibility was collective. But of course the knight, like the hundred, was a notional or fiscal concept, rather than an actual knight, in real armour, riding a living horse. Once this distinction is understood we can dispense with the word feudal, which is merely confusing in English terms. The purpose of Norman, as of English, administration was primarily to raise money. The English did this on a territorial, the Normans on a personal, basis. Thus the information for Domesday was first gathered territorially, then rearranged for each shire under

* They may have known the word *feodal*, meaning the tenant of a fief (in Latin, *feudum*), but 'feudalism', had such a concept existed, would have been a meaningless abstraction to them. Phrases and ideas such as 'feudal England', 'the feudal army', and so forth, are the inventions of antiquaries. The term 'feudal' does not occur in print until 1614; since then historians have used it freely, to the confusion of innocent schoolchildren. It is more appropriately employed, if at all, for purposes of indiscriminate journalistic abuse, as in 'the feudal magnates of the Jockey Club', etc.

tenants-in-chief; for it was on persons, not communities, that the Normans placed the responsibility.

But this change was more a matter of habit and attitude than a fundamentally different way of doing things. The English instinct was that the army should consist of the local men of each region fighting side by side as conscript territorials: the Norman was that great landowners should produce the men in respect of the property they held under the King. But neither system, or course, ever worked in practice. The English and Anglo-Norman States both created armies, when it came to the point, in more or less the same way: by hiring mercenaries, and by adding to them such local elements as were fit for battle. The two rival conceptual systems were thus, in reality, two slightly different ways of raising money to pay for professional troops. There was no other means of getting an effective army into the field, or keeping it there long enough to serve its purpose.

This purpose, of course, was the maintenance of the integrity of the State from its external and internal enemies. The English might grumble at royal taxation, the weight of which they attributed, quite wrongly, to Norman innovation. But William and his successors, by virtue of their ability to command the military situation, itself dependent on the continuous flow of cash, set very high standards of government. They grasped the full potential of the royal institution they inherited, and gave it new vitality. The Old English State had been running down in the eleventh century. The growth of regional earldoms, the relative decline in royal revenues, were accompanied by an immense and depressing conservatism which finds expression in innumerable charters and documents. The vigorous impulses of Alfred and his immediate successors had been wholly expended. It was as though Old English society was turning its back on the real world, and looking inwards on itself. Edward the Confessor himself set the tone:

> I Edward, by divine mercy King of the whole English nation, counting the perishable things of this world as worth nothing, and, with all creatures of passage, desiring to obtain these things that last for ever, hasten to grant a fugitive and doubtless transitory little estate in order that I may obtain in the kingdom of Christ and of God an everlasting dwelling-place.

This was all very pious; but an attitude fatal to strong, central government of the type the English want and need. Bede would have disapproved. Excessive Continentalism has always been a danger to the English; but so has excessive isolationism. The advent of the Normans was a necessary corrective. The territorial convulsion which followed Hastings not only arrested but reversed the erosion of the royal estates.

The Conqueror was a man born into the harsh Continental world of incessant warfare and fragile security. He appreciated, perhaps better than the English themselves, how difficult it was to achieve the internal stability they enjoyed, how easily it could be jeopardised, and how vital therefore to maintain and strengthen the institution of centralised monarchy in all its plenitude. He was a king in the English tradition, but a much more effective one than his immediate predecessors. And his sons built upon his work, blending Continental innovations, in the Exchequer and Chancery, with the structure of the Old English administrative machine, to produce the most formidable instrument of royal government in Europe.

The result was a progressive rise in English living-standards. Other factors certainly helped: the period 1050–1300 was one of the warmest and most favourable climatic periods in historic times;* international trade was reviving. The area under cultivation was steadily expanding. The towns were growing; so was the population as a whole. With all deliberate speed, the Anglo-Norman State abolished agrarian slavery, which had persisted in conservative England long after it had virtually disappeared on the Continent. Under the old regime, exalted members of the royal house had openly engaged in the slave-trade.† The Anglo-Norman monarchs, by contrast, would not give it countenance, and their courts were readily available to terminate servitude. They were also at the disposal of the ordinary villein, or peasant tied to his lord's land, unless the proof of his status was explicit. Improvements in legal administration, springing from a reinvigorated central government, tended to sharpen definitions and define obligations more closely; in that sense of course the villein found it more difficult to wriggle out of his dues. But there is no evidence that the freedom of the ordinary villager was greatly curtailed as a result of Norman rule; the damage had already been done in the tenth and eleventh centuries. It is more likely that he saw the law, if he could afford it, as his road to freedom. In any event, the restructuring of English landed society in the late eleventh century, by emphasising exactions at all levels, tended to increase agrarian productivity – the only way in which the medieval world could escape from its economic prison of subsistence living. The coming of the Normans thus gave a salutary forward-impulse to the progress of the English; but it did not change the direction.

The Conqueror evidently did not believe it possible, or desirable, that

* See H.H. Lamb: *The Changing Climate* (1966).

† They were not alone. William of Malmesbury, a patriotic Englishman, says that the English nobility were accustomed 'to sell their female servants, when pregnant by them, after they had satisfied their lust, either to public prostitution or to foreign slavery'.

the English State should be linked permanently, under the same head of government, to large Continental possessions. To his most responsible son, William, he gave the kingdom; his heir by primogeniture, Robert, got the Duchy of Normandy. In the long run events were to prove William I right. But in the short run the folly of Robert led to the reunification of the two territories; and for over a century England was politically a part of the Continental system.* This had a marked effect on the development of the English language. Had the Conqueror's will been enacted, the Channel would have constituted a political as well as a linguistic barrier, and English would rapidly have become the language of the Anglo-Norman ruling class. But Normandy was not lost until 1204, and in the meantime French had become – and would long remain – the vernacular of administration, at any rate at the highest levels. Of course the mass of the English people never learnt French. The evidence for the Domesday survey, for instance, was presented by English and French sworn juries, before being recorded in Latin; to some extent administration had to be bilingual. But government itself was French-speaking: one reason why English notables played no effective part in it was that they could not participate in the lengthy political debates, conducted in French, which took place at the King's court. Englishmen who rose in society thus became French-speaking; the linguistic division followed social rather than racial lines; a French literature sprang up in England, and the French dialect spoken there showed early and marked divergencies from Continental French.

But many members of the ruling class were bilingual. It was clearly a great practical advantage. Henry I certainly knew some English; Henry II understood it perfectly, though he preferred to speak in French or Latin. Some Continental ecclesiastics learnt not only to speak English but to read Old English texts and documents. It depended, of course, on whether a man was born and raised in England; if so, English was likely to be his first tongue. The historian Odericus Vitalis, born in England under the Conqueror of a Norman father and an English mother, had to learn French from scratch when he went to Normandy at the age of ten. By the end of the twelfth century a man occupying an important administrative position was open to censure if he spoke no

* Robert finally lost Normandy to Henry I in 1106 and spent the remaining 30 years of his life in English gaols. He may have learnt Welsh while shut up in Cardiff Castle; a Welsh poem attributed to him laments the fate of those 'who are not old enough to die'. He is buried in Gloucester Cathedral, under a splendid effigy of coloured wood. He thus fared better than his father, whose magnificent tomb in Caen was rifled by Calvinists in 1562; a single thighbone was preserved, and reburied under a new monument; but this, in turn, was demolished by the revolutionaries in 1793. Only a simple stone slab now commemorates England's greatest king.

English at all. Even before this all obvious racial distinctions had disappeared, and the language division was increasingly functional. Richard Fitz-Nigel, treasurer of the Exchequer, writing about 1179, stated: 'With the English and Normans dwelling together and alternatively marrying and giving in marriage, the races have become so fused that it can scarcely be discerned at the present day – I speak of freemen alone – who is English and who is Norman by race.'

After the loss of Normandy, when men who held lands on both sides of the Channel were forced to chose a single allegiance, the dynamic behind the continued use of French rapidly disappeared. Long before this we hear of complaints from men of the highest rank that castles should not be entrusted to aliens, and that English heiresses should not be married to men whose birth would disparage them, 'that is, to men not of the nation of the realm of England'. French lingered on as a class distinction. Paradoxically, it was the sheer conservatism of the English, especially of lawyers and civil servants, which kept French alive as the administrative vernacular. By 1400 we find English textbooks purporting to teach French to the upper classes. Evidently by then no one learned it from birth, but it was still used, in a fossilised form, for many purposes of law and government. But even in the thirteenth century some statutes were written in English as well as in French and Latin. In the fourteenth century, war, nationalism and racism completed the destruction of French. In 1356 the mayor and aldermen of London ordered that proceedings in their courts should be conducted in English. Six years later Parliament was opened for the first time by a speech in English; and it enacted the Statute of Pleading, laying down that henceforth, as French 'is much unknown in the said realm',

> . . . the King . . . hath ordained . . . that all pleas which shall be pleaded in his courts whatsoever, before any of his justices whatsoever, or in his other places, or before any of his other ministers whatsoever, or in the courts and places of other lords whatsoever in the realm, shall be pleaded, showed, defended, answered, debated and judged in the English tongue, and that they be entered and enrolled in Latin.

This conquest by the English language, it should be noted, took place against fierce resistance from authority, for French was the spoken, as Latin was the written, language of international culture: in particular, the lawyers and the universities fought a vigorous rearguard action against English. In the last decade of the thirteenth century the monasteries at Canterbury and Westminster adopted regulations forbidding novices to use English and requiring all conversation to be in French; a fourteenth-century Oxford statute ordered construing in French 'lest the

French language be entirely disused'; the complaint was made that at Merton College the fellows talked English (and wore 'dishonest shoes'). But by 1400 the battle was over; French was wholly, and for all except some legal purposes, a foreign language.

Meanwhile English had derived enormous benefit from this process. As it began to emerge again as the language of business, it became necessary to transfer to it a large number of key French words used in administration, justice and the general preoccupations of the ruling class. It is notable that most of these words were adopted by English during the thirteenth and fourteenth centuries, when the use of French was in rapid decline. The total number of French words absorbed during the Middle English period was slightly over 10,000, of which 75 per cent are still in current use. Over 40 per cent of the 10,000 came in during the period 1250–1400; in the years 1350–1400 twice as many French words were adopted than in any other half-century.* This great enrichment in vocabulary was accompanied by internal changes in the structure of the language which were even more important. The Wessex dialect, the language of government in the Old English State, was both clumsy and highly conservative. The Conquest dealt it a death-blow. The English which emerged as the official (and therefore, eventually, as the universal) form in the later Middle Ages was essentially based on the Mercian dialect as spoken in the south-east Midlands, and above all in London. It was far more flexible and capable of dynamic growth. Old English, as written and spoken before the Conquest, is essentially a foreign language to us; the so-called Middle English, as we read it in Chaucer, is merely an archaic version of our own. The Norman invasion thus made a crucial contribution to the development of English as the international language of government, culture and commerce – in which role, by a supreme irony, it has decisively displaced French.

In the year 1153 two of the greatest territorial magnates in England, the Earls of Chester and Leicester, sat down at a table with their clerks to draw up a treaty between themselves. Its text has fortunately survived, and has a curious interest. It lays down in considerable detail how frontier disputes, and other matters of contention between the two earls and

* Many were duplicates of Old English words; in the later Middle Ages, members of the upper class, and still more those aspiring to such status, used French derivatives by preference. This process has recently been reversed. Use of French derivations, as opposed to 'honest' and 'earthy' English words, is said to be 'non-U', and an affectation denoting suburban gentility. Thus, one should say 'looking glass' not 'mirror'. But such distinctions lead to confusion: 'lavatory' is claimed to be socially preferable to 'toilet', though both are of French origin.

their dependants, were to be resolved. The King is mentioned only once, and then merely by implication. At first glance, it is as though the English State did not exist, and any form of stability was entirely dependent upon the personal exertions of independent local sovereigns, and the arrangements they made between each other. Yet first impressions are deceptive. This document, and the motives which inspired it, are not a testimony of anarchy; on the contrary, they are a tribute to the intense longing in England for some kind of system of law and order. These two great Anglo-Norman princes were trying, as best they might, to devise a set of club rules to fill a legal vacuum. They were not renouncing the State: they were endeavouring to reinforce its weakened authority. The treaty was written in Latin; its terms were debated and settled by arguments conducted in Norman-French; but the instincts of its authors were those of responsible Englishmen. It was only 80 years or so after the Conquest, but already the men who mattered among the Anglo-Norman ruling class were behaving with the reflexes of true offshore islanders.

The alleged 'anarchy' of Stephen's reign is one of the most interesting, and instructive, episodes in English history. It was the deviation which supplies the true key to the norm. It aroused a sense of outrage, among Englishmen of all classes, out of all proportion to the weight of the facts as they can now be discovered. Of course it was unfortunate that Henry I's son and heir should have been drowned, thanks to a drunken crew, in the *Titanic*-like accident of the *White Ship*.* It was still more unfortunate that Henry should then have tried to force, on a reluctant populace, his daughter Matilda as his heiress. She had been brought up at the German imperial court, and her intolerable Germanic manners were resented by all classes; moreover, Henry promptly remarried her to the heir to the House of Anjou, whom the Anglo-Normans

* The distinguished passengers were all drunk too. When Stephen saw this he refused to travel on the ship, and thus lived to be an unhappy king. Drunkenness was not supposed to be a Norman vice, but an English one. Perhaps by this time many Normans had adopted English habits. William of Malmesbury says of the English: 'Drinking in parties was a universal custom, in which occupation they passed entire days and nights . . . they were wont to eat until they became surfeited, and drink until they were sick.' He blames the loss of the Battle of Hastings on drink. By contrast, the Normans were abstemious, though they liked delicate cuisine. William the Conqueror drank little, and mixed water with his wine; Henry I 'drank simply to allay his thirst, and he deplored the least lapse into drunkenness both in himself and others'. Another account says he never 'drank more than thrice after dinner'. But Walter Map, in his account of Henry's court, said there was a standing order for a carafe of wine to be put in Henry's bedroom at night. As he never called for it, the servants drank it; one night Henry asked for his wine, and it was not there. Royalty does not vary much. At Balmoral, a whole bottle of whisky was put out for Queen Victoria's use every night; she never touched it, and for decades it became a perk of the servants. On his accession, Edward VII discovered the practice and ended it.

regarded as hereditary enemies. In the circumstances, it was only natural that Stephen, one of the Conqueror's grandsons, the richest landowner in north-west Europe, with half a million acres on this side of the Channel alone, should snatch at the throne, and get it. He was not, as it turned out, a suitable choice. He was, says the *Chronicle,* 'soft'. He lacked 'a hearty voice', and could not give orders on the battlefield himself, using a spokesman instead. He was indecisive. There is an apologetic note in some of his charters; in one, issued to Worcester Cathedral, he admits having made a wrong decision by failing to take proper advice. On the other hand, he could be arbitrary, and act without due legal procedures. He was good-natured, especially to children, chivalrous and open-handed. He certainly tried hard. The sheer physical demands made on early medieval kings were always extraordinary; Stephen, in particular, was a martyr to duty. In 1139, for instance, he made at least 34 major journeys, covering virtually the whole kingdom, and took part in five major sieges; he was in the saddle in all weathers and at all seasons. He died exhausted and disillusioned with power, and it is not surprising that his surviving son renounced his claims for a financial settlement. But despite his efforts Stephen failed to maintain a unitary kingdom. Matilda's claims were put forward; there was a certain amount of fighting of an inconclusive nature; and the English, with one voice, cried 'Anarchy!'

Now this was a monstrous distortion of events. The famous passage in the Peterborough version of *The Anglo-Saxon Chronicle,* our most striking authority for this reign, paints an appalling picture of chaos and savagery. It speaks of 'nineteen terrible winters' when 'Christ and his saints slept', and describes in devastating detail the dreadful tortures and wickedness inflicted on the people when the State abdicated and desperadoes took over. It has become the received version of history, much quoted and enjoyed by generations of law-abiding Englishmen ever since, and cited as a warning of what happens when central government breaks down. Yet the passage is certainly a gross exaggeration; its details may be pure fiction. Peterborough was one of the few areas where government had, in fact, ceased to operate effectively. But immediately following the famous description of chaos is a long account of how prosperous and wealthy Peterborough Abbey was throughout this period; and this in turn is followed by a disgusting anti-semitic story, which is devoid of foundation. As an account of Stephen's reign, this final section of the Peterborough version of the *Chronicle* is almost worthless.

What are the facts? The actual fighting in the 'civil war' was of little importance. It took place chiefly in Wiltshire and Gloucestershire for a few years in the middle of the reign. Matilda never controlled anything which could be called a government. Those baronial thugs who operated outside the law, such as Geoffrey de Mandeville, had short careers and met violent ends. Their illegal castles were flimsy affairs, which have left little trace. In many shires the laws of King Henry I were administered effectively from start to finish. The Chancery continued to function, though with a diminished staff and a smaller volume of business. So did the Exchequer, though its shortcomings under Stephen were naturally exaggerated by the men who claimed credit for getting it back into full working order under Henry II. Stephen did his best to maintain the currency; some of his seven issues were small, and others light-weight, and many irregular coins were minted; but there was no calamitous devaluation, as the chroniclers imply. Even where central government failed, local great men kept their territories at peace, like the two earls. Few Englishmen, or indeed Normans, were killed. Apart from the personal retinues of certain great barons, the fighting was conducted by Welsh mercenaries (for the Empress) and Flemings and Bretons (for Stephen). It is remarkable that even in the most disrupted areas, abbeys such as Malmesbury and Tewkesbury carried out ambitious and costly building schemes at this time; more monastic buildings were started or finished, and more religious houses founded, during Stephen's reign than ever before.

The hullabaloo, indeed, was set up not because the English experienced anarchy, but because they came close enough to it to sense how appalling it might be if the State really did abdicate its functions. England was more stable, and better governed, during King Stephen's reign than any other territory in north-west Europe. What frightened the English was the way in which their country was slithering towards the Continental norm. They over-reacted, in what seems in retrospect a hysterical fashion. But this sprang from a sound instinct. Conditions in Stephen's reign were sufficiently disturbing to convince great masses of people, of all classes, that the decline must be instantly and dramatically reversed. Thus Henry II came to the throne with an overwhelming mandate for strong government, which he was delighted to exercise. The English, once again, had engaged in a skilful exercise in the re-writing of contemporary history to suit their own purposes. William I's work in rebuilding the Old English monarchy was therefore continued by an Angevin who became a thorough offshore islander in his turn. The universal satisfaction which greeted his accession showed that the English put effective central government, operating on traditional lines of law and custom, before any other consideration. The Conquest was now a distant memory, the Old English line mere folklore; racial pride found greater satisfaction in the tales of King Arthur and his knights than in the consciousness of a more recent past. 'Normalcy' meant the good old days of Henry I. The new King was judged by his ability to bring them back. The aristocracy was no longer alien; just upper-class. Stephen's reign gave a glimpse of the terrifying prospect if they were called upon to assume a Continental role. Henry II, by restoring the prestige of the Crown, by refurbishing and improving the machinery of government, once more emphasised the distance between the King and all his subjects, irrespective of class, which is the foundation of equality before the law, and so the precondition of national and racial unity. It was as subjects that Normans and English came together; and since the framework of this subjection was essentially the Old English State and law, it was the Normans who became Englishmen. The connection with the Continent was maintained; but its nature and limits were laid down firmly and exclusively on this side of the Channel.

PART THREE

'This Realm is an Empire'

[1154–1603]

IN January 1308 King Edward II, at the request of the Pope, ordered the arrest of all the Knights Templars in England, sequestered their property, and appointed commissions (which included certain papal inquisitors) to sit at London and York to try them on charges of heresy and moral depravity. The Templars were largely, if not wholly, innocent of the charges against them, which they hotly denied. Their fault was that they belonged to a rich order; from their original function as custodians of Solomon's Temple in Jerusalem, they had become wealthy bankers and the owners of substantial estates throughout Christendom. Their French possessions were particularly valuable, and had attracted the greedy attention of Philip IV of France. He wanted their money and their lands, and he had sufficient leverage over the Pope to force him into ordering a general dissolution of the order. The Pope complied, and the trials were held to provide moral justification for an act of blatant theft.

The English had no particular animus against the knights. But like any other wealthy churchmen, they were not popular in England, and their lands were tempting targets. So the commissions were set up. But then came, for the English, a difficulty. How could the knights be made to confess? The only conceivable way was to torture them. This was what was done, with little hesitation, in France and elsewhere. The bishops of the southern province, meeting in London, were eventually persuaded to seek permission to use torture; but the inquisitors who attended them complained they could find no one in England to do the work. At York, the northern bishops were outraged: they said that torture was unknown in England, and quite illegal; that they did not employ torturers, and had no idea where to find them. Were they supposed to import them from abroad? Let the Pope, or the King of France, do his own dirty work. A strong note of English indignation – a thrill of horror at the wickedness and barbarity of foreigners – runs through this curious episode. In legal matters, in respect for certain inalienable principles of decency in the administration of the law, the English already placed themselves on a different, and higher, plane than men across the Channel: there were some things not done in England, things which Englishmen could not be brought to do. But at the same time the English, then as now, were pragmatists, with a streak of what their critics would call hypocrisy. Having made their protest, they set to work. After all, the Templars were a foreign order; their wealth did no good to England; if the Pope, in his wisdom, wanted it disposed of, then who were the English to refuse? So the trials took place. Exactly how the confessions were obtained is not clear. But they were eventually forthcoming. English honour was satisfied; and so, in time, was English avarice. Most of the Templars' property – which, according to the Pope, should have gone to the Knights of St John – was quietly absorbed by the English Crown. Leaving the issue of torture aside, a remarkable precedent had been set, of sinister implications for the Pope, for the religious orders as a whole, and for Roman Catholicism. Its significance was not lost on Englishmen: it passed into the national memory, for convenient use at some future time.

But the precedent of using torture was not followed. More than two centuries later, the great judge and legalist Sir John Fortescue, writing on the laws of England, stated emphatically that torture was not permissible under English law; it was, he said, one of the respects in which English law was superior to foreign systems. That Fortescue reflected not just legal opinion but the overwhelming sentiment of the English people was shown, at the time, by the popular fury aroused by John Tiptoft, Earl of Worcester. Tiptoft was the most accomplished lay scholar of his age. He had travelled widely in the Near East and Mediterranean, and had spent two years at Italian universities. Returning to England, he was a harbinger of the Renaissance and, in this context, an important figure in English cultural history. But Tiptoft had brought certain other ideas and practices back with him, which seemed to the English altogether more significant, and wholly evil. He had absorbed rather too much of the Continental 'culture'. As Edward IV's justiciar, he found the English common and statute law inadequate for his purposes of suppressing the Lancastrian cause. In 1462 he had the Earl of Oxford and Aubrey de Vere tried and condemned by what the English wrongly called 'law padowe' (Paduan law), an outlandish and intolerable importation, which denied the Englishman his traditional rights of defence. The trial was no more unfair than many others of the period conducted according to the customary process. But it caused deep anger, and the execution of the two men was regarded as simple murder. In 1468 Tiptoft had two Lancastrian agents, Cornelius and Hawkins, examined under torture – the first time such a practice had been permitted under the aegis of the law. Then, two years later, he introduced the punishment of impalement for traitors: a form of execution no more barbarous than those hallowed by English tradition, but Italian in origin and (it seems) utterly repugnant to the hardened and bloodthirsty London mob. When Tiptoft himself fell the same year, and was carried to execution at Tower Hill, a mass of heavily armed soldiers had to protect him from a lynch-crowd, which screamed out: 'The butcher of England.' Few judicial murders (it was little better than that) in English history have given more general pleasure, and Tiptoft lingered on in popular folklore as a cultured and alien monster. The irony is that he almost certainly saw himself as a civilised legal reformer, providing the benighted and unruly English with the benefit of the latest Continental ideas.

The extraordinary attachment of the English to their system of law (if indeed it can be called a system), the positive affection it inspires, the awe-inspiring confidence, often unwarranted, which they repose in its ability to do justice, the tenacity – indeed ferocity – with which they resist attempts to modify it with foreign importations, is one of the most enduring national characteristics. In a sense, the law is the only true English religion – the only body of doctrine in which the great mass of ordinary Englishmen have consistently and passionately believed. It is impossible to turn to any period of English history, where written records survive, without finding striking evidence of a huge and dogged conviction in the adequacy of the law if only, and this is the vital qualification, it is administered according to tradition and custom. Complaints about the law are purely conservative in nature. It is not being observed. It has fallen into disuse. It is being obscured and perverted by innovation. Grievances are strident and incessant: but they are invariably directed against agents – kings, justices, sheriffs – not against the law itself. It has a pristine virtue which will always shine through, provided modern accretions are periodically removed.

Now this English attitude to the law poses a delicate problem to any English government which wishes to improve it, or even to make it work at all. The concept of an ancient and perfect legal framework is, of course, an illusion. Such a thing has never existed, could never exist. But the English conviction that it does and must exist is so strong that any approach to change must be made from a conservative standpoint. It must be introduced under the guise of putting the clock back to an imaginary period in which the law flourished in all its majesty. The only form of progression is to move backwards into the past – but a past so imaginatively reconstructed that, in reality, it contains the necessary elements of novelty. This is the essential principle behind the development of English law, indeed of English constitutional history as a whole. The present is reformed by rewriting the past in such a way that it becomes the future. Thus continuity is maintained, no one's prejudices are disturbed (they are seemingly endorsed), and a forward motion is achieved under the appearance of regression. It is a process which requires a contempt for logic, a degree of self-deception, and often barefaced hypocrisy, with all of which the English are richly endowed.

It is with this principle in mind that we should look at the development of English law during the Middle Ages. Change is occurring all the time: but at no point can change be isolated from the body of custom, identified as novelty, and so objected to: on the contrary, it is disguised as reaction. The greatest English rulers had a positive genius for performing this conjuring trick. Consider the case of Henry II, perhaps the most gifted of them all. No one could conceivably call him a radical. He was the richest and most heavily endowed monarch in Europe, the lord of half of France as well as England: no one had a bigger stake in the established order. His programme was ostensibly one of pure reaction. William I had confirmed the laws of Edward the Confessor; Henry I had done the same. Henry II inherited a kingdom which had come close to breakdown, because (it was universally believed) these laws had been ignored, or broken, or tampered with. Henry II's coronation charter thus promises to erase the 'nineteen terrible winters' of Stephen's reign by restoring conditions as they were when his grandfather was alive. Thus, in a sense, the imaginary golden age of the Confessor is to be recreated.

But what was this golden age in precise terms? No one knew. The corpus of English law, such as it was, was obscure and bewildering in its contradictions. The written versions of the Confessor's laws were rambling documents, often specific only on points already, in the 1160s, irrelevant; the laws of Henry I were also largely useless for many practical purposes. The Normans had inherited three different systems of Old English law, in Wessex, Mercia and the Danelaw; they had added elements of their own; the Church was now energetically striving to insert the wedge of its own canonical system; and the resurrected principles of Roman law were attracting the enthusiasm of professional lawyers. How could these various elements be fused into a common system, so that everyone knew where they stood? And how could the law be modified, as and when required, to meet the needs of a society which was beginning to change with some speed?

The solution which Henry II in fact adopted was, in a sense, no solution at all: he simply embarked on a vigorous policy of law-enforcement, using himself as the principal instrument, but recruiting and employing a growing number of able men who acquired the expertise of professional judges, and who were sent on regular expeditions through the country to try cases. None of this was precisely an innovation; it was new only in its scope and thoroughness. The principle behind it was that if the law were consistently enforced, it would codify and rationalise itself by usage; the ubiquity of a centrally administered system would itself erode regional variations; and the experience of routine would automatically encourage judges to devise improved procedures which, in the guise of mere aids to efficiency, would in fact radically alter the law. All of this happened. Henry II carried out a legal revolution. But no one was aware of it. It was impossible to point to any one element in it which was not legitimised by earlier usage. Yet the law was fundamentally changed.

Henry II, in fact, legislated by stealth. In 1166, at the great Assize held at Clarendon, he carried out a comprehensive inquiry into all crimes and suspicious happenings, all legal commissions and omissions, which had occurred, or were alleged to have occurred, since the beginning of his reign. It led to furious activity, on his own part and that of scores of professional justices, which virtually turned the kingdom upside down. Many copies of the document recording the Assize were drawn up and published throughout the country; it was in effect reissued in 1176 and 1195, thus replacing Old English law-codes and their confused Anglo-Norman successors by a uniform system of common law applied to the whole country. It was, in reality, a statute which should, by rights, replace Magna Carta in the honoured place as the first of the Statutes of the Realm. Yet what, strictly speaking, was new about it to contemporaries? The King had simply held an important court, as all his predecessors had done from time to time, and given detailed instructions for the law to be enforced, 'with the consent', as the document says, 'of all his barons'. There was nothing revolutionary here: the only novel element was the scope of the action, and the vigour with which it was carried out.

Again, take Henry II's famous 'petty assizes'. What was fundamentally wrong with English twelfth-century society, as he found it, was the terrifying uncertainty which surrounded rights and property. Economic relationships were becoming far more complex; land was changing hands rapidly by death, inheritance, subdivision, gift and sale. Many titles were in dispute, records non-existent, suspect, stolen or forged. The law was a clumsy apparatus. Often it might not work at all. Sometimes it took years to achieve any decisive result. A curious document has survived in which a landowner called Richard of Anstey noted down in immense detail the wearisome steps he had to take to secure from his uncle an inheritance which was also claimed by a bastard niece. It is a profoundly depressing account. The cost to Anstey was enormous: over £330, probably more than the total value of the estate; and the money, almost certainly, had to be raised on mortgage at rates of over 50 per cent. The temptation to resort to force as an alternative to the law was always strong. A widow, a young heir, an heiress, anyone not in a position to defend their rights against swift and violent dispossession, might never get redress. It was the chief remediable cause of human misery. Where the dispute was complex, of course, there was little reform could do. But the vast majority of cases were open and shut; all that was required was a simple legal device to make their settlement swift, simple and cheap. So Henry II and his advisers produced a series of short writs, applying to the commonest types of case, which could be bought from the Crown. If a man had been turfed out of his property, he bought a writ of '*novel disseisin*', took it to the local authorities, went before a jury, and if they swore he had been ejected, he automatically got possession in the King's name and could claim damages. To get inheritance of his father's estates he applied for the writ '*mort d'ancester*'; to prove his land was held by clerical or secular fee – a vital financial point – the writ '*utrum*'; to demonstrate his right to an advowson the writ '*darrein presentment*'. In the last case the procedure, typical of the new technique, was simplicity itself. The jury was merely asked: 'Who presented last to the benefice?' When they gave their answer, the verdict was: 'The same or his heir should present again.'

These writs appealed strongly to every instinct of the English. Where was the innovation? There was none, or so it could be argued. King's writs had been issued for centuries, if not precisely for this purpose. Juries of local people had been summoned to establish facts for almost – if not quite – as long in such cases. The very notion of establishing a verdict simply on what had gone before was a profoundly conservative and satisfying principle. The three elements were all old; only the conjunction was new. So a precedent was successfully established, on a sound basis of ancient tradition, and a momentous revolution in English law – perhaps the most important in its entire history – was carried through without anyone noticing. The precedent set, and hallowed, the way was open to further progress. The use of the jury in such cases was seen to be such a neat, equitable and popular instrument that it was soon applied to other types of case. In 1179 it was substituted for the judicial duel in property suits, to the general satisfaction. This led to the rapid demise of the superstitious element in all cases. And the use of the jury in civil suits opened the way, where neither party could be quite confident of the verdict, to the neat device of out-of-court settlements. The parties composed, paid a fine to the Crown, drew up an agreement, kept one copy each and deposited a third with the Treasury as record. From the reign of John onwards this procedure was adapted to cases where there was no dispute, but simply a need for an absolutely sound title: the fine became a conveyancing fee for a right to deposit a copy in the official records, giving the Crown's perpetual sanction to possession. The Crown was delighted by these developments, which brought it a reliable and growing income. The public welfare was enormously assisted. The law was set upon a new, radical and fruitful course, capable of infinite elaboration. Somewhere along the line a revolution had occurred, but before the point was noted the elements which composed it were already encrusted with the reverence of centuries. It was a very English operation.

Yet Henry II, the man who gave this powerful and skilfully judged impetus to the development of a just and effective legal system in this country, has a slight and insecure position in the English pantheon. The historians of law regard him with profound respect – in the case of Maitland with real affection – but to most Englishmen he is simply a rash and intemperate monarch whose inconsiderate words led to the brutal murder of England's greatest medieval saint. For this the monastic chroniclers and the hagiographers of Becket are chiefly to blame; and even a shrewd and original writer like Gerald the Welshman presents a hostile portrait of the King, for Gerald had a personal grudge: Henry's refusal to make him bishop of St David's. The image of the King as a man born to greatness and ruined by unbridled passion, the tyrannical head of a bawling, screaming family of incompatibles, is almost wholly false. In fact Henry was a man peculiarly well suited to rule medieval England. The English expected their King to be a chief executive in every sense of the word, to be a man of *gravitas* and dignity, learned in the customs of the country, scrupulous in observing their spirit, who listened to and noted the views of the magnates, was attuned to popular opinion, but at the same time willing to accept full responsibility for the initiation and execution of policy. He had to be kingly – he had to look and behave regally – but he had also to possess a dedication to the minutiae of official business rarely found even in the most industrious civil servant.

The English in fact expected too much, and they very rarely got complete satisfaction. But Henry II must have come close to their ideal. He was a very professional king. He took a deep interest in the proceedings of the Exchequer, and may have presided over its sessions: 'Where the King's treasure lies,' quoted the *Dialogue*, 'there lies his heart also.' He ran the finances not only of England but of his wide French territories in a highly capable manner. Estimates of the surplus he left at his death range from £60,000 (over a year-and-a-half's State income) to the enormous sum of £600,000, and this was achieved not by abuse but by careful and efficient management. He was the last king for 300 years to leave the state in a creditor position. He was not just a legal innovator but an assiduous and greatly respected judge. A large portion of his time was absorbed in presiding over court cases, some of which lasted from eight in the morning to nightfall and beyond; all really important cases the King handled personally – this was sometimes a cause of delay; and there is no doubt that litigants were anxious to get their cases settled before Henry, not simply because his verdicts carried the highest authority but because they were reached in a convincing and impressive manner. Wherever he went, and he travelled during his reign to practically every corner of the kingdom, he was besieged by immense crowds of people: his face was familiar to a very large proportion of his subjects. Henry was shy and diffident; but he forced himself to move among the crowds, and never lost his composure when they pulled him about. He had an impressive capacity to remember faces and names; once he looked hard at a man's features, he never forgot them. He was an accomplished linguist. A talk with the King was a memorable experience, for he quickly seized on the heart of the matter, and had a gift for the lapidary phrase. Some faint echoes of the great political debates in which he participated reach us from the documents, revealing a formidable marshal of arguments. A financier, a judge, an administrator, a public relations expert, an articulate politician and diplomat: all these things Henry was expected to be, and was; but hardest of all, he had also to be, at frequent intervals, a professional soldier. There is something pathetic in the spectacle of Henry, at the end of an arduous life, in his late fifties, buckling on his armour and preparing to take part personally in energetic and ferocious hand-to-hand combat.

But these duties could not be avoided. A professional medieval king had to be not only omni-competent but ubiquitous. His personal presence on the battlefield was mandatory. His active supervision of all aspects of government was essential if the machine was to function at all. Any attempt to innovate or reform required intense and relentless exertions on his part. Henry had some good servants; but no official could be trusted beyond a limited point. The ablest ones were often the most corrupt and suspect. Any delegation of authority was a risk. The King travelled incessantly on business: his consumption of horses was enormous. On one occasion he covered 140 miles in two days of riding; 50–80 miles a day was not unusual for him. Speed of movement was the key to successful kingship; so, also, was secrecy. Some of Henry's most important movements escaped the knowledge even of the best-informed chroniclers, and are revealed only by his charters and pipe-rolls. The King of France marvelled at Henry's celerity, and thought he must travel by some supernatural means.

Such a life was ultimately intolerable. Henry's only relaxation was hunting. He never took a holiday, and can rarely have passed a night without worry. He had great nervous energy: his courtiers resented the fact that he never sat down except to eat and sleep. But the strain evidently told. Henry's frantic attempts to diet sprang from the ominous knowledge that a King who could not ride a horse long distances, at great average speed, would rapidly lose control of events, and might in the end forfeit his crown and his life: an inactive monarch was always at personal risk. In many ways this was a brutal and merciless society where the punishment for political failure was death. The last months of Henry's life were clouded by the despair engendered by the physical and nervous exhaustion of decades. It was the fate of all the great medieval English kings who did not die young.

Henry was motivated only in a superficial degree by personal ambition. What made him a great and characteristic English statesman was a passionate regard for public order; and it was to this that the English people responded. No race on earth has such a consistent and rooted hatred of unauthorised violence. Extremely violent by nature and instinct, their political capacity for self-knowledge has always placed the highest premium on the control and subjugation of these terrible forces within them. From Anglo-Saxon times to the present, English history is the long record of the struggle for self-mastery, the remorseless, often unsuccessful, attempt to release themselves from the drug of violence. It has been, on the whole, a remarkably successful struggle; but for this drug there is no such thing as a wholly complete cure, and constant vigilance will be needed so long as the English race lasts. At any rate, Henry II was unusually well attuned to this English preoccupation. He had violent instincts himself; equally, he was a passionate self-disciplinarian. His love of order was an intellectual concept which he ruthlessly superimposed on his own chaotic nature. His kingdom was the macrocosm of himself. How could some degree of respect for the law – some alternative to habitual violence – be imposed upon it? This was the salient object of his public life.

The volume of violent, serious crime in twelfth-century England was enormous. When court records begin to appear, as they did shortly after Henry died, they present a picture of viciousness which would appal even the most pessimistic American police commissioner today. The justices who visited Lincoln in 1202, for instance, found 114 cases of homicide, 89 of robbery, usually with violence, 65 of wounding, 49 of rape, and a great many others. Moreover, many violent crimes never came before the court, for want of evidence or unwillingness to lay charges. Against this tide of perpetual lawlessness, Henry struggled with only partial success.* But to some extent he was able to involve the more public-spirited elements of society in the process of law-enforcement. Having clarified both civil and criminal law, he enlisted ordinary freemen (as jurors) and local gentry (as administrators of justice) in the business of getting it observed; his more responsible barons could always find regular and well-paid employment as judges. He invested the royal courts with a salutary measure of terror and majesty, and drew to them a growing volume of business from the ramshackle private courts of the baronial honours. It was very limited and piecemeal progress, but progress all the same; under Henry a murderer stood a growing chance of apprehension, and a judge could rarely be defied with impunity. The State thus moved perceptibly closer to the people, sometimes in the most ominous and uncomfortable manner. Clause 6 of the Assize of Northampton (1176) obliged everyone, even villeins, to take a personal oath of allegiance to the monarch, under pain of arrest; Clause 2 went even further:

> Let no one either in a borough or a village entertain in his house for more than one night any stranger for whom he is unwilling to be responsible, unless there be a reasonable excuse for this hospitality, which the host of the house shall show to his neighbours. And when the guest shall depart, let him leave in the presence of the neighbours, and by day.

Yet however much such measures might be resented by some, it is clear that Henry's campaign for law and order met with the enthusiastic approval of the overwhelming majority of the English people of all classes. It responded not only to a national need, but to a popular demand.

There was, however, one element in society which not only refused to cooperate but actively resisted the Crown. This was the Church, or rather an influential section of it. Here we come to a significant watershed in English history, an episode which tells us a good deal about England and the English. The Church of England, though an importation from Rome, had played a very important state function in Anglo-

* William I had a conscientious objection to capital punishment and substituted mutilation. But hanging was restored by Henry I; women were burnt. At a single court session in Leicestershire, his judge Ralph Basset had 44 thieves hanged. Under Henry II hanging alternated with mutilation; thus in 1166 at the London-Middlesex assizes, 14 men were hanged and 14 mutilated. Henry II's legal reforms made it more difficult for the authorities to secure a conviction, but on the other hand made it far more likely that criminals would be brought to trial. Men caught in the act were often executed without trial; as late as 1603, James I, travelling south to London for his coronation, had a red-handed thief hanged on the spot.

Saxon times. From its earliest existence, it had been closely associated with the spread of civilisation, the concept of law, especially of written law, the development of a fair system of taxation, defence in war, administration in peace, above all with the proper functioning of a powerful central monarchy. Next to the Crown itself, it had played a bigger role than any other element in the evolution towards a stable, unified civil society. With scarcely an exception, bishops and kings had worked together in constructive harmony. Indeed, it was impossible to separate the functions of Church and State. Now this is a very English concept: the idea that the spiritual authorities should underwrite the operations of civil government. It is also a brilliant formula for domestic tranquillity. No English writer of the Dark Ages expressed it in words; it was taken for granted, an obvious and pragmatic contribution to the problem of maintaining order.

But the Church of England was also linked to an international organisation, based on Rome. This link, as a channel for ideas and culture, was welcome to the English, in the same way that the English Channel itself served to transmit the controlled importation of Continental goods and notions. But the stress is on 'control'. It was never tolerable that the international links of the Church should be used to transmit orders which in any way limited the sovereignty of the English people. Ultimately the direction of the Church of England had to lie in English hands: the links had either to be controlled from this side of the Channel, or snapped. The history of England's relations with Rome over five centuries is a series of variations on this unwavering theme.

The late eleventh century introduced a period of papal expansion. Under the guise of spiritual reform, Pope Gregory VII (Hildebrand) elaborated a programme of political action which would have absorbed the whole of western Christendom into a centralised theocracy under the sole and absolute direction of the Pope. He set down his aims in a series of propositions which he dictated to a secretary and which was placed in his letter-book. The world was Christ's kingdom: the Pope his vicar, exercising all authority whatsoever, with the Church as his spiritual and the State his secular arm: bishops and kings alike would be mere functionaries of Rome, with the priest-emperor transmitting to them the commands of the Deity. Not until the communist manifesto of 1848 did the governments of the world face such an audacious programme of international subversion. It was a threat to the established order of all countries. But it was in England that it was most bitterly resented and most fiercely resisted. Here, for the first time, England was to lead the struggle against a deliberate attempt to set up a European tyranny.

The opening phase of the papal plan was to elevate its international agents, the priesthood, into a separate and privileged class, exempt from the normal processes of local administration, and above all of justice. The Church sought, in the first instance, to set up separate courts to try cases solely concerned with spiritual matters; then to use these courts as the exclusive instrument for clergy accused of secular offences; finally to embrace within the spiritual ambit an ever-widening variety of secular crimes and disputes. Ultimately, then, the Church would take over the judicial function of the State, and from thence it was only a small step to take over administration and everything else.

Such a plan might be carried out in Tibet; it was inconceivable that it should succeed in Europe, above all in England. The only questions were: could a compromise be reached, and if so what form would it take? The English are gifted at finding such pragmatic solutions, in which theory is not pressed too hard on either side. It says a lot for the English genius that the break with Rome was delayed so long. William I set the pattern, with a combination of firmness on essentials and gracious gestures (chiefly money) to keep the Pope happy. Despite Anselm and his like, there is plenty of evidence that by the mid-twelfth century arrangements satisfactory to both sides could usually be reached on most points in dispute, such as the selection of bishops and the overlapping of jurisdictions. In most cases the State got its way, as was inevitable: for it was the State which had the responsibility of actually running the country. Then came Thomas Becket: and his story illustrates what can happen when a powerful and violent personality tries to overthrow a characteristically English way of doing things.

Henry, as we have seen, was preoccupied with the problem of violence, and especially violent crime. One very important aspect of this was the rising volume of crimes committed by clergymen. There were perhaps 100,000 people in the country who could make some claim to possess clerical orders. Only a minority possessed one of the 9,000 or so available benefices, or could rely for their support on a religious house. Some lived by crime; a large number were attracted to felony from time to time in order to live. The King's mother, the old Empress, could recite from memory a long list of outrageous clerical offences. Some were of national notoriety; and what most angered the public of all classes was the use of the clerical courts to enable these men to escape punishment. There was, for instance, the case of the Archdeacon of York, who was alleged, in 1154, to have murdered his archbishop by slipping poison into his mass-chalice. The archdeacon may have been innocent; what angered all laymen, from the King down, was that he was never brought to trial in a royal court. There was a more recent case of a canon of Bedford, charged with the murder of a knight, who had then insulted the sheriff and pleaded benefit of clergy. (Becket eventually had this man scourged for contempt of court, but he never stood trial on the capital charge.)

It was this growing problem, almost certainly, which led Henry to appoint Becket Archbishop of Canterbury. Becket was a Londoner, the son of middle-class Norman parents. He had received a good education, had travelled, and had picked up some of the new papal ideas in the household of old Archbishop Theobald. But he was a worldly man, a pluralist; the King had found him an energetic and single-minded public servant, and had made him Chancellor. He had fought alongside the King and acted as his ambassador in Paris. He was fond of money, still more of spending money; but always willing to fight the King's battles against the Church. Henry did not want a war with the Pope; on the contrary, he wanted a workable compromise, of the type Lanfranc had achieved. He thought that this could be brought about if Becket combined the role of Chancellor with metropolitan. This was why he appointed him, over the claims of several more qualified churchmen.

Now to an exceptionally busy and serious-minded head of government, like Henry, Becket's behaviour after his installation must have seemed criminally irresponsible. Not only did he immediately resign as Chancellor, but he set himself to sabotage the royal efforts at law-enforcement on precisely the issue of criminous clerks which Henry, as he knew, was determined to resolve. Henry seems to have completely misread Becket's character: perhaps he knew him much less well than popular tradition supposes. Becket was the dangerous type: a man of concentrated energy, with second-class brains and no sense of proportion. It was beyond his nature to balance the claims of Church and State in one judicious personality. Becket patronised actors; in a sense he was one. He could only play one role at a time, but into that he threw everything he possessed. He had acted the Chancellor; now he acted the archbishop; when that role palled he would act the martyr.

Becket's provocative attitude led the King into the tactical error of setting down his definition of Church–State relations in writing – the Constitutions of Clarendon. This was bound to lead to a formal debate with the Pope, and indeed with the English hierarchy, who in other respects were behind the King and willing to concede that he had tradition and custom on his side. It is a useful lesson of English history that, if you have the advantage of tradition, it is a mistake to prejudice it by putting it in words. But Henry was doubtless angered by Becket's flat refusal to compromise. The Archbishop took no steps to curtail crime among the clergy by improving the methods of selection, by a judicious programme of unfrockings, by an increase of severity in clerical courts.

There is no evidence that he took any interest in the moral aspects of the problem; or indeed, that he had the slightest concern for his primary duties as archbishop – the pastoral care of two-thirds of the English people. He did not in fact perform any strictly clerical duties, beyond saying mass. He paid no attention to purely moral or religious questions; whether he was a Christian in any meaningful sense is open to doubt. What mattered to him was power, authority, jurisdiction and the privilege of his caste. It is not surprising that Henry, who had a very definite layman's sense of right and wrong, found Becket's spiritual pretensions intolerable.

The course of this famous dispute – Becket's condemnation by Henry's court at Northampton, his flight and long exile, his apparent reconciliation with the King, his defiant return and his murder – is one of the best-documented episodes in the whole of medieval history; 11 contemporary biographies of Becket survive, together with over 700 letters from the interested parties, and there is a mass of other material. What strikes the modern reader is the extraordinary violence of Becket's attitudes, and his gradual loss of balance. When Becket was condemned at Northampton, he turned on two members of the King's court, Henry's illegitimate half-brother Hamelin and Ranulf de Broc, and shouted: 'Bastard lout! If I were not a priest, my right hand would give you the lie. As for you [to de Broc] one of your family has been hanged already.' Men were killed for less than this in the twelfth century. The correspondence on all sides is pretty vituperative; but Becket's side of it is distinguished by the ripeness and variety of his abusive language. The English aristocracy, who were absolutely united behind Henry, had some grounds for their view that Becket was a vulgar and mannerless upstart; their hatred of him was coloured by a strong sense of class solidarity. Even Becket's natural supporters found him an embarrassment. No pope was more anxious to enforce clerical claims than that accomplished litigant Alexander III; but he had got himself involved in a disputed papal election – the fatal weakness of the medieval papacy – and needed Henry's support against the Anti-Pope. Lesser papalists were dismayed by Becket's tactical blunders and by his growing bitterness and intransigence. The senior English clergy generally sided with the King. The wisest of them, Gilbert Foliot, Bishop of London, a man distinguished for his learning and pastoral work, took the view that Church and State had to live together in harmony, thus expressing (perhaps unconsciously) the broad-minded Anglicanism of Bede. As he said to Becket: 'If the King were to wield the temporal sword with the audacity with which you wield the spiritual one, how can we ever have peace?' The trouble with Becket, he added, 'is that he has always been an ass, and always will be.' The Empress Matilda, speaking for the older generation of the English establishment, gave some sensible advice: the Constitutions should never have been written down. All documents and oaths should be withdrawn on both sides. After that, royal justices should be told to be careful, and bishops to be reasonable. Henry himself seems to have been willing, at all stages, to accept any workable compromise which allowed him to get on with his job of running the kingdom: the expenditure of time and energy on the controversy was prodigious, and he (unlike Becket) had plenty of other things to do.

But Becket was an extremist and a doctrinaire; he did not want a compromise but a royal humiliation, of the sort Gregory VII had inflicted on the Emperor at Canossa. Towards the end, he seems to have lost sight of the original issues in dispute, and was searching for new grievances. Before he set foot again in England, on 1 December 1170, he found one. Earlier that year the King had had his eldest son crowned; Becket being unavailable, the Archbishop of York had officiated, assisted by six other prelates. No doubt Canterbury usually had the right to crown a king; on the other hand, this kind of rivalry between the two provincials was ecclesiasticism at its most disreputable and vulgar, something which, if tolerated at all, was best left to semi-literate monks from the two chapters, who could (and in 1176 actually did) beat each other with clubs in support of their claims. But Becket was by now so obsessed with his wrongs that he seized on this trivial point to excommunicate the offending bishops. It was their natural complaints which exasperated Henry beyond endurance, and led to the tragedy.

It is evidence of Becket's state of mind that, in his sermon on Christmas Day, a few days before the murder, preached to the text 'Peace be to men of goodwill', he dwelt angrily on the wickedness of the King's men who, he said, had cut off the tail of a horse belonging to one of his servants. He excommunicated the culprits there and then. Thus the momentous conflict between Church and State came down, in the end, to the simple matter of a horse's tail. The archbishop was in danger of dissolving his cause in ridicule. Perhaps he realised that he was now at the end of the road, and that the time had come to abandon the tattered role of embattled cleric and embrace the new one of martyr. At what point does the quest for martyrdom become a matter of suicide? To get himself killed was now the only way in which Becket could damage the King. There is no evidence that the four men who travelled from Henry's court to question Becket were bent on murder. They were not riffraff, but senior barons, substantial landowners; one had been a royal justice. They came to Becket, in the first instance, unarmed, intending remonstrance, perhaps threats; their object may have been to take him into custody, for trial, for Becket had undoubtedly broken the law. It was the angry argument which then developed which led the King's men to rush off for their armour, and made killing inevitable. Becket's behaviour brought a sad protest from his ablest adviser – the only one who was not a sycophantic monk – the famous writer and canonist John of Salisbury. Why, he asked, did Becket always refuse to take advice? Why bandy furious words with these wicked men, and exasperate them still further? Why had Becket followed them to the door shouting at them? Becket's reply was that he had done with taking advice. Even now, he could have sought refuge. His subsequent actions that evening, of which we have various minute-by-minute accounts from eyewitnesses, make it plain he deliberately chose to be assassinated.*

The murder threw Henry into a state of nervous prostration; and it is not hard to see why. He was a man of law, a sworn enemy of violence, whose whole life and policy were devoted to the establishment of order and the due processes of the courts. This atrocious crime, for which he must bear some responsibility, was the negation, the denial, of all his principles. It was a disaster for the English Crown, a humiliation for the English people; and it was also a personal tragedy for himself.

But when Henry recovered his composure, he in fact acted with remarkably sagacity. He drew heavily on the deep wells of English hypocrisy, on the English capacity to muddy hostile waters and confuse inconvenient issues. First he suddenly found it necessary to go campaigning in Ireland (then and for many centuries to come a refuge for English grandees in temporary disgrace) until the immediate storm blew over. On his return he did humble penance. Substantial sums of English taxpayers' money were transferred to eager palms in the Lateran. By now the first miracles at Canterbury had taken place; and, when all was said and done, they were English miracles. A murdered archbishop was a threat to the English Crown; an honoured martyr could be made into an English national asset. The English establishment moved into action. Becket got his martyr's accolade with remarkable speed. Alexander canonised him only two years afterwards: Rome is always grateful to clergymen who make life in England difficult. But he misjudged the resourcefulness of the English character. With shameless effrontery, the Crown took over the new saint. Henry was heard to mutter appeals for St Thomas's assistance in moments of crisis. His three daughters, comfortably married to European sovereigns, enthusiastically spread the new English cult throughout Europe. His justiciar, Becket's mortal enemy, built a chapel in his honour, without abating

* For a recent, and far more charitable, view of Becket's conduct, see David Knowles: *Thomas Becket* (London 1970).

by one jot his campaign against clerical privilege. Canterbury became an international shrine to rival the much-envied tomb of St James in Compostella, a financial godsend for centuries to the citizens of the town, monks and laymen alike, and a rich source of foreign exchange for the English economy.

As for the causes for which Becket died, the Crown carried on much as before. The Clarendon Constitutions were, in practice, enforced. Only on the matter of criminous clerks was the State, for some time at least, chary of pushing its claims. Becket died so that a few clerical murderers might go unpunished. Or did they? There is evidence that, from time to time, they met rough justice, not in the King's court, or in the Church's, or in any court at all: they were simply (so clerical spokesmen complained) 'hanged privily at night or in the luncheon hour'. The Becket affair changed English history in only one respect: it gave birth to English anti-clericalism, a smouldering national force which was to grow in depth and volume until it found expression in the Reformation. There was more than gruesome symbolism in Henry VIII's treatment of Becket. In 1536 he instructed his attorney-general to institute *quo warranto* proceedings against St Thomas. The corpse was assigned counsel at public expense, but found guilty of 'contumacy, treason and rebellion'. The bones were scattered and offerings made at the shrine confiscated by the Crown. It was declared illegal to call Becket a saint, and it was further ordered 'that all images and pictures of him should be destroyed, the festivals in his honour should be abolished, and his name and remembrance erased out of all books, under pain of His Majesty's indignation, and imprisonment at His Grace's pleasure'.*

English anti-clericalism was, of course, merely one important branch of English xenophobia. Hostility to foreigners is one of the most deep-rooted and enduring characteristics of the English; like the national instinct for violence, it is a genuine popular force, held in check (if at all) only by the most resolute discipline imposed, against the public will, by authoritarian central government acting out of enlightened self-interest. Racialism has always flourished in England when government has been weak, and the sophisticated governing minority have lacked the will to resist public clamour. The claim, sometimes advanced today, that England

* There is no authentic account of this 'trial', and it may be an invention by sixteenth-century Catholic propagandists. There is also a tale, impossible now to prove or disprove, that Henry VIII's body met a similar fate: his daughter Mary, it is alleged, had it resurrected and burnt as a heretic. Certainly, Henry's tomb was never finished; the screen was taken down and the ornaments sold by Parliament in 1646; the sarcophagus was eventually used for the body of Nelson, and is now in the crypt of St Paul's.

has an internationalist outlook, and a talent for promoting inter-racial harmony, is spurious and lacks historical justification, at any rate so far as the great mass of the English are concerned. Tolerance has only been imposed in the teeth of their resistance. The evidence on this score is overwhelming; the only difficulty is to determine precisely where English racialism begins. The area of racial respectability, centred on London, has often appeared to extend no further than the lowland zone, bounded by Severn and Trent, and not invariably to all of this.* Henry VIII, admonishing the men of Lincolnshire who had participated in the rebellious Pilgrimage of Grace, told them that they hailed 'from one of the most brute and beastly shires of the whole realm'.† English governments could usually cope with insurrections from outside the zone, which lacked a pure English centre of gravity; but a rebellion in the south-east was always a serious matter, and usually fatal. It was the south-east which determined the course of the Reformation, the Civil War, the Glorious Revolution and the Hanoverian settlement. Not until the nineteenth century, with its dramatic shifts in population and economic resources, was the north-west able to assert decisive political influence: the repeal of the Corn Laws in 1846 was its first regional victory, and in a sense its last, for the balance has since swung back to the south-east: no English government has been formed in recent decades without majority support in this area.

But the real racial frontiers were fixed in the Welsh and Scottish marches. Beyond these limits even Roman military power had encountered difficulties which ultimately proved too expensive to resolve. It is true that the Romans established a form of military occupation in Wales. But the normal processes of economic colonisation could not operate there; the Welsh economy remained pastoral; the people could not be effectively disarmed, indeed in the closing stages of Roman rule they were recruited as auxiliaries; their tribal organisation, laws, language and customs remained intact; only Christianity made any impact. In Scotland even the Roman military presence was fugitive and ineffectual, and the Roman economy made no progress north of the Tyne–Tees, or indeed for many miles to the south of it. This pattern was repeated during the Germanic settlements, which made no substantial penetration beyond the line now known as Offa's Dyke, and the Old Roman Wall. The racial and cultural frontiers began to solidify in the

* Hence the ancient phrase 'the Home Counties' (i.e. the non-foreign counties); these are the counties bordering on London, plus Hertfordshire and Sussex. The people of these areas spoke the East Mercian dialect which became the basis of modern English.

† Henry VIII never felt safe in the north, and went there only with the greatest reluctance, and under heavy escort. Elizabeth, though on constant progress, never went north of the Trent or west of the Severn; she was a Home Counties queen.

eighth century and have never changed by more than a few score miles. The Normans, as the residual legatee of the Old English State, became the dominant landowning element only within the areas of effective English occupation. Thus the relationship between England and its Celtic neighbours began to assume its modern form from the beginnings of the twelfth century.

This relationship was, and remains today, essentially ambivalent. The English could never establish a cultural ascendancy in Wales or Scotland, or destroy in their peoples a sense of separate nationhood based on race, even though the English language became predominant and eventually triumphant. On the other hand, England was inevitably the paramount power, in a military and political sense, in the British Isles. To what extent should this preponderance based on greater resources of wealth and manpower be expressed in direct political sovereignty? This question has never been finally answered; perhaps there is no answer which all the parties can find fully acceptable. The claims of the Old English kingdom, as expressed for instance in Edgar's coronation, were theoretically limitless; and they were inherited by the Anglo-Norman monarchy. But to enforce them was a different matter: here the pattern of development is ragged and contradictory. Wales was just close enough to the English centre of gravity to permit conquest; Scotland just too far away, and the modern relationship was established by diplomacy and agreement rather than force. But all this took time: throughout the Middle Ages the English confronted their Celtic neighbours in an atmosphere of mutual and often violent hostility. Moreover, racial fears were intensified by geographical factors. Both the Welsh and the Scots quickly learned to synchronise resistance to the English with the hostile efforts of England's Continental enemies, above all the French monarchy. The domestic divisions of the British Isles became an integral part of the struggle for supremacy in north-west Europe. Thus the racial antagonism, based on an arrogant sense of cultural superiority, which the English felt for the Celts, was sharpened by the fear that England was always the potential victim of a conspiracy of encirclement.

The Welsh were the earliest victims of this terror-psychology. It is interesting that the Welsh initially saw the Normans as their natural allies, in more than just a military sense, against the hated English. Gerald the Welshman, writing in the late twelfth century, described the English as a people born to slavery: the noble Normans, and the free-born, fearless Welshmen, were the racial types to be admired. But a hundred years later Norman and English interests and stock had coalesced and united in a common anti-Welsh racialism; Edward I determined on a final solution to the problem of the Welsh frontier by an outright and permanent military occupation, underwritten by a colossal infrastructure of castles, roads, ports, and towns colonised by Englishmen; and this political aim was reinforced by a racial ideology. The Church of Rome had always favoured the outright English conquest of the Celtic fringe – a policy inaugurated in the seventh century by the Synod of Whitby – as the only means whereby the Celtic Churches could be brought within its unitary system of discipline and administration. Thus the Archbishop of Canterbury, Pecham, was instructed to be the fugleman of Edward's armies; all Welsh who resisted were automatically excommunicated; and one of the archbishop's clerks wrote a racist diatribe which became the moral manifesto of conquest. The Welsh were 'Trojan debris', swept into the wooded savagery of Cambria under the guidance of the Devil. Their sexual promiscuity was notorious; they spent their lives in theft and rapine, or sloth; they were so depraved that only a few had learned to till the soil. Only the mild forbearance of the English kings had prevented the English from long ago blotting out the existence and memory of this 'detestable people'. Edward's military architect, James of Savoy, put the matter more prosaically: so long as war with Scotland and France was possible, Wales would always constitute a threat to the English, for, he said, 'Welshmen are Welshmen'. More than a century later, continued Welsh resistance to the English Crown led to petitions to parliament for the enactment of racial legislation: privileges enjoyed by Welshmen residing in England should be withdrawn, and Welsh purchase of land in England forbidden; Welsh tenants should be automatically obliged to give securities of good behaviour, and Englishmen in Wales should be given special legal protection against the malice of Welsh juries. In 1403 one of Henry IV's officers told him that 'the whole of the Welsh nation in these parts are concerned in the rebellion', and he pleaded with the King 'to ordain a final destruction of all the false nation aforesaid'. No such extermination took place; it was beyond the capacity, and perhaps the desire, of the English; it proved impossible to carry out even a general policy of colonisation; but the Welsh ruling class and aristocracy were largely destroyed, and Wales wholly absorbed in the English system of administration, a process completed by Henry VIII.

The English undoubtedly wished to impose the same fate on Scotland. At all times they hated their northern neighbours even more than they hated the Welsh, because they feared them more, because a Scottish–French alliance was a more dangerous combination, and because Scottish treachery (as they saw it) was more expensive and difficult to punish. It was always much easier for English governments to recruit

English armies against Scotland than against France, though the chances of profit were far smaller. To fight the Scots was often a pleasure as well as a duty. The English attitude was summed up by Henry VIII's envoy in Scotland, Ralph Sadler, who complained to his master: 'Under the sun lives not more beastly and unreasonable people than here be of all degrees.'* The Scots were saved by geography, by timely resistance, perhaps most of all by Elizabeth I, the first modern-minded English monarch, whose Scottish treaty of 1560 prepared the way for a political solution based on consent.

By contrast, the relationship of the English with the Irish is a saga of unrelieved tragedy, from the mid-twelfth century to the present day. Any theory that the English have a natural capacity for governing other races cannot survive even the most cursory examination of Anglo-Irish history: English policy-makers committed every conceivable error from the first moment of contact, then sought to retrieve their blunders by savagery. The chief trouble, ironically, was that Ireland never sufficiently occupied the centre of England's political consciousness. It was a marginal threat, a marginal problem, and a marginal asset. The English have never been able to let the Irish wholly alone; on the other hand they have never given Ireland a high priority in their national schemes. Until the great labouring Irish migration of the nineteenth century, the mass of the English had had no direct contact with the Irish; racialism, on this side of the Irish Sea, was a matter of hearsay, distant rumours of an unsatisfactory people. Contact between the races devolved upon a small group of Anglo-Norman settlers, invested by the English State with plenary powers of conquest, but lacking the means and numbers to achieve it. They were beyond the range of effective supervision from London. They were active and aggressive enough to arouse Irish antagonism and provoke periodic resistance; but physically incapable of suppressing it without help from England. Thus Anglo-Irish relations became a succession of episodes, following a dreary and repetitive cycle of misrule, rebellion, suppression, and then malignant neglect, leading again to misrule and rebellion. The English settler class could not complete the conquest; neither would they adopt a thorough-going policy of assimilation: they oscillated uneasily between the two. Irish society became stratified on a racial basis: the kind of relationship which existed between English and Normans immediately after 1066 was, in Ireland, frozen into permanent antagonism. English policy in

* On the other hand, ambitious Scots had been travelling south to make careers in England since at least the thirteenth century. Robert Bruce, Earl of Carrick, and claimant to the Scottish throne, served Henry III as a judge for nearly 20 years, ending up as chief justice of the King's Bench. A King of Scotland served in Edward III's army, and was paid £2 10s a day. Welshmen, too, served in English armies, as archers.

Ireland, right from the start, failed to create a viable state: it was never more than an occupation, and a precarious and partial one. Its political motivation was fear, its instruments invariably force. The English could not conquer Ireland; but they would not relinquish it. They could not administer it; they could only, from time to time, subdue it. To complete the tragedy, the English in England, viewing Ireland from afar, conscious indeed of its existence only when misery erupted into violence (which they then felt in duty bound to put down), came to regard the Irish as wholly unreasonable people, who could not be fitted into any known scheme of government, a society *contra naturam*. The blame for English failure was complacently shifted on to the heads of the victims, and the English closed their eyes to the true nature of the Irish problem. Their eyes are still closed: though Ireland can still attract English attention by violence, it cannot command an understanding. Booksellers and publishers agree that it is not easy to sell serious books on Ireland to the English public; and not one Englishman in a hundred has ever heard of the Statutes of Kilkenny.

Yet these English laws, passed in 1366 and retained on the statute book until well into the seventeenth century, were the keystone of a policy which turned Ireland into the South Africa of the Middle Ages. They were wholly racist in inspiration, and their object was a crude form of apartheid. The attempt to govern the whole of Ireland was abandoned. The English colony was to be limited to the 'obedient shires' which constituted the Pale. Those who lived beyond the Pale were officially designated as 'Irish enemies'. As early as 1285 complaints had been made that Irishmen should never be appointed to bishoprics, 'as they always preach against the King'. Now the Irish – the custodians of an ancient church, which flourished when the English were still pagans – were to be excluded by law from all ecclesiastical office. The English were not to enter into negotiations with the Irish; or to marry them or sell them horses or armour. The English settlers were to be protected from 'degeneracy' by a variety of prohibitions. They must not employ Irish minstrels, poets or story-tellers; they must use English sermons, the English language and English customs. They were forbidden Irish sports, such as hurling and quoits, and commanded to learn the use of the bow 'and other gentle games'.

Thus religion had very little to do with the origins of the Anglo-Irish problem; or, rather, it did so only in a sense which later history made richly ironic. The original English invasion of Ireland, in the 1170s, was carried out at papal request, and with papal authority, by the bull *Laudabiliter* (1155). It is true that the bull may be spurious; true also that its supposed author, Hadrian IV, was the only Englishman ever to sit on the throne of St Peter. But Hadrian was more a cosmopolitan clerical careerist than an English nationalist; and in any event the real authority for the conquest was contained in letters written to Henry II by Pope Alexander III, a resolute Hildebrandine pontiff, with no love for the English. The English were encouraged to brutalise the Irish in the name of papal supremacy. England's title-deeds to Ireland were inscribed not in London but in Rome. The popes had always hated the Irish Church since the Synod of Whitby, and even before it.

By the time Richard II visited Ireland in the 1390s, the racial mould had set. He classified its inhabitants into '*irrois savages, nos enemis; irrois rebelz; et les Englois obseissantz*'. By '*irrois rebelz*' he meant 'degenerate' English, who had 'gone native'. English-born and Irish-born English settlers were forbidden by statute to shout racist expressions at each other: they were to stand together, in racial solidarity, against the 'Irish enemies'. Now from Richard II's classification it is only a short step to Cromwell's more famous one: 'English protestants (loyalists) and Irish papists (rebels)'. And from Cromwell's it is an even shorter step to the Unionist-Nationalist division in present-day Ulster. Religion, indeed, is not the root of the problem in Ireland; it is merely the colouration of an underlying racist division which is much more ancient.* The identification of Irish nationalism with Roman Catholicism was largely accidental. It was the fanatical Catholic sovereign Queen Mary who began the systematic plantation of English settlers in the confiscated lands of Irish rebels. Hence, when England turned to Protestantism under Elizabeth, the older Anglo-Irish landed class clung to the Church of Rome more as a protest against the newer English plantations than for doctrinal reasons. Catholicism and the Pope became an expression of Irish nationalism, and the Papacy, which had given Ireland to England, now exhorted the Irish to resist. No doubt Celtic conservatism helped to stiffen Irish resistance to reform; but the main impulse in Ireland's religious choice was political and racial. If England had remained Catholic, and France had turned Protestant, there is little doubt that Ireland would have turned Protestant too.

As a final irony, it is arguable that England might have accorded

* Dean Swift, writing from Dublin to Alexander Pope, said he was 'grieved to find you make no distinction between the English gentry of this kingdom, and the savage old Irish (who are only the vulgar, and some gentlemen who live in the Irish parts of the kingdom)'. Edward Carson, the founder of modern Ulster (though himself a Dubliner), made no bones about his racialism. In 1933 he wrote: 'The Celts have done nothing in Ireland but create trouble and disorder. Irishmen who have turned out successful are not, in any case that I know, of true Celtic origin.' (H. Montgomery Hyde: *Carson* [London, 1953], p. 491.)

religious toleration to the Irish, while still clinging to her political sovereignty. Henri IV's Edict of Nantes in France had set an important, and at the time successful, precedent. Its adoption was discussed in ruling English circles. Bacon noted: 'A toleration of religion (for a time not definite) except it be in some principal towns and precincts, after the manner of some French edicts, seemeth to me to be a matter warrantable by religion, and in policy of absolute necessity.' But such a policy broke down on the rock of racialism, particularly now that northern Ireland had become the theatre of Scottish Presbyterian settlement. Ireland thus became, and remains, the victim of a racial aggression, masquerading in the trappings of a religious controversy; and faith became the emblem of race, and thus of allegiance.

The growing aggressiveness of the English towards their Celtic neighbours, reflecting a new consciousness of their nationhood and racial unity, also found expression in hostility towards alien elements within the English community. The ruling class of the early Anglo-Norman kingdom had a distinct cosmopolitan flavour: we hear of no complaints against Lanfranc and Anselm, both Italians, on the grounds of race. The Jews, too, came to England for the first time in the wake of the Conqueror, and formed substantial and flourishing communities in many of the chief towns. For seventy years or more they appear to have lived unmolested. But as the twelfth century progressed, the native population, including the largely French-speaking aristocracy, began to draw fierce distinctions between themselves and 'those not of the nation of the realm of England'. It has always been in the economic interest of the English State to protect foreigners and allow them to go about their business to our mutual profit; equally, it has always been the popular desire to persecute and if possible rob them. When the medieval State was strong, foreigners were safe; the moment the Crown relaxed its grip, their lives and property were at risk. Alien trading communities had always to be placed under the personal safeguard of the King.

The Jews were a case in point. They were, in a legal sense, the property of the Crown, which systematically 'farmed' them. They alone, in theory at least, were allowed to lend money at interest (at rates usually around 50 per cent, but sometimes up to 66⅔). The King could tallage them at will, and at death their property reverted to the Crown, or could be possessed by the heirs only on payment of a heavy percentage fine. The Crown made it possible for the Jews to enforce the law against their debtors so that, in turn, it could take its cut. At Westminster special justices of the Jews and a separate exchequer were set up to administer the community. Their dealings were vast and played a crucial role in the money economy: they financed the development both of agriculture and the arts, and made possible the very rapid advance in English standards of life and culture which was such a marked feature of the twelfth and thirteenth centuries. Aaron of Lincoln, perhaps the most successful Jew in English history, operated in 25 counties; among his clients were the King of Scotland, the Archbishop of Canterbury, a score of bishops, abbots and earls, and innumerable lesser fry. He financed the building of Lincoln Cathedral, Peterborough and St Albans Abbeys, and at least nine Cistercian houses. When he died in 1185 the Crown set up a special exchequer, the *Scaccarium Aaronis*, to collect his debts, a process which took 20 years; it was probably the biggest financial windfall ever received by an English government.

But rising English racial consciousness made it increasingly difficult for the Crown to guard its protégés. The Church, heavily in debt to the Jews, fed the racial flames by manufacturing tales of Jewish ritual murders: the first and most notorious, of the child 'St William of Norwich' in 1144, eagerly spread by monks whose splendid estates and edifices the Jews had financed, led to an ugly rash of anti-Semitic riots. It is significant that this occurred during the worst years of Stephen's reign, when central government was near breakdown. Henry II's re-establishment of law and order allowed the Jewish communities to flourish once more – indeed reach the height of their prosperity; but soon after his death anti-Semitism again took to the streets. There were pogroms in London, Norwich, Lincoln and Stamford; and in York 150 Jews, who had taken refuge in the royal castle, were massacred. The ability of the Crown to protect the Jews was a faithful index of its general authority: when John was brought to his knees at Runnymede, the victory of the 'constitutional' forces was symbolised by the insertion of three anti-Semitic clauses in Magna Carta. Archbishop Stephen Langton, one of those who helped to draw it up, celebrated in English history as one of the architects of the constitution, was a notorious anti-Semite: he had an archdeacon, who married a Jewess and apostatised, burnt as a heretic, and he tried to enforce regulations compelling Jews to wear distinctive signs sewn on to their clothes.

Magna Carta undermined the economic basis of English medieval Jewry, though the communities struggled on. In 1264, when the Crown was again humiliated, there was a further wave of pogroms: part of Simon de Montfort's popular appeal (he was, ironically, a French racist who despised the English) was his aggressive anti-Semitism. By the time Edward I, a most magisterial exponent of monarchical authority, took over the government, English Jewry was near to ruin. In 1275 he enacted a Statute of the Jews with the object of transforming them from usurers into artisans. But this aroused the fury of the city tradesmen; moreover, it took from the Jews their unique role of service to the Crown. In 1290 Edward washed his hands of the problem and expelled the entire community, which was now destitute. In the fourteenth century English agriculture suffered grievously from the absence of Jewish finance, and the failure to provide a native substitute.

The departure of the Jews created a new role for the Italians, both in the economy and in the English racial consciousness. They became the new hate-objects in the towns. The Italians were unpopular because they were bankers; because they were the chief beneficiaries of the system of papal provisions to English benefices which developed during the later Middle Ages, and because they were successful tradesmen, with regular emporia in the chief cities. With the decline of Crown authority under the Lancastrians, and especially under Henry VI, their lives and property were increasingly vulnerable, with Parliament naturally leading the xenophobic pack. In 1440–2 Parliament passed measures which, in effect, sanctioned piracy against merchant ships owned by foreigners. In 1456 and again the following year, there were anti-Italian riots in London. The Italians fled in terror to Southampton and made it their operational base; but in 1460 an anti-alien faction captured control of the town, and the Italians left. One of the objects of Edward IV's restoration of royal power was to make England safe for foreign communities, and to end the legalised piracy conducted against foreigners in the Channel. On the whole he and his Tudor successors made steady progress; but hatred of the Italians remained a strong English characteristic. It was, of course, reciprocated. The prevailing Italian view seems to have been that England was a rich country, inhabited by barbarous fools, ripe for plucking by the civilised and the sophisticated. England, said the papal envoy Piero da Monte in 1436, was 'a very wealthy region, abounding in gold and silver and many precious things, full of pleasures and delights'. Silvestro Gigli, Henry VIII's agent in Rome, put the point more crudely (or so the King was told): 'Let the barbarous people of France and England every one kill another. What shall we care therefore so we have the money to make merry withal here?' The English saw the Italians as greedy cosmopolitan adventurers, with no sense of nationhood, loyalty or patriotism. As one Elizabethan writer put it: 'The Italians serve all princes at once, and with their perfumed gloves and wanton presents, and gold enough to boot if need be, work what they list and lick the fat even from our beards.' Anti-Italian feeling was one of the great popular engines of the English Reformation.

It was, however, the French who above all crystallised, and then for centuries symbolised, the xenophobia of the emerging English nation. From the middle of the twelfth century until the middle of the nineteenth, the external history of England is very largely the history of Anglo-French enmity. Sometimes the hostility is expressed in open war; sometimes in diplomacy or commerce; sometimes in all three simultaneously. From time to time a different enemy – the Spanish or the Dutch – flits briefly across the stage of history, as it were to separate the combatants; but always, and inexorably, the great brooding conflict between French and Englishmen seizes control again, and re-establishes the pattern of cross-Channel hatred. It is one of the great tragedies of mankind, this senseless aggression and rivalry between two well-endowed and immensely civilised peoples; and who can swear that it will not break out again? Perhaps it is inevitable that the English should view with suspicion any power which occupies the southern shores of the Channel. We first hear of rabid anti-French feeling in England in the 1050s, when a French-speaking faction formed at the court of Edward the Confessor. But neither the Channel nor language provide a satisfactory explanation for the origins of the quarrel. The Normans had no sooner established themselves in England than they began a policy of relentless hostility towards the French Crown. The matrix of Anglo-French diplomacy was quickly established, with the French encouraging and financing anti-English factions in Wales, Scotland and Ireland, and the English financing and arming anti-French coalitions in the Low Countries and Germany. Henry I inaugurated England's anti-French diplomacy in Germany by marrying his daughter to the Emperor, and later taking a bride from German-speaking Lorraine. Henry II began the new policy of subsidising elements hostile to France on her northern and eastern frontiers, and even in Italy; it was continued by Richard I and John, who built up an immense anti-French alliance in Flanders and along the Rhine. Over seven centuries the amount of English gold which has flowed overseas for this purpose is beyond computation. At one time or another, every independent territory – kingdoms, principalities, duchies, palatinates, counties, archbishoprics and city-states – within military striking distance of France, has been in the service of the English taxpayer, as can be seen from the exchequer pipe-rolls stretching from Henry II to William Pitt the Younger.

This enmity transcended language and culture. The English aristocracy hated Frenchmen when they still spoke their language from birth, when indeed many of them still owned broad estates in France. The loss of Normandy in 1204 enormously widened the cleavage, because it meant that men who owned lands on both sides of the Channel had finally to choose their national allegiance; it meant also that the southern side of the Channel was now a hostile shore; it was thus a decisive event in English history.* But it did not only begin the struggle between the French and English States: it served to give that struggle an increasingly strident nationalistic flavour. The collapse of the French language in England in the fourteenth century was both a cause and a consequence of that intense phase of the struggle we know as the Hundred Years' War; it brought a cultural separation between the French and English ruling classes which made reconciliation more difficult.

Yet the English attitude to French culture was curiously ambivalent. It was something the English – particularly the educated and leisured classes – felt they needed. French was the international language of culture, as Latin was of scholarship. The first flowerings of English national literature in the late fourteenth century were accompanied by strenuous efforts to keep up the study of written and spoken French. French was the vehicle by which new cultural elements reached this country; it was the hallmark of the fashionable and the pretentious to punctuate their speech with French words and expressions. In the intervals of hostilities, fourteenth- and fifteenth-century Englishmen travelled widely in France, just as Englishmen would rush to Paris in the brief Peace of Amiens in 1802, and again after Waterloo. From the late fourteenth century we get the first French-conversation manuals for the use of English travellers. One, entitled *La Maniere de language qui t'enseignera bien a droit parler et escrire doulx françois*, and dating from 1396, tells the Englishman what to say while on the road or at an inn. It unconsciously gives the English racial view of the French: how to instruct lazy, incompetent and venal French hostlers in their duties; how to tell French innkeepers to clean up their filthy and vermin-ridden bedrooms, and serve food which is wholesome and not messed-about; how to take advantage of the lascivious French habit of supplying girls to travellers, and how to avoid being cheated in consequence. It differs only in detail – certainly not in fundamental attitudes – from the phrase-books supplied to the English Grand Tourists in the eighteenth century, or even the patrons of Mr Thomas Cook. The English were already beginning to attribute to the French all kinds of undesirable habits and

* Among other things, it forced the English to maintain regular naval forces; John's military incompetence thus gave him a place in history as the founder of the Royal Navy. By 1205 there were 51 royal galleys, grouped in three commands; about this time we first hear of Portsmouth as a naval base, the use of a mariner's compass, and of a code of maritime law; by John's command all vessels had to strike their colours when passing a king's ship. The sailors were paid 3d a day, masters 6d – in advance. But press-gangs were soon necessary. John's first 'keeper of the King's ships' was an archdeacon, thus reviving a connection between the Church of England and the navy first established in the ninth century.

attitudes and, with more justice, political customs which the English found abhorrent.

Of course this racialism was based partly on fear: France was four times England's size, with many times its wealth and population; it was universally assumed that the French were aggressive, predatory and malevolent towards the English: an early English proverb had it: 'When the Ethiopian is white the French will love the English.' Torture was believed to have had its origins in France, and to flourish there. Fortescue drew rabid distinctions between English and French law, entirely in England's favour; the English King, he wrote, must rule in conjunction with the commonwealth while the French King was essentially an uncurbed tyrant. A fifteenth-century Frenchman related with horror that he overheard two citizens of London say that they would go on dethroning or executing kings until they found one who suited them. In 1460, the Duke of York and the Earl of Warwick, fomenting from Ireland a revolt to overthrow the Lancastrians, accused the Crown in their manifesto of seeking to introduce the abominable and servile French custom of conscription. Henry VII, exasperated with Parliament, said he would never summon one again, and would rule 'after the French fashion'. Even at moments of national reconciliation, the English racial hatred for the French festered beneath the surface. At the Field of the Cloth of Gold, the Venetian ambassador overheard a snatch of conversation between the Marquis of Dorset and one of his friends: 'If I had a drop of French blood in my body, I would cut myself open and get rid of it.' 'So would I.'

To some extent all English medieval governments were under pressure to make war against the French. While the tradition that the aristocracy owed some form of unpaid military service to the Crown persisted, there was a natural reluctance to cross the Channel for royal wars which brought no obvious profit to leading landowners; after the loss of Normandy this became the chief bone of contention between John and his barons, as we shall see. But with the development of purely professional armies in the late thirteenth century, a huge segment of English society began to acquire a vested interest in perpetual warfare with the French. The armies, from top to bottom, were well paid so long as the Exchequer could continue to ship sacks of sterling silver across the Channel. The Black Prince got £1 a day for active service, an archer 4d, which was the annual rent for an acre of fertile arable land. Military success brought enormous profits, for all prisoners were ransomable, some for sums running into tens, even hundreds of thousands; there was an elaborate system for the distribution of this livestock booty among the various ranks, and a regular market in captives, operated from Calais and other centres of trade. This system had the merit of keeping down the slaughter; but it was also a prime motive for renewing the conflict at the slightest excuse. 'By reason of these hot Wars,' wrote a contemporary of Edward III, 'many poor and mean Fellows arrived to great riches.' The Duke of Gloucester complained to Richard II that peace 'was disheartening to the poor knights, squires and archers of England'. Professional soldiers were not the only profiteers; a huge segment of the English economy had a vested interest in supplying the wartime commissariat; in the mid-fourteenth century, for instance, 2,000 bales of English cloth were supplied in a single year for the use of the navy. There was a popular impression that the wars brought a net profit to the English nation as a whole. It was assumed that war, if properly conducted, would pay for itself: this was emphatically laid down by Parliament in 1376.

For the King, as personal head of the government, the wars brought both great opportunities and appalling risks. It was shown time and again that the King could establish the ascendancy of the monarchy in popular esteem by successful operations in France, even if these were expensive; equally, failure was bound to bring political retribution at home. Richard I, for instance, was one of the most irresponsible monarchs ever to occupy the English throne. He was interested solely in the professional business of warfare and treated England (which he visited only for six months in a reign of ten years) purely as a bank for his expeditions. To get money, he auctioned off the government. As his biographer says: 'Everything was for sale – powers, lordships, earldoms, shrievalties, castles, towns, manors and the rest.' Richard joked: 'I would sell London if I could find someone to buy it.'* Yet all this was redeemed by his dazzling reputation as an international commander; a besotted and bellicose public made him a folk-hero even in his lifetime, and has honoured his memory ever since. John, by comparison, was a conscientious sovereign, but was ruined by military failure. The wretched Henry III, whose real interests lay in religion and the arts, who

* c.f. the alleged remark of the Chancellor of the Exchequer, Harold Macmillan, to John Foster Dulles during the 1956 Suez crisis: 'We would rather sell the National Gallery than surrender to Nasser.' Richard I's finances were engulfed by the revolution in military technology caused by the Crusades, which brought an enormous increase in the cost of war. The pay of a knight went up to 1s a day; men-at-arms cost 4d if mounted, 2d on foot. In a single year Richard spent £49,000, more than the regular income of the English State, on fortifying part of Normandy; most of it went on a single castle, Chateau Gaillard. This expenditure was outrageous: a century later, Edward I managed to build first-rate castles in Wales for £10,000 each (but then he understood accountancy and cost-control). Hardwear, especially armour, became vastly more complex and expensive, and larger horses had to be bred to carry the increased weight. By the fourteenth century, a charger could cost £100 or even more; Richard II paid £200 for the horse he rode at his coronation. (For comparison, a first-class hunter cost up to £300 in the 1860s.)

had no aptitude for government and still less for warfare, nevertheless felt it incumbent on himself personally to conduct expeditions to France; their inevitable failure, compounded by his preposterous scheme to make one of his sons King of Sicily, brought about the political crisis which made him the prisoner of the parliamentary party. Even his masterful son, Edward I, got into trouble towards the end of his reign through an unsuccessful French war. Edward III inherited a disgraced and humiliated throne; his father had been shamefully murdered, and he himself placed under the tutelage of a rapacious oligarchy. Yet all this was erased by the splendour of his French victories, ephemeral though they proved. His reputation as a warrior-king even survived long years of dotage, marked by economic distress and political chaos. Richard II, who was mad, would probably have destroyed his dynasty in any case; what made his destruction certain was his attempt at authoritarian rule on the basis of peace and alliance with France, under which the French Crown pledged its support for Richard 'against all manner of people who owe him any obedience, and also to aid and sustain him with all their power against any of his subjects'. It was thought intolerable that an English king should conduct a frontal assault on the liberties of his subjects with the aid of England's natural enemies; when Richard was brought to trial, 58 magnates, lay and ecclesiastical, were each asked to give their opinion separately: they were unanimous that he should be deposed and placed in perpetual imprisonment.

The question of war with France continued to have a direct bearing on the fortunes of the English monarchy until the middle of the sixteenth century. Henry V reconstructed a strong central government – of a type denied to Richard II – entirely on the basis of his successful French campaigns; and it was the military failures which followed his death which destroyed the Lancastrians. A monarch was under no compulsion to get embroiled in France; both Edward IV and Henry VII declined to take the risk, and their prudence made possible the restoration of political stability. But it is significant that Henry VIII, anxious to recreate the glamour of the English monarchy, embarked on unprovoked and senseless aggression against France. Polydore Vergil states that Henry 'considered it his duty to seek fame by military skill'; he commissioned an English translation of a French biography of Henry V, and in the introduction the translator calls on him to emulate his illustrious predecessor. Henry, a conservative in all things, responded to the appeal with the same mindless improvidence of a Richard I or an Edward III; but maybe also he had a shrewd instinct that a victory over France still exercised a mesmeric appeal over the English public of all classes. He was, in this respect as in others, the last medieval king of

England. Something of this old-fashioned and backward-looking potentate lingered on in his daughters. Mary felt the loss of Calais more than any other of her sorrows, though it was the most expensive and useless colony England has ever possessed. Even Elizabeth, as a young sovereign, felt the old hankerings, and tried to grab Le Havre. But experience brought wisdom. As the first modern-minded English monarch, she brought firmly to a close the long and fruitless history of England's efforts to acquire possessions in France.

How was it possible that the English persisted in these atrocious and consciousless wars of aggression? Apart from Henry v, a very un-English monarch, whose fanatical religious zeal convinced him that he had a mission from Almighty God to occupy the French throne and who attributed his remarkable victories to the enthusiastic intervention of the Deity, the English did not take their French claims as anything more than a pretext. Violent chauvinism, I fear, was the biggest single impulse throughout. When, in 1295, an Englishman, Thomas de Turberville, was discovered to be working as a spy for Philip the Fair, the event created a national sensation; such treachery was regarded as unprecedented, and a crime against nature; as the spy was dragged to a horrible death, the Londoners tried to tear him apart with their bare hands. The Edwardian victories bred in the English a violent arrogance. Milan's ambassador to Burgundy wrote to his Duke in 1475: 'The English are a proud race, who respect nobody, and claim a superiority over all other nations.' The wars brought grave economic difficulties to England, but little direct physical suffering. The chief victims were French peasants.* Elaborate rules, on the whole well observed, governed

* In theory medieval rules of war protected the clergy, and peasants going about their lawful business. But the evidence shows the rules were worthless. Edward III and the Black Prince sought to avoid pitched battles and used scorched-earth tactics to bring economic pressure on the French monarchy. Henry v alone imposed sufficient discipline to prevent excesses, at least against clerics, so that during his campaigns Norman peasants donned monks' cowls to escape slaughter. But as a rule the English had no respect for Church property. In 1373 an eyewitness said he saw over 100 mass-chalices robbed from churches being used as drinking-bowls at a supper given by Sir John Harleston and his men. Both sides murdered and tortured French peasants to extract money. A report on the excesses of the Dauphin's soldiers in Luxeuil and Faucogney in 1439 contains marginal references to '*femme violée*', '*gens crucifiez, rotiz et penduz*', '*homme roty*', etc. Even theoretical writers on war made no bones about the realities. In *The Tree of Battles*, Bonet writes: 'In these days all wars are directed against poor labouring people.' Paris de Pozzo admits in *De Re Militari*: 'A man may not torture a prisoner to extort money from him by way of ransom, but it is different in the case of peasants, at least according to the custom of the mercenaries.' Thus medieval chivalry, whose fundamental principle was the protection of the weak by the strong, proved meaningless. See Maurice H. Keen: *The Laws of War in the Later Middle Ages* (Oxford, 1965) and H. J. Hewitt: *The Organisation of War under Edward III, 1338–62* (Manchester, 1966); recent work is summarised in Kenneth Fowler (ed.): *The Hundred Years War* (London, 1971).

relations between the combatants, but there was no protection whatsoever for civilians, who were invariably robbed by both sides and murdered in their thousands. Whole communities starved to death in the wake of voracious armies. Pestilence was the only impartial leveller. It debilitated and then destroyed the Black Prince; the arch-aggressor Henry v met a miserable end from dysentery. The Church, the one international institution which commanded some kind of respect, made periodic efforts to arbitrate. On several occasions the popes were able to arrange truces; but equally often they egged on the combatants, especially in times of papal schism, with rival popes backing the opposing sides. All the armies were blessed by the national hierarchies. Wyclif, in this as in other respects ahead of his time, denounced 'the sin of the realm in invading the kingdom of France'; but his was a lonely voice.* Very few people questioned the morality of anti-French aggression, merely its expediency.

Indeed, the only restraints on English militarism were financial. But these restraints were important. The medieval English were exceptionally violent, aggressive, xenophobic and racialist; they were also greedy, parsimonious, business-minded and pharisaical. They applauded aggression; they were much less anxious to finance it. They thought war was a business, which should turn in a profit. In fact it never did so. From the reign of Richard I, when the French wars opened in earnest, the English Crown was heavily in debt, and at times actually bankrupt, for three centuries; and the biggest single drain, by far, was the aggression in France; the retention of Calais alone cost a fifth of the regular revenues of the State. Only the action of Edward IV, in renouncing his French claims in return for a regular pension, brought the English Government back to solvency; and it was Henry VIII, by resuming the war, who once more toppled the English Government into debt, a position from which it has never since recovered. Of course these facts

* There was a strong anti-violence, even pacifist, strain among the Reformers. Bishop Pecock, an Establishment maverick of the mid-fifteenth century, denounced the war with France as immoral; he also argued that the Lollards should be fought with reason, not the stake, and ended his days in close confinement in a monastery. Roger Ascham called chivalry a licence to plunder and murder. The Pope, on the other hand, egged on Henry VIII to invade France, and promised him the Crown if he succeeded. To the anti-French nationalism of the English Church in the fourteenth century we owe English perpendicular architecture, developed in preference to later French Gothic. The design of the choir of Gloucester Cathedral, its first masterpiece, coincided with the decade of Crécy and Sluys. The Scots, in alliance with France, clung tenaciously to 'flamboyant' Gothic in the French style. The English clergy 'did their bit' to help the war-effort. The English had long been jealous of the French kings, who were anointed, at their coronation, with oil presented by the Angel Gabriel to Clovis. In 1399, just in time for Henry IV's coronation, the Canterbury monks 'discovered' a jar of oil presented to St Thomas Becket by no less a person than the Virgin Mary.

are more apparent to modern historians, who can analyse the State accounts over long periods, than they were to the English ruling class at the time. But whenever the English grasped the point that the war was losing money, as from time to time they did, they were abruptly overcome by a rash of pacifism. One might say that much of the history of England has been a conflict between xenophobia and avarice, with the latter usually, in the end, getting the upper hand. The irresistible force of the English desire for war meets the immovable object of the refusal to pay for it. The English love to inflict violence on foreigners; happily they love money more. This is the hammer and anvil which forged the structure of English political society.

Consider, for instance, the history of Magna Carta, which is an example of this process in operation; it also illustrates the English political genius for transforming the muddles in which they involve themselves into triumphs of the national spirit. It is worth examining in some detail because it tells us a great deal about our national character. The crisis which led to Runnymede really began in 1204, when John lost Normandy in circumstances which suggested he lacked nerve, resolution and energy, and was quite possibly a coward. The English ruling class wanted the duchy back, for personal as well as national reasons; but they did not think John was the man to recover it, at least by a frontal invasion; at all events they were not prepared to help him do so. It was easy to make life difficult for the French King, by financing his enemies, and by using the navy to protect them. This John did; and so far so good. In 1213 his ships demolished the French fleet at Zwyn, in Flanders. He carried on financial warfare on a considerable scale. The anti-French emperor, Otto, got 1,000 marks a year; the Count of Boulogne a pension of £1,000; large sums went to the Dukes of Limburg and Brabant, the Counts of Flanders and Holland, and to hundreds of Flemish knights kept permanently on John's payroll. The anti-French coalition thus created was not as formidable as it looked on paper, but it had considerable nuisance value. The English did not resent the role of paymaster. They were quite prepared to defend their own territory. Great preparations were made to resist invasion. Every male over 12 took an oath. Constables were appointed to organise the urban communes 'for the defence of the kingdom, and the preservation of the peace against foreigners and other disturbers'. These steps were not unpopular; on the contrary. But what aroused increasing opposition was John's evident determination to make the English fight in France, under his command. They thought he was incompetent, treacherous and unreliable in a crisis. They knew he was cruel, and a liar; most of them believed he had murdered his nephew Arthur with his own hands, in a drunken rage.* Though quite prepared to fight the French by other means, they would not serve under him across the sea; nor would they finance such an expedition. John's advisers warned him against such a policy in 1205; but he returned to it again and again, as if his honour could not be satisfied until he personally beat Philip Augustus on his own soil. In 1213 he conducted a formidable inquest on the military service owed to him by his tenants-in-chief. Many of the barons, particularly from northern and eastern England, made their opposition perfectly plain. But the next year John sent an army to bolster up his motley Flanders coalition, and himself landed in La Rochelle to attack the French from the south. This venture, flying in the face of growing national sentiment, might have undermined criticism by success. But John was deserted by his Poitevin allies in the south, and his stipendiaries in the north met complete disaster at the Battle of Bouvines. He returned to England a conspicuous and humiliated failure; but almost his first act there was to impose a provocative three-mark (40s) scutage per knight's fee on all the barons – the vast majority – who had not come to France with him. This was the immediate, as well as the fundamental, cause of the baronial revolt.

Now all Anglo-Norman and Angevin administrations had faced baronial conspiracies, which were ill-formulated and narrowly based affairs, promoted by personal frustrations and ambitions. But John, like Stephen, was gravely weakened by a headlong conflict with the Pope. This was no novelty either; but always, in the past, the majority of bishops had closed ranks behind the King. It was John's misfortune that he not only had to contend with the masterful Philip Augustus in secular matters, but in Church affairs faced the most audacious and relentless of all the medieval popes, Innocent III. In 1205 Canterbury fell vacant. The monks of the chapter, without royal permission, elected their prior; then, terrified by John's angry reaction, accepted his nominee. The Pope quashed both elections, summoned the chapter to Rome, and forced on it an English cardinal, Stephen Langton, who was

* Some of the stories about John's cruelty are exaggerations, or at least unproven. But he certainly starved to death the wife and son of William of Braose in Windsor Castle; and 22 of his prisoners died of starvation in Corfe Castle. He enjoyed watching judicial combats, which were often horrible affairs; English public opinion was already sharply opposed to them. Apart from his cruelty and military incompetence, it is difficult to see why he aroused so much dislike. He was well educated (Glanvill had been his tutor) and fond of reading; in 1203, 43s 10d was paid 'for chests and carts for carrying the King's books beyond the sea'. He took a bath on average every three weeks, and was the first English king to wear a dressing-gown. He had a huge collection of jewels. Although he had at least five bastards, and a mistress called Suzanne, he founded Beaulieu Abbey, gave donations to many religious houses, and regularly fined himself for breaking fast or going fishing on feast-days. John was exactly 5 ft 5 ins tall, as was confirmed when his tomb, in Worcester Cathedral, was opened in the late eighteenth century.

a reliable exponent of his canonical views. There was no precedent for such a brutal infringement of the royal prerogative, and John naturally refused to confirm the appointment: from 1207, when he was consecrated, until 1213, Langton was unable to set foot in England. In 1208 Innocent imposed an interdict on England, which in theory at least meant the suspension of all Church rites; and the next year he excommunicated the King. By itself, such a breach with the Church need not be disastrous: John began by taking the interdict calmly – it was not widely observed – and even threatened to hang anyone who insulted papalist clergymen; but after his excommunication, more and more of the bishops turned against him, and his anger and exactions increased. Royal agents seized Church property, and diverted its revenues on a growing scale to the Exchequer: over £100,000 was thus obtained. These funds largely paid for John's successful expeditions in 1210–12 in Scotland, Ireland and Wales; and to that extent his conflict with the Church was popular with the laity. But a kingdom at odds with Rome was the potential object of a crusade: and Philip laid claim to John's throne in the name of the Deity, rather as the Conqueror had done 150 years before. In 1213 John abruptly turned Innocent from an enemy into an ally by resigning the kingdom to him, and receiving it back under an oath of fealty; all charges against him were withdrawn, the bishops returned, and Innocent now exerted his considerable diplomatic powers on behalf of his new subject.

This desperate act of *realpolitik* might have solved all John's troubles. But the military disasters of 1214, and his ill-judged reaction to them, turned many of the barons, who had stood by him against the Church, decisively against him. The Church, in theory – and in practice, so far as Innocent's orders were obeyed – was now on his side; but many of the senior clergy had bitter financial grievances against the Crown. In any case, the submission to Rome had placed the Crown in a new legal context, opening the gates wide to constitutional opposition. It was now possible, for instance, for the barons to lay their grievances against John in the papal court. How far opposition to an anointed king was lawful had been for half a century a subject of debate. In 1159 John of Salisbury had published his *Polycraticus*, the first attempt to expound a philosophy of politics: this declared it obligatory to dethrone, even assassinate, an evil king; but of course John of Salisbury was thinking purely in terms of a conflict between Church and State. An alternative view, on behalf of the royal administration, was put in the *Dialogus*, written by Henry II's treasurer about 1178: 'Though abundant riches may often come to kings, not by some well attested rights but . . . even by arbitrary decisions made at their pleasure, yet their deeds must not be discussed or condemned by their inferiors.' How far all, or indeed any, of John's barons were familiar with such arguments is not known; but the legal revolution of the twelfth century had forced all landowners to take a much more sophisticated interest in the law, and to adjust their ideas accordingly; and it is significant that John's most resolute critics came from the younger, better-educated, generation. In any case, the barons were perfectly aware of what might be called the constitutional tradition of the English State. They might not know of Æthelred's concessions of 1014; but they certainly knew of the coronation charters. John had given one himself, in which he had promised to end an abuse of Richard's of particular concern to the barons – the royal device of forcing tenants-in-chief to get their charters re-sealed on payment of a stiff fee. Versions of both the laws of Edward the Confessor and of Henry I's coronation charter were circulating at this time. Magna Carta was not a bolt from the blue but the expansion of a long-established, if intermittent, practice; it was called great simply because it was so long.

Nevertheless, though the background to the crisis is clear, everything else about the charter, except its basic chronology, is surrounded by mystery. Far from being simple, it is one of the most puzzling and complex events in English history. As we have seen, a baronial party had been collecting even before John went on his disastrous French expedition. After his return, he met these barons at Bury St Edmunds in November 1214, at London at the end of the month, and in the new year in London again: discussions were angry and inconclusive, and a truce was arranged until April. Both parties appealed to the Pope. In March the Pope instructed the barons to abandon their conspiracies, but on the other hand told the King to meet their just demands. In April, the King met certain barons, probably loyal to him, and later in the month they acted as intermediaries with the rebels; on 26 April the truce expired. In May the King offered to provide a charter himself, saving his right of appeal to the Pope; he also offered to go to arbitration, with the Pope as 'superior'. The barons occupied London, and further truces were arranged. On 15 June the King met the barons at Runnymede, and in the next five days the charter was negotiated, agreed, signed and sealed. On 17 July the King met the barons at Oxford to arrange for the execution of the charter; but the meeting broke up in disagreement, and immediately afterwards the King wrote to the Pope asking him to quash the charter. The barons defied the King in August, the Pope excommunicated them, and civil war broke out.

This is the chronology; but what of the objects and motives of the parties? This was a protracted and multiple negotiation. There were the northern and eastern barons; there were other barons who supported them; there were some barons who supported the King; there were some who regarded themselves as intermediaries; there was the royal administration, including substantial lay and clerical landowners; there was a Church party loyal to the King and another, under Langton, which was intrinsically hostile to John, or at least neutral, but under pressure from the Pope to support him; there was the Pope himself, represented by his legate; and finally the King. All had different objects and motives; no wonder there was confusion and cross-purposes, with letters and envoys travelling to and fro between Rome and England, and up and down the country.

The dissident barons were clear, or thought they were clear, on their objects. They did not want to serve abroad, or pay for others to do so, except by consent. They wanted an end to a variety of exactions they felt were illegal; trial by the customary processes; and affairs of State to be settled in the King's great council, with their right of attendance (in the case of great barons) or representation (in the case of lesser barons) guaranteed. The first demand – service abroad – was obviously what mattered most to the more obdurate rebels. It is conceded in a preliminary draft, called 'The Unknown Charter of Liberties', probably drawn up in May. But it does not appear at all in Magna Carta itself. As for the exactions – as regards wardship, marriage, and so forth – kings had always readily conceded them before; it may have been a positive advantage to the administration to have them codified, as Magna Carta does; in any case, many of these dues were already obsolescent. The demand for formal councils is still more mysterious. The strongest kings had always been anxious to conduct important business surrounded by as many tenants-in-chief as possible. That was the object of the solemn crown-wearings, three times a year. At one of them, for instance, the Conqueror had settled the plan for Domesday Book. Rufus had insisted that all his magnates should attend him from time to time, and regarded those who did not as potential rebels. A king liked to do business with the great personally, look them in the eye, discover what they were really thinking. Again, no great act of State could be carried through effectively without the general approval of the aristocracy. All Henry II's solemn assizes had been attended by a multitude of barons. John had not only continued this practice, he had formalised it. In 1213 he had held an assembly at St Albans, to which he had summoned not only tenants-in-chief but four men and the reeve from each township on the royal lands; a few months later, at Oxford, he had again summoned the barons, plus four lawful men from each shire, to discuss, as the writ said, 'the affairs of our realm at our colloquy'. Why should the King now be forced to concede what he had already expressly practised?

We cannot make sense of the charter if we regard it as a baronial document. Nor was it simply a version of a baronial list of demands, sifted and softened down by an establishment party, under Langton. Langton was a feeble nonentity, unfamiliar with English politics, personally unknown to the barons, important only in so far as he possessed the Pope's confidence; when he lost this, shortly after the charter was signed, the Pope snuffed him out, and he ceased to count at all. What, then, was the charter? The only real answer is that it was a muddle, a spatchcocked compromise which did not represent the attitudes of any of the parties – or, rather, represented bits of all of them – and was therefore unworkable as a political settlement. Its very spirit was confused. Baronial demands in the past, and indeed on this occasion, were essentially conservative. They wanted the clock put back to where, in their view, it had always been: this was what the coronation charters meant. All revolts against the King were in the strict sense reactionary. It was the King who, traditionally, had the reformer's role. Henry II's great assizes were essentially innovatory. It was Henry, Richard and John, prompted by the demands of the administration and its growing expertise, who had brought in changes, most of them beneficial. The barons stood for stability, the Crown for movement. But the charter, as eventually signed, stands for both. It is a document without a unifying viewpoint. Nor is this surprising, granted the circumstances of its creation. The parties involved were too numerous, the physical forces behind them too evenly balanced, to produce any other result. The barons dropped the demand about overseas service. But they successfully inserted a 'security clause', appointing a committee of 25 barons as watchdog on the King's behaviour. The Church got in a clause about its own rights. London got in a clause protecting its liberties. Virtually all the parties were in debt, so three clauses were accordingly inserted, the onus of which fell largely on the Jews. The drafting of the treaty, as of the earlier barons' demands, was in the hands of the administration, for the Chancery clerks alone could do a job of this kind. So many clauses were inserted to suit the convenience of bureaucracy, and others were tidied up in a manner approved by royal officials. Some radical fellow even succeeding in putting in a bit to protect the rights of villeins. During these five days of argument and re-drafting, the charter grew and grew. It became enormous, in its totality quite beyond the comprehension of any one of the parties. After the solemn agreement, when the exhilaration died down, men began to read the small script again. All of them found something they disliked. To the northern barons, the overseas service clause was crucial; to omit it was to yield the whole issue. Some of them left Runnymede in anger before the charter was even signed. To

John, the inclusion of the security clause was equally intolerable; when, a month after Runnymede, the barons made it clear they intended to enforce it, he denounced the whole document. But if the King had adhered to the charter, in the sense he placed on it, the barons would have denounced it in their turn. The story of Magna Carta, in fact, is not of a negotiation which succeeded, but of one which failed.

Happily the genius of the English for rewriting history while it is still happening turned an acrimonious disaster into a triumph of constitutional good sense and moderation. As Cnut was transformed from a Scandinavian ruffian into an English Christian gentleman, as the disaster of Dunkirk was transmuted into the prelude to victory, so by a process of constructive national myopia the confusion and muddle of Magna Carta was canonised as the bedrock of the English constitution. Innocent III conveniently died the next year, closely followed by John himself; the removal of the two chief actors cleared the way for creative fiction. With a new, young king on the throne, the charter, suitably amended, could be represented as a solemn concordat, to which all the community subscribed; what was actually in it mattered less than the consensus it inspired. It was reissued in 1216, and again the next year; it took its final form in 1225, was confirmed by the King in 1237, in 1297 and on many subsequent occasions – at least 32 times in all. It was entered as the first document on the statute book, thus ousting from the honour the more important and deserving acts of Henry II. It became a national institution, a symbol not of the civil war it provoked, but of the constitutional peace it was supposed to have established. Few read it; everyone quoted it.* Archbishop Pecham flourished it against the King in the defence of the rights of the Church; Edward I flourished it against the Pope in defence of the rights of the State; Parliament cited it against the Crown and the Crown against Parliament; unlettered peasants used it against their masters, masters against townsfolk, townsfolk against rural lords. To appeal to Magna Carta became the one, great, unanswerable argument which any and every section of society could employ. Within a generation its provisions became largely incomprehensible – some of them remain enigmas even today – but it was none the less the written embodiment of the golden English past, a massive monument of constitutional rectitude. For the first time it made politics

* Even today Englishmen frequently cite Magna Carta without knowing what is in it; in 1970 I heard Lord Wigg, Chairman of the Race Course Board, claim on the BBC that off-course betting was a unique English right, guaranteed by Magna Carta. The Americans have inherited this characteristic. All swear by the Declaration of Independence; few know what it says. In 1970, in Cleveland, Ohio, its text, without attribution, was shown to passers-by, who were asked to sign it; only one in 50 did so; the rest declined, on the grounds that it was 'commie stuff', 'written by hippies', etc.

respectable, because it made them old. So the English came to see compromise, consultation, the settlement of dispute by argument as opposed to force as their outstanding national characteristics; and in time shaped their habits to conform with this image. The history of Magna Carta is a triumph of English hypocrisy – always one of our most useful assets.

The development of English political society in the long shadow cast by Magna Carta is rich in irony and paradox. It cannot successfully be analysed in terms of conflict between the classes. Men believed that such conflicts were, or ought to be, unnecessary, for each section of the community had its ancient and predestined role to play, and conflict was a sign of malfunctioning, to be corrected by a return to the past. What everyone wanted was continuity; all men were, or believed themselves to be, conservatives; political progress was thus in fact achieved only by what might be termed constructive self-delusion, by the use of conservative instruments to achieve radical reform. This applied both to acts and to institutions. Thus Edward I, for instance, who undoubtedly regarded himself – and was so regarded – as an ultra-orthodox conservative, inaugurated a social revolution by statute under the impression that he was putting the clock back in the soundest possible manner. His two great acts of *Quo Warranto* and *Quia Emptores* were designed, as he saw it, to curb and redress illegal innovations over the whole field of tenurial rights, to stop landowners from acquiring privileges for which they had no warrant, and from creating new social structures in the disposition of their lands. He intended them to be thoroughly reactionary pieces of legislation, which would have met with the warm approval of such illustrious predecessors as William the Conqueror and Henry II. They were indeed popular for this reason; but their net effect, in the long term, was to destroy the tenurial basis of society, to undermine the system – which went back to the origins of English society, and which the Conquest had merely reinforced – under which political, military and jurisdictional power sprang directly from the ownership of great landed estates. It was one of the decisive events in English history, for henceforth men would have to seek power increasingly through the formal institutions of the State.

Yet these institutions were seen not as instruments of change but, on the contrary, as the tenacious guardians of custom, a guarantee that the past could always be conjured up to buttress the present against the future. Parliament was essentially a development of the later Middle Ages; not until after 1325 was it established that it must include representatives of the shires and boroughs; not until 1376 do we find a Parliament angrily and self-righteously taking to task a corrupt and unsuccessful administration; not until the mid-fifteenth century do we find a desperate government, *in extremis*, submitting a form of national accounts to Parliament, to prove that it simply could not carry on without more money. Yet nobody regarded Parliament as in any sense an innovation, still less a revolutionary instrument; they believed it had always existed and had always exercised, more or less, its current functions; its customs had been honoured, in the phrase parliamentarians used repeatedly, 'since time out of mind'. There were no historical or constitutional textbooks; such relevant literature as existed always stressed the immemorial antiquity of everything. In any case, institutions were seen not merely as ancient but as natural and God-ordained. In the twelfth century John of Salisbury had described the State as a body, with the King the head and other sections of society as the limbs and organs. The image persisted, though sometimes with variations. In the mid-fifteenth century, Sir John Fortescue, in a sophisticated discussion of how England was governed, saw the King 'as a stomake which dystrybuteth the mete that it receyveth to all the members and reteyneth no thynge to hym self but only the neuryssynge'. Such natural arrangements could not be changed; they could only live, and grow. Even where commentators had a distinct political viewpoint, they presented it as fact, not programme. Thus, a mysterious document called the *Modus Tenendi Parliamentum,* now known to date from 1321, is clearly written from the point of view of a parliamentarian anxious to enlarge its powers and privileges. But he does not say: 'This is what Parliament ought to do;' he says: 'This is what Parliament does, and has always done.' He was lying, or rather exaggerating. But his tract was popular, and was reissued and added to many times over the next century; eventually the practice of parliament – influenced no doubt by the tract – did indeed come to correspond with its theories.

Parliament, in fact, did not become an open political issue until long after it had established itself as an indissoluble part of the political fabric. It spoke for the realm; that was its job; the more accurately it reflected opinion – that is, the more representative it was – the more effectively that job could be performed. Broadly speaking, Parliament was there to help the King to perform some national task which was beyond his own unaided powers. Its duty, says an official document of 1300, is 'to hear and do what is necessary for the common convenience of the realm'. Walter Burley, commenting on Aristotle's *Politics* in 1340, says: 'The King convokes Parliament to deal with hard matters.' Two hundred years later, we find Henry VIII telling the Commons that his power, majesty and dignity is never so great as when Parliament is sitting and the Crown is operating through it. Parliament is the servant of the executive government, though a servant which enjoys considerable (and increasing) trade union rights. Its primary function is to raise emergency revenue. The King is always in debt, for his normal sources of income, through which in theory he should conduct the business of the kingdom, tend to be static, and are continually overtaken by the creeping inflation which is characteristic of all dynamic societies, and is a constant motive-force behind political change. Only Parliament can provide the money, because only those who attend it can, in practice, ensure that it is collected. But Parliament does other things to help the King. His sworn duty is to prevent undesirable change. Society is moving forward under its own impetus; from time to time the King must call a halt, by some resounding statute; such acts often have the opposite effect to that intended; they invariably have unforeseen effects; but this point is never grasped. Parliament is called in to give the King's acts authority and to make them work. It shares the job of dealing with petitions and grievances, and eventually takes over the whole business. There is nothing new in this. Grievances have always been presented to the Crown for redress. They call not for change and innovation, but for reform – for the restoration of ancient rights, and the original, mythical justice. Parliament thus comes to be identified with liberty because it is seen as the most effective means by which new oppressions can be removed, and the past restored. Even Milton, writing *Areopagitica* during a political revolution, had no higher ambition for Parliament than this: 'When complaints are freely heard, deeply considered, and speedily reformed, then is the utmost bound of civil liberty that wise men look for.'

Hence it is wholly mistaken to see the origins of Parliament as an attempt to challenge, let alone usurp, the power of the executive. In the fourteenth and fifteenth centuries, Parliament was employed to ratify the deposition of kings, but it always did so on the instructions of the ruling group and the new monarch, to vest with legality a *fait accompli*, and usually on the grounds of restoring justice. This power, said Parliament, was based 'on ancient statute and modern precedent'. Again, Henry VIII used Parliament to provide a legal framework for the religious revolution he had ordained; but again the object was to restore what all agreed, sincerely or not, to have been the original situation. That was what the Middle English word 'reform' meant; not until the late eighteenth century did it assume the modern connotation of introducing change without precedent.

Indeed, the idea of a dynamic conflict between the executive and institutions representing the public was, and is, wholly alien to the English mind. The English have always gratefully accepted strong central government, based on ancient custom. What they wanted – what they have always wanted – was the kind of regime provided by Edward I, one of the greatest of the English kings. Edward was 35 when he came to the throne, an immensely experienced administrator and soldier, with a European reputation. He was tall – over six feet – wore his clothes and armour well, moved with grace and dignity, had a fine, powerful voice, a wide vocabulary, a talent for impressive phraseology. His manners were exemplary, with a touch of gravity. He always looked and behaved like a king. He understood perfectly, and shared, the assumptions, tastes, likes and dislikes, prejudices and emotions of the ruling territorial aristocracy. It did not occur to him that a king who knew his business need fall out with those whom God and nature, logic, precedent and ancient custom, had appointed to act with him in the affairs of the community. He put his theory of the constitution neatly in 1280 when he said he would always 'according to God and justice do what the prelates and magnates of the realm shall advise, especially as no one supposes that such prudent men will give the King advice dissonant with or contrary to reason'. He was a thoroughly professional king, active, industrious, well informed, able to discuss details of law with his judges, to draft a statute, to preside over the King's Bench, and the Court of the Exchequer, to draw up a line of battle and supervise the construction of a fortress. He was the last English monarch to possess a detailed grasp of the whole range of government activities. But even he found this exacting role a strain. He sought to conserve his time and energy by the deliberate and methodical conduct of business, regularly summoning Parliament in the late spring and late autumn, to coincide with the busiest periods of the financial year, thus concentrating administrative work at a fixed time and place. But it is significant that all the troubles of his reign occurred during the last decade, when he was an old man and his mental and physical energies failing. No medieval king was safe once he had passed the prime of life: kingship was a pitiless and ultimately thankless career.

No later monarchical head of government came within measurable distance of Edward as a chief executive. Some could not even act the part: his own son, Edward II, whom he despised, was a man of common tastes, who enjoyed digging, rowing and village sports – harmless enough to us, but fatal to the authority of medieval kingship.* Some tried to act the part without the substance: Richard II's theatrical dressing-up of the throne, his instructions that men should grovel before him, and genuflect when his eye alighted on them, his flourishing a sword at his archbishop – these empty gestures carried no weight and eventu-

* But even Edward has found an apologist. See H. F. Hutchison: *Edward II, The Pliant King* (London 1971).

ally aroused contempt and hatred. The big men of the realm could see that he was not the genuine article, that he did not know his business, and he was snuffed out without compunction or pity. As for Henry VI, even the commonest people could see he would not do: the Londoners despised him for being seen always in the same old blue surcoat, 'as thowth he hadd noo moo to channge with'.* His virtues were disastrous handicaps. Loathing warfare, he sang hysterically during the battles which his supporters insisted he should witness. He could not even recognise the stinking quarter of a convicted traitor, as it hung in the London street; he had to be told what it was, and recoiled in horror. He was prudish: 'Fie, fie, for shame!' he said to a troupe of topless dancers, and he objected even to seeing naked men taking the waters in Bath. The English did not want a monk on the throne, and only the vigorous efforts of his ferocious wife, a 'she-wolf of France', kept him on it so long.

Edward III could, and did, perform the physical functions of monarchy with success. He got on well with the grandees, was successful in battle, enjoyed the military theatricals of chivalry – as in the Garter ceremonies he devised. But unlike his grandfather, he never mastered the less spectacular side of government business. His debts were beyond remedy, his administration was always in chaos, and his regime, such as it was, quickly dissolved into warring factions when, at the age of 60, his powers began to fail. His was a façade of professional kingship. Perhaps, as the complexities of administration grew, this was all anyone could reasonably expect. Henry V made a deliberate effort to grasp again all the reins of power; hugely self-confident, industrious, clear in his objectives and determined to have his way in all things, he was a frightening and much feared figure among the ruling class; but he simply did not have the time to supervise directly the administration of justice and finance, while engaged on a war of conquest.

A king, increasingly, had to choose those aspects of government on which he preferred to concentrate. Edward IV restored the authority, and solvency, of central government by deliberately renouncing his foreign claims. Even so, there was something of a juggling-act quality about his highly successful period of office. He had to stoop to things

* But Henry VI could be extravagant in the cause of culture; like other English kings who patronised the arts (Richard II, Charles I, George IV) he thereby added to his financial difficulties, and died owing over £350,000, nearly seven times his annual regular income. His foundation at Eton was very costly, and when Edward IV took over he wanted to wind it up, but was dissuaded, no doubt by wily Old Etonians. The Yorkists favoured Wykehamists. One of them, John Russell, Bishop of Lincoln, preached the sermon at the opening of Richard III's 1484 parliament; he said its duty was 'to harken to the commyn voyce grownded in a resonable presydent' – a typical Wykehamist sermon.

which his predecessors would have found repugnant. He sold his birthright in France for a (very welcome) pension from the French Crown; he personally engaged in trading operations, which brought him huge profits; he ran the Crown lands with the sharp eye of an estate agent; he was the first king – the first English politician, indeed – who engaged in a deliberate policy of fostering good public relations, especially with the mercantile community of London. Six London aldermen were made Knights of the Bath, not for any particular services, but merely to mark Edward's coronation; a very early example of the honours list. In 1474 we find his wife, Elizabeth Woodville, no doubt on the King's instructions, writing to one of her bailiffs, and commanding him

... to deliver to our trusty and well beloved the mayor and brethren of my Lord's city of Coventry and their wives ... twelve bucks of this season to be evenly distributed amongst them; that is to say, six of the said bucks to the said mayor and his brethren, and the other six of them to their said wives ...

In 1482 Edward invited the Lord Mayor of London, the city aldermen, and 'a certain number of such head commoners as the mayor would assign' to meet him in the royal forest at Waltham. There they were given an excellent morning's sport, and afterwards they were

... brought to a strong and pleasant lodge made of green boughs and other pleasant things. Within which lodge were laid certain tables, whereat at once the said mayor and his company were set and served right plenteously with all manner of dainties as if they had been in London, and especially of venison both of red deer and of fallow and ... all kinds of Gascon wines in right plenteous manner.

Twice during the meal Edward sent the Lord Chamberlain 'to make them cheer', and did not sit down to his own dinner 'till he saw that they were served'. Afterwards the King went hunting with them again, and a few days later he sent the wives of his guests 'two harts and six bucks with a tun of Gascon Wine'.* For this shameless type of PR exercise Edward was well suited: a large, fleshy, handsome, carnal man, always smiling, friendly to all, a great teller of dirty stories.† He boasted

* The tradition of royal PR hospitality lingers on anaemically in the Buckingham Palace garden-party. Edward IV's magnates also engaged in politically motivated hospitality. When the Earl of Warwick was in London six oxen were consumed for breakfast in his house; anyone with a suitable introduction could attend, and take away with him as much meat as he could carry on the point of a dagger. Territorial bigwigs maintained this practice until the end of the nineteenth century in some cases; such PR feasts were called 'ordinaries', and one, given by the Duke of Omnium, is described in Trollope's *Dr Thorne*, Chapter XIX.

† Edward's brother, Richard III, was also tall and good-looking, but thinner. His hunchback was an invention of Tudor propaganda. Nicholas von Poppelau, who met him

of his womanising, claiming he had 'three concubines, which in diverse properties diversely excelled, one the merriest, the other the wiliest, the third the holiest harlot in the realm'. Edward's smooth good looks he inherited from his mother, Cecily Neville; but he had brains, too, – one of the cunningest men who ever sat on the English throne – and every kind of accomplishment. His handwriting was superb. He was a beautiful and enthusiastic dancer. He understood accountancy, and his reform of the royal household led to a dramatic cut in expenditure. But above all he was a glad-hander. It was said that he knew the names, faces and incomes of every man of importance in his kingdom.

By the end of the Middle Ages, indeed, the monarchy was beginning to lose some of its executive functions and was becoming, to some extent, a show put on for the benefit of the public. A king in the high Middle Ages had to keep friendly relations with the great territorial magnates to be reasonably secure; the circle of mandatory approval widened dramatically in the fifteenth century, and in the next it embraced a significant proportion of the entire nation. A king had to devote himself to certain aspects of government business, and spend laborious days in the details of administration; his real power depended to a great extent on the actual amount of time he was prepared to spend exercising it; this principle held good until the days of George III, the last sovereign who attempted to be a professional; it was George's eventual incapacity through madness, and the idleness of his sons, which led to the final collapse of monarchical authority. But the ability to concentrate on administration had increasingly to be supplemented by showmanship. It is significant that Henry VII, a true (though incomplete) professional, who was far abler and more industrious than his son, was never so secure on the throne, and could not risk putting his authority to the test, because he lacked regal glamour. Henry VIII was idle, irresponsible, ignorant, lacking in judgment and totally oblivious to any sense of duty to the community. But he knew how to beat the big drum of monarchy, and the nation trouped in his wake. Through all the vicissitudes and miscalculated adventures into which he led the realm – disastrous foreign wars, state bankruptcy, the debauching of the currency, change of religion, government by confiscation and judicial murder – the great mass of the people obediently followed. The English have always

in 1484, said he had very delicate arms and legs. The Countess of Desmond, who lived to be over 100, told Walter Ralegh that she had often danced with Richard, and that he was the handsomest man at court, apart from his brother Edward. There is no conclusive evidence that Richard killed the Princes in the Tower; he probably believed, as did many others, that Edward's marriage had been irregular, and that they were bastards. But he would not have scrupled to murder them. Between Henry VI and Elizabeth, all the reigning sovereigns of England were killers.

responded to strong central government, invested with majesty and colour, and operated by a self-confident will. Henry's last speech to Parliament was an astonishing performance. His government had nothing to report but failure, but the King subjected the assembly to a magisterial harangue, in which all sections of the community were in turn soundly rebuked for their shortcomings. He contrived to give the impression that the nation was entirely to blame for any evils which had befallen it, and that it was exceedingly fortunate he was still prepared to remain at the helm and protect it from the worst consequences of its folly. The speech was heard in breathless admiration, and was never forgotten by all those present. Many of them, we are told, actually wept tears of love, penitence and gratitude.

This gift of royal showmanship Henry passed on, in all its plenitude, to his dazzling daughter Elizabeth. To be sure, she supplemented it with an enviable range of qualities and accomplishments: a subtle intelligence, industry and self-discipline, prudence and deliberation, a warm heart and a virtuous mind. But without it she could not have kept her throne, let alone given a divided, weak and desperately vulnerable nation the strength which comes from unity and a common purpose. No woman had ever presided successfully over a medieval court, whose function was to associate the chief landed proprietors with the business of government, and determine a fair division of its spoils. Their animal energies found natural expression in violence, whether civil or international; such energies could only be diverted into more useful channels by the cynosure of the throne, whose authority sprang from its tenant's ability to epitomise and transcend the ruffianly virtues of a military aristocracy. A woman's sex was thus a daunting handicap. Elizabeth's political genius consisted in turning it into an asset. She did not attempt to disguise her sex; on the contrary she emphasised it. In her great speeches, she always reminded her hearers that she was a woman. But she was a woman *sui generis*. They could turn her out in her petticoats, she said, and she would make a living anywhere in Europe. She had a woman's body but 'the heart and stomach of a king'. She was careful not to say 'of a man'. She was not an emancipationist; she did not believe in woman's liberation. She did not seek to play a masculine role, and so injure the men she had to control in their pride. The English had burnt Joan of Arc for precisely that mistake. Elizabeth vaunted her sex. Her weapons were an astonishing wardrobe, a collection of jewels which even the popes envied, false hair, paint and powder, and the universal knowledge that behind these trappings lay a resolute and imperious spirit which it was perilous to challenge. Elizabeth did not need men, unlike her wretched half-sister Mary, and her still more unhappy cousin Mary of Scotland. She was chaste by choice, and virtuous by policy and inclination. The mystique of her court – the cult of the Faerie Queen, the sexual favourites, the pretend love-affairs, the political minuets she danced with the popinjays who surrounded her – was an elaborate and calculated exercise in royal diplomacy, designed to replace the licensed gangsterism of masculine chivalry by a non-violent system which a woman could manipulate. It seems to us in retrospect shameful that this noble and virtuous queen, whose intelligence soared above her courtiers', and whose ability and sense of responsibility rivalled that of even her most devoted and accomplished advisers, should have felt it necessary to demean herself to this masquerade. But there was no other way.

Behind this public-relations façade, Elizabeth did her best to supervise the actual operations of government. But she was obliged to be selective. Thanks to her knowledge of languages (English did not become an accepted vehicle for diplomacy until the late eighteenth century) she was able to negotiate with ambassadors, and correspond with their masters, directly; she thus kept foreign policy firmly under her control, and all the decisions were ultimately hers. Dynastic and religious policy, too, she settled herself, though with great difficulty. But in finance her touch was less authoritative, because although she decided how and when funds could be allocated, her supervision effectively ceased once the Exchequer issued the money – as she was painfully aware, for it was at this point that the corruption and incompetence began.* Nor (beyond a little genteel piracy) could she carry through the revenue-raising operations by which Edward IV and Henry VII restored the solvency of the State. Elizabeth inherited a bankrupt government during a period of raging inflation, and maintained the credit of the State only by the desperate expedient of selling the royal estates, and creating monopolies for purchase. In fact throughout her reign she covered the gap between revenue and expenditure by living on capital. On the other hand, she always finally managed to meet her obligations; England's credit was better than that of any other state. The truth is that Elizabeth handled as much business as any one person reasonably could. She was a hard-working, devoted public servant, rapid at her paper-work, abstemious in food and drink, a professional whose pleasures were essentially functional and constitutional. But the economic problems of her reign – and

* For Elizabeth's angry, but unavailing, efforts to straighten out the finances of her army in the Low Countries, see Sir John Neale: 'Elizabeth and the Netherlands, 1586–7', *English Historical Review*, xlv (1930), pp. 373–96. She made a far more successful intervention, against determined and self-interested opposition from Burghley, Leicester and Walsingham, in the administration of the customs; see Howell A. Lloyd: 'Camden, Carmarden and the Customs', *EHR*, lxxxv (1970), pp. 776–87.

the frantic efforts of her Government to cope with them – were beyond her ability to supervise, or perhaps even comprehend. In the last resort the Tudor monarchy, which revived the ancient notion of the omnicompetent sovereign, proved inoperable because the widening scope of government made the constitutional sharing of responsibility inevitable. Even Elizabeth faltered. Her last years were ones of rising difficulties, and increasing inability to face them. Like all her great predecessors, she died tired and dispirited, worn down by a system of government which placed too great a burden on a single body, and by the insatiable demands of a people which always expects too much of its ruling servants. The old English monarchy, the one-man, one-woman show, founded by Alfred, endorsed by the Conqueror, died with her. It was a cruel system, which murdered its failures and killed its successes by overwork. Three of the greatest of its practitioners, William I, Henry II and Edward III, had been stripped to the skin and robbed of their rings the moment they breathed their last. It was hard for a monarch to die in dignity; even Elizabeth lost her hold on the creatures she had elevated. The English evolved a new type of constitution, and thus made their unique contribution to the history of human society, not from any love of change but because their old system was beyond the physical strength of those called upon to operate it.

I have made the point that the idea of government by a process of conflict between Monarchy and Parliament was wholly alien to the English mind. The Monarch was, in fact, a member of Parliament, its supreme member. The English did not want self-government. What they wanted was authoritative government, operating under the law, in a highly conservative manner, and in the national interest, with the King taking the decisions and answerable for them. In local administration they assumed responsibilities because there was no one else; but even here it was self-government by the King's command; and the 1,000 or so civilian knights, who took on these roles in ever-increasing measure, felt themselves to be overworked. No one wished the King to escape from his responsibilities; but it was a tall order to expect him everywhere and always to discharge them to the general satisfaction. When, as frequently happened, he proved incapable of doing so, there arose the question of who was to help and advise him. The role of Parliament was not at issue. What was at issue was the composition of the King's council. Here was the central political problem of the Middle Ages, and it was one the English never solved. Beyond a vague assumption that the King ought to be advised by his 'natural' counsellors, the great magnates and prelates, there was no consensus, and even this vague assumption was eroded in time. The English were never prepared to accept any written set of rules, for which there was no precedent (only in our own time, and with reluctance, have they been prepared to give formal acknowledgment to the existence of a prime minister, and they have never given legal definition of his rights and duties). The security clause of Magna Carta, which placed 25 barons in supervision over the King, and gave them the right to resist him if he broke it, was the one important clause never to be put into operation, though to the baronial party it was the core of the document. In the mid-thirteenth century various formulae were put forward for a committee of management, laid down by statute. They may have derived from the precedent of the college of cardinals in Rome, or the imperial electors in Germany, or more likely from the experiment in the Latin Kingdom of Jerusalem, a limited monarchy conducted on the crazy theoretical principles of Continental feudalism. But it is significant that their most enthusiastic advocate was an alien, Simon de Montfort; and he finally snarled in disgust that 'the English always turned tail' when it came to placing formal restrictions on the Monarch's choice of ministers. Such ideas were periodically revived: during the reigns of Edward II and Richard II, during the minority and dotage of Edward III, and throughout the reigns of Henry VI and Edward VI, all kinds of proposals were made for some kind of formal oligarchy to supplement the manifest incompetence of the Sovereign. None was generally accepted, or operated with success. The English managed by a series of expedients, usually born and terminated in blood, and punctuated by judicial murders.

Oddly enough, the desire of the aristocracy to get formal protection from arbitrary proceedings by the King – expressed in the famous Clause 39 of Magna Carta – was frustrated not by the Monarchy but by their own clumsy efforts to usurp its authority. The life of a magnate who played politics was more seriously at risk in the fourteenth and fifteenth centuries than under the Norman, Angevins and early Plantagenets. William I did not believe in capital punishment, least of all for the aristocracy; nor, on the whole, did Henry II; and even a magnate caught in *flagrante delicto*, in open and violent rebellion against the Crown, usually got a full and fair trial. It was the King who sought, in 1352, to define and limit the categories of death for treason. But as the practice of changing the government by force developed, judicial standards in political cases declined. Not only did Parliament invent the disgraceful practice of impeachment, but the rival oligarchies proceeded against their opponents with a ruthless disregard for the substance, and often the forms, of law. Magnates who played the political game and lost were adjudged 'guilty by notoriety' and often got no real trial at all. The judicial massacres of public men were sufficiently numerous to produce structural changes in English society. The habit of members of the English ruling class of executing each other, often on trumped-up charges or even without trial, and sparing neither friend nor close relatives, became internationally notorious. A foreigner at Elizabeth's Court noted with disgust that the savage English aristocracy openly boasted of their ancestors or kinfolk who had gone to the scaffold, swearing, in their horrible, godless way, that no one could claim true nobility without a history of treason and judicial murder in his family. By the mid-sixteenth century the old medieval clans had virtually wiped themselves out on Tower Hill. Government by landed oligarchs had, indeed, already discredited itself a century before. Sir John Fortescue wrote that the King's advisers should not, on the whole, be drawn from the magnates, who were corrupt, incompetent and selfish, but from the prudent professional men, who would devote themselves to the State service. On the whole this advice was followed, and it became increasingly common for the council to consist of the professional managers of Parliament. Thus Parliament was eventually drawn into the executive role of government; but the process was unconscious and haphazard, and no one ever attempted to justify it on a theoretical basis, or indeed on any basis except convenience. When Elizabeth died there was no general agreement as to how the country ought to be governed. That was still entirely up to the King, and his methods would not be challenged until he got himself into trouble over money. If he managed to keep solvent he would not be challenged at all.

There can be no doubt that this consensus among the propertied classes in favour of strong central government, preferably conducted by one man who knew his business and respected the need to take advice from the proper quarters, was fully shared by the great mass of Englishmen. The common English people are not docile – they have never been docile – but they are deeply conservative, respect the hierarchies of society, reverence existing institutions and customs, and as a rule are provoked to protest only when menaced by change. It is true that, with the development of the cloth-manufacturing industry in the last decades of the fourteenth century, a different spirit begins slowly to show itself. Parliament noted sombrely in the middle of Elizabeth's reign: 'The people who depend on making of cloth are of worse condition to be quietly governed than the husband men.' The rise of industrialism coincided with the demand for the spread of political responsibilities. But until the last part of the eighteenth century the overwhelming majority of the English people were directly engaged in agriculture; and agricultural workers in England are notoriously reluctant to press demands of any kind. (Hence, even today, though their productivity has risen, and is rising, faster than in any other industry, they have the lowest average wage-rates of any male occupational group.) Moreover the very structure of society was based firmly – and until the nineteenth century, it seemed, irrevocably – on the land. The native inhabitants of Roman Britain firmly declined to adopt the city-civilisation Rome proffered them; even the wealthiest, while accepting the towns as administrative centres and economic instruments, made their homes in the midst of their estates. The Old English were an incorrigibly rural people, who saw with indifference the Roman cities crumble into ruins, and whose own towns were built as markets and fortresses. The Normans and their successor-rulers encouraged urban growth as a source of revenue, not of culture, and their few experiments in deliberate town planning (such as Winchelsea) were conspicuous failures.

It was the land that mattered: its ownership was universally regarded as the ultimate source of satisfaction, the criterion of worldly success, and the only sure index of status. The English respect hierarchies, but only those based on real estate. The Old English wergilds, a system of classification which governed not only social status but legal and political rights, and military and financial obligations, were strictly related to landed property, with the exception of clergy and substantial merchants. The wergild system fell into disuse as the development of the common law abolished monetary compensation for criminal offences; but it was, *pari passu*, replaced by a social hierarchy based, not on a hereditary caste system, but on the actual, current occupation of land.

The English, indeed, are acutely but also realistically – perhaps even cynically – conscious of status. A man, whatever his origins, who did not have the land was nobody; a man, whatever his origins, who did have the land, could take his place in the hierarchy, with all its public rights and duties, to which his possessions entitled him. From the fourteenth century onwards, efforts were made by statute to oblige men of a certain estate to take up the distinctions and burdens of knighthood. From the mid-fifteenth century, the right to elect knights of the shire to Parliament was limited by statute to freemen holding real estate to a minimum value of 40s annually, and with some exceptions this remained the basis of the suffrage until 1832. The test of nobility, or peerage, became restricted to those who received an individual writ of summons to Parliament. Hence a man was noble not by right of birth but by ownership of land. Even an eldest son of a peer was a commoner until he succeeded his father; younger sons remained commoners all their lives, and drifted down the social hierarchy, unless they acquired land by marriage or industry. The formal status of peers even of the highest rank was in peril if their estates dwindled. The ruthless landed snobbery of the English was often enforced with outspoken brutality. It is worth quoting the statute passed in 1478 depriving George, Duke of Bedford, of his titles, on the grounds that his estates were insufficient:

> ... for as much as it is openly known that the same George has not, nor by inheritance may have, any livelihood to support the said name, estate and dignity, or any name or estate, and often it is seen that when any lord is called to high estate, and has not livelihood conveniently to support the same dignity, it induces great poverty and indigence, and often causes great extortion, embracery and maintainence, to the great trouble of all such countries where such estates shall happen to dwell.... Wherefore the King, by the advice of his lords spiritual and temporal and the commons assembled in this present parliament, and by authority of the same, ordained ... that from henceforth the same ... naming of a duke and all the names of dignity given to the said George, or to the said John Neville his father, be from henceforth void and of no effect. And that the same George and his heirs from henceforth be no dukes, nor marquesses, earls or barons.

Until the end of the nineteenth century, insufficiency of estate was considered absolute, or at least adequate, grounds for refusing a peerage to a man otherwise entitled to it by virtue of public services. A successful general, admiral or politician usually had to demonstrate that he had a sizeable property and the income to support it before his claims were acknowledged. Even in 1918, when Field-Marshal Sir Douglas Haig was negotiating, through Sir Philip Sassoon, with Lloyd George for suitable recognition of his wartime services, such as they were, the point was made that if he accepted an earldom Parliament must vote him £250,000 to justify the title; eventually Haig settled for £100,000.

Land carried with it direct political power, and still more political influence; it was universally acknowledged that it must be linked to status. Those who owned land must be brought into a loyal relationship with the Crown. Conversely, those whose loyalties were suspect must not own it. From the beginning of the thirteenth century, there was a general feeling that foreigners whose allegiance lay elsewhere should be debarred from ownership of English land. This eventually found statutory expression, and remained the law of the land until 1870. As late as 1864, justifying his opposition to the repeal of the statutes, Lord Palmerston wrote to the Lord Chancellor:

> I do not think we ought to alter the long established law of our land to suit the private purposes of a Foreigner however respectable or entitled to consideration.... According to our social habits and political organisation the possession of Land in this Country is directly or indirectly the source of political Influence and Power and that Influence and Power ought to be exercised

exclusively by British subjects and not to pass in any degree into the Hands of Foreigners. It may be said that the possession of landed Property by a few Foreigners would produce no sensible effect on the working of our constitution, but this is a question of Principle and not of Degree and you might on the same ground propose a law to allow Foreigners to vote at elections, as well as to allow them to purchase the means of swaying the votes of other Persons at elections.

Within the harshly realistic English system which related status and power to land it follows, as a corollary, that there was infinite room for mobility. The English class system has always been severe but never exclusive. If even dukes could drift down the hierarchy, assisted in their declension by a tremendous statutory boot from Parliament, the newly rich could always scramble up.* There were always plenty of vacancies. The turnover at the top was very rapid. Very few great landed families held together for more than a century or so. The chief cause of their eclipse was not death in battle or on the scaffold but the failure to beget male heirs. In the fourteenth century 13 earls died without any legitimate children at all, and four left only girls. This led to a sharp contraction in the numbers of the great proprietors, but a corresponding increase in the size of their estates by inheritance. The fundamental cause of the Wars of the Roses was the narrowing of the gap between the Crown itself, whose revenues were falling, and those of the few great proprietors, which were rising – the King became a mere *primus inter pares* of half a dozen heads of families, who alone or in combination could replace him almost at will.† But the wars and their aftermath virtually wiped out these super-aristocrats, and the nobility of the late sixteenth century was, for all practical purposes, a new creation.

The English aristocracy, greedy and realistic, never showed much compunction in allying itself with its social inferiors, if the price was right. In the fourteenth century, one-third of the daughters of London aldermen married into the nobility; in the next century this proportion rose to over 50 per cent. All the professions, law, warfare, public

* The English are a pushful people. In Anthony Trollope's remarkable study in social ambition, *Is He Popenjoy?*, the hero, the Dean of Brotherton, gives a classic statement of English social philosophy: 'It is a grand thing to rise in the world. The ambition to do so is the very salt of the earth. It is the parent of all enterprise, and the cause of all improvement. They who know no such ambition are savages and remain savage. As far as I can see, among us Englishmen such ambition is, healthily and happily, almost universal, and on that account we stand high among the citizens of the world.' The Dean's father had kept stables; his grandson becomes a Marquess.

† Richard II's regular income had been about £120,000; Henry VI's fell from £75,000 in 1422–32 to £54,000 in 1442–52. The Lancastrian monarchy has been described as 'a pauper government ruling with the consent of its wealthier subjects' (A. Steel: *The Receipt of the Exchequer, 1377–1485*, Cambridge, 1954).

service, the Church, offered to the man of meanest origins a speedy entry into the upper echelons provided he acquired enough money to purchase land. Of the richest and most magisterial medieval prelates, one, Wykeham of Winchester, had a peasant father, another, Wolsey of York, was the son of a butcher. Such men could acquire wealth and status very rapidly indeed – often in a mere decade – but they were not resented provided they conformed equally quickly to the manners and prejudices of the men they joined. All kings created new grandees from lowly origins, through the Church or the civil service, but it is significant that the two kings who were most successful in managing the aristocracy, Edward I and Edward III, were careful to select for promotion men who were circumspect in their social behaviour, and who showed an early aptitude for assuming their traditional responsibilities as landed proprietors. New men who behaved arrogantly risked judicial murder or (in the case of one imprudent bishop) being torn apart by a London mob. The children of such upstarts invariably conformed, and their parents' birth was rarely held against them, for there were few whose family origins would bear prolonged investigation. When an angry London mob screamed at John of Gaunt, Duke of Lancaster, that he was the son of a Flemish butcher (his mother was widely believed to have been an adultress), they were motivated not so much by snobbery as by xenophobia.

The English landed class was thus (and has remained) flexible, because all suitable persons were admitted; rich, because its financial standards of admission were high; and tiny, because its unsuccessful members were expelled. The income tax returns for 1436 reveal 51 magnates with an annual average income of £865; 183 greater knights, averaging £200; 750 lesser knights, averaging £60; 1,200 esquires averaging £20–35; 1,600 property-owners averaging £10–19 and 3,400 with £5–10. (Some of the men with the largest incomes, who also supervised the returns, deliberately underestimated their incomes, in some cases by as much as 50 per cent; thus the Duke of York's real income was at least £7,000, not £3,231, as stated, and the Earl of Warwick's was at least £6,000, not £3,116; ever since the eighth century, the English aristocracy had been engaged in various forms of tax-evasion on a massive scale, and of course still does so, though it is nowadays called tax-avoidance.) Thus less than 7,000 people owned virtually all the property in the country, and controlled a nation of three million.

It may be asked: how was this astonishing social and economic imbalance preserved? What prevented the great mass of the propertyless English from rising and sweeping away this minute ruling class? Certainly there were no physical barriers to stop them. There was no standing army. Even Richard II's amateurish efforts to recruit a permanent force of Cheshire archers was violently resented by the aristocracy, and held as one among many good reasons why he should be deposed (and murdered). There was no professional police force. The ruling class was armed, and to some extent trained to warfare, but then so was virtually everyone else. It is true that in theory men of villein status were not permitted to bear arms; but they in fact did so; moreover villeinage was in rapid decline in the fourteenth century, and soon after became obsolete. And in many parts of England, particularly in the east and north, the great majority of the peasants had always been free. In any case, the supposed defence needs of the realm led the Government positively to encourage large numbers of peasants to become proficient in the use of arms, especially of the long-bow. This formidable and characteristically English weapon required a great deal of training to master; but in the hands of the skilled man, it was lethal up to 600 yards and at a distance of up to 100 yards could penetrate chain mail. Most able-bodied Englishmen knew how to handle it. The villages often trained together, as an operational unit; from Oxfordshire in 1355 we have an echo of their brutal war-cry, the signal for general pillage: 'Havok, havok, smygt faste, gyf good knok!' No medieval English Government could possibly have withstood a peasant uprising which aspired to take over the State.

Yet this motivation was lacking. It was never difficult, in those times, to assemble a crowd, often of thousands, sometimes of tens of thousands. There was huge misery everywhere; a large proportion of the population suffered from chronic ailments, beyond the skill of doctors, even if they had been available to the poor; many were incurable cripples; the merest rumour of some miraculous cure, of a travelling miracle-worker, was sufficient to set up a stampede which the authorities were powerless to control. These hysterical crowd-movements were a recurrent feature of the Middle Ages, more common on the Continent than in England, but nevertheless to be found here; the chroniclers superstitiously paid them as little attention as possible; there was, as it were, a conspiracy of silence about the fearful potential of the mob. There is no doubt that agitators could have diverted this uncontrollable energy into directly political channels. But in England the elements of political conspiracy were lacking. There may have been good reasons for this. Slavery itself had been killed by the Normans. Villeinage was never as oppressive, or as widespread, here as across the Channel. It was already in decline in the twelfth century, and the slow declension was never, as on the Continent, successfully reversed, or even arrested. Moreover, many of the villeins themselves were men of substantial property, with a stake in the existing order; there are cases where we know they declined a change of status, for sound economic reasons. There is evidence of a very widespread distribution of property, if only on a small scale, in the countryside. Many local riots took place; but they were the medieval equivalent of a strike, organised and carried through to achieve specific, limited objectives, not to overthrow society. The aim of most of them was to put the clock back. None sought to impose an alternative government. The English medieval peasant was deeply conservative, as were, many centuries later, his industrial progeny. In 1926 the General Strike was wholly successful, in that it brought transport and industry to a standstill; but, having achieved this, the workers could go no further, as they lacked the desire – though they had the opportunity – to take over the State; so they went back to work. It was the same with medieval uprisings; the furthest their political aspirations could reach was to present petitions to the authorities for the redressing of their wrongs, in the belief that the King, if informed of their grievances, would not hesitate to restore the ancient law. They were usually deceived; but their faith in the virtue of a divinely inspired kingship never faltered. Even in the time of Henry VIII, a sovereign who had introduced dramatic changes in the religious structure of the country – changes which the northern peasants bitterly resented and which, indeed, were a chief motivation of revolt – the Pilgrims of Grace put implicit trust in the King's word that they would get redress, dispersed quietly, and were then massacred in detail. The truth is, rebellious peasants were not violent revolutionaries, but violent reactionaries.

The attitude of the English masses can be seen in its most striking form during the fourteenth century, a period of fundamental economic change. The agricultural system of early medieval England had been based on 'high farming'; that is, large estates farmed directly by their great secular and ecclesiastical owners, and based to a considerable extent on labour services. Some of these estates were big business, by any reckoning. In 1322 the cathedral church of Canterbury farmed 8,373 acres of arable and owned 13,730 sheep; the Priory of St Swithin's had 20,000 sheep. The standard of living of all had been rising steadily for two centuries, as more land was brought into cultivation; population had increased steadily too. By the early fourteenth century, only marginal land remained; it may be that the population had reached saturation point, at any rate in terms of the existing methods of exploitation. The large estates were often grossly inefficient. Compulsory work-service was no more effective in England than compulsory knight-service. High farming required large injections of capital to work at all, especially when harvests were bad. After the expulsion of the Jews, it was difficult for big landowners to get bridging finance, even at crushing rates of interest; and increasing taxation often pushed them deeply into debt, and so destroyed what credit they had. In 1315–17 there were three appallingly bad harvests, caused by torrential rains and floods. It brought the last general famine in English history.* But if thousands of peasants died of starvation, the great landlords were badly hit too. Indeed high farming in England never really recovered from this catastrophe. Many of them began to lease out major portions of their estates for fixed rents. With the decline of the old manorial system, the labour market for agricultural workers, paid by the day, grew rapidly. Large-scale farming for the market ceased to be attractive; marginal land fell out of cultivation. The fall of money incomes from farming was one reason why the English ruling class was so attracted to the French wars, which offered the only other occupation they knew which would bring cash profits. The peasants were keen to get land, because they were less affected by agricultural recession; the estates were keener to sell or lease it, because they had to raise money for war taxation. Moreover, the effects of *Quo Warranto* proceedings were diminishing the social value of large estates.

Into this already critical situation, the Black Death introduced a new and revolutionary element. This mixture of diseases, in which pneumonic and bubonic plague were predominant, arrived in August 1348 and lasted until the end of 1349; a second wave struck in 1361–2, attacking mainly children, and a third in 1369. The second attack was perhaps the most important from the economic point of view, affecting the labour market in the 1370s. The cumulative impact on the population was dramatic. Over the country as a whole it fell by about one-third; in many areas by a half or more. There was an immediate and rapid increase in wage-rates, which rose 30 per cent in the decade 1340–50, 60 per cent in the next decade, and continued to rise. There was also a steady upwards movement in what we would call 'wage-drift', with employers forced to concede a range of amenities to get any labour at all. Boon-labourers at harvest expected a midday meal of bread, ale and pottage, plus either beef, pork and mutton, or a fish dish and five herrings. For the first time in English history, the ordinary man had the possessing class at his economic mercy.†

* C. Britton: *A Meteorological Chronology to AD 1450* (London, 1937), p. 133. But see also Barbara Harvey: 'Population Trends in England, 1300–1348', *Transactions of the Royal Historical Society*, 1966.

† Landlords made desperate efforts to counter the manpower shortage by labour-saving devices; use of the scythe, instead of the sickle, and of the butter-churn became common at this time. See B.H.Slicher van Bath: *The Agrarian History of Western Europe* (London, 1961).

The rich and the well-to-do sought desperately to reverse the trend by legislation. Ordinances put out in 1349, revised and enacted as statutes in 1351, were the first of many measures designed to hold wage-rates and force men to work for them. They were accompanied by sumptuary legislation which sought to underpin the crumbling class-structure by denying the masses the right to exercise their new purchasing-power. These acts recall the despairing economic legislation of the late Roman Empire. They were almost wholly ineffective because they could not be generally enforced; and the spasmodic attempts to enforce them provoked anger. The clock could only have been put back by bringing about an actual and massive cut in the standard of living, the first for centuries. This was beyond the power of any medieval government, or indeed of any modern one; it was something only war or natural disaster could achieve. A statutory freeze of wages was unacceptable to the peasants, who preferred to withdraw their labour; it was unacceptable, in the end, to the landlords, who preferred to evade it and pay high wages simply in order to stay in business. Rents fell and land declined in value. Villeins ran away in large numbers because the economic opportunities now open to them greatly outweighed the small risk of capture. Parliament tried to enforce work-services, but the thing was impossible without an expensive apparatus of repression, which did not exist. Where action was taken it was too weak to secure compliance, but irritating enough to provoke violence. Moreover, the King's courts were available to free peasants; even lowly elements of society could, and did, seek legal protection. The peasants were capable of pleading the Great Charter in their support, even if it did not strictly apply. They were litigious, and if the law failed them could be relied on to resort to arms to resist what they saw as a novel and revolutionary invasion of their ancient rights. There were numerous uprisings, most of which were successful in attaining their short-term object of frustrating Parliament's attempt to control the economy.

By the 1370s the Government, reeling from military disaster and divided by rival factions, had virtually abandoned the hopeless attempt to freeze wages; instead they sought to tap the new wealth of the masses by resorting to poll-taxes, that is by a direct, fixed levy on every adult, irrespective of income. This monstrous device, of Continental origin, was the perfect formula to provoke civil commotion. Attempts to take a preliminary census, in 1377, led to riots in the big cities, such as London, and resistance almost everywhere. In 1381, the actual imposition of a shilling poll-tax was the signal for revolution, particularly in the heavily populated areas of the south-east. A peasant had to pay 2s for himself and his wife in ready cash; in many cases he did not possess it. His lord might pay it for him, advancing the money against future labour. But in many villages there was no longer a lord – he had contracted out – and the peasant, for the first time, came up against the revenue officers directly. He could not pay, and so he took to arms.

Yet the Peasants' Revolt was essentially an exercise in English conservatism, or rather in two kinds of English conservatism. The representatives of the propertied classes wanted to put the clock back to the early years of the century, before the peasants acquired their present economic bargaining-power. The peasants also wanted to put the clock back, but to an even more distant, and largely imaginary, period. It was thus a conflict between two reactionary forces, operating under different time-scales. The authorities saw the peasants as revolutionaries, in that they offered violence against due forms of law. The peasants saw the authorities as revolutionaries, in that they used the instrument of the statute to demolish ancient customs. They turned first against the Church, for clerical landlords were the most unscrupulous and tenacious in their employment of new legal devices; and anti-clericalism had been a burgeoning English tradition for over two centuries – the earliest organised peasant riots, of which we have records, in the 1230s, had anti-clerical objectives. There were assaults on monasteries and even on Cambridge University. But the rioters, if anti-clerical, were not anti-Christian; on the contrary, they sought a return to a primitive Christianity. Lawyers were also objects of attack, as the agents of new and vicious forces. In many cases recent legal records were seized and burnt; lawyers were captured and beheaded in the name of the King. But the ancient law itself was not challenged; on the contrary, it was exalted, and quoted in support. The men of Kent said they would pay no new taxes; no taxes at all, 'save the 15ths which their fathers and forebears knew and accepted'. One of Wat Tyler's demands was that there should be no new law – only 'the law of Winchester', a reference to the legislation, real or imagined, of Edward I, already a century old. The peasants demanded an end to attempts to restore villeinage, a return to the ancient free market in land, and the right to rent land at the old price of 4d an acre.

This programme was, in effect, a revolutionary one. But it was presented in the customary English manner; that is, the sanctification of change by dressing it up in the guise of a return to an earlier order. The rebels delved deep into their capacious folk-memories, and came up with some surprising symbolism – though they were thinking in real terms, not symbols. The Kentish rioters called themselves 'men of Kent and Jutes'. In East Anglia, there were demands for 'county kings', of the Northfolk and the Southfolk, each of whom would issue 'county charters', modelled on Magna Carta. There were evidently, among these people, living memories of pre-Norman times: not even of the great unitary Wessex state but of the kingdoms of the heptarchy which had preceded it. Indeed, Norfolk and Suffolk had already been merged in the Kingdom of East Anglia by the mid-seventh century, at the time when the great Sutton Hoo treasure was buried. The rioters of 1381 were thus going back over 800 years, to a form of territorial tribalism.

The riots were put down without much difficulty. Some of the leaders were murdered or hanged, but there was no general repression. The State was in no position to carry one out; and the English have always quailed before the prospect of class war. The rioters got their way on all essential points. On the Continent similar – and much more bloody – uprisings led to a genuine reaction, in which the villeinage system was brutally reimposed, and was to endure in some respects until the French Revolution and beyond. But in England the agricultural revolution, of which the Peasants' Revolt was an illogical by-product, went on almost unhindered, to the great benefit of the community. By the end of the century high farming was dead, and the new pattern of freeholders, tenant farmers and landless labourers, which was to last until the end of the eighteenth century, was firmly established.

But the revolt undoubtedly struck terror into the English ruling class. Its one real consequence was to turn that class decisively against the new religious movement associated with Wyclif. There was here undoubtedly a confusion in ideas, for Wyclif was in some respects a conservative, upholding the ancient rights of the State against the clerical encroachments of Rome; he had the support of what would later be called the Whig element in society, such as the Duke of Lancaster, John of Gaunt, and the Black Prince's widow. But he also, through his use of the vernacular, had a great and growing popular appeal; he challenged the established order in the Church, and a wide range of religious assumptions. His views were shared by many poor clergymen, in revolt against their ecclesiastical superiors. It was not easy to make a distinction between those religious reformers and a priest like John Ball, who had been released from the Archbishop of Canterbury's prison to take a leading part in the peasants' uprising. At any rate, the authorities were in no mood for such hair-splitting. All innovation, reform, change – call it what you will – was dangerous. Hence in the 1380s there was a series of moves to suppress Lollardy. The prelates recovered their nerve; the State moved to assist them, for the first time hunting out and burning heretics on a considerable scale; the kings, Lancastrian, Yorkist, even Tudor, took on a new role as the custodians of religious orthodoxy. Fear of economic and political subversion sent the ruling class back to the old, discredited altars. The one clear result of the Peasants' Revolt was to delay the Reformation in England by 150 years.

The English eventually approached the business of changing their religion, if that is a correct description of what happened in the middle decades of the sixteenth century, in a characteristically haphazard and confused manner, and were later to congratulate themselves on the constitutional propriety with which it was done, and the admirable compromise which they eventually evolved. Yet the breach with Rome, and indeed the three centuries of growing hostility to the papacy which preceded it, had comparatively little to do with religion as such; its principal dynamic was anti-clericalism, which was itself a form of English xenophobia. The papal aggression of the twelfth century had ended the old easy relationship between Church and State which had been such a striking and constructive feature of Anglo-Saxon society. The Becket affair made it clear that henceforth the two powers, one national, the other international, would be in a permanent state of tension and often of conflict, with public opinion inevitably moving in support of the national position.

The thirteenth century, it is true, saw the universalist claims of the papacy come near to triumph. The English King became a vassal of the Pope. For the first time the Pope had a major voice in senior clerical appointments. Even a dominant personality like Edward I could not get his own man made Archbishop of Canterbury. A large number of ecclesiastical benefices were made subject to the system of papal provisions, under which nominees of the Pope, most of whom were Italians, enjoyed the revenues of English bishoprics, canonries and rectories, without in most cases ever setting foot in the country.

Yet the power of the Pope was more theoretical than real. Successive kings found their relationship with the Pope convenient, chiefly because it enabled the State, in the name of the Pope, to impose heavy taxes on the clergy. Throughout the later Middle Ages, vast sums were raised in this manner; the Pope on average got about 10 per cent of the proceeds, if he was lucky, the State took the rest. If papal action at any point constituted a real challenge to the Crown, the King could immediately turn to Parliament for assistance; and Parliament always faithfully reflected the growing anti-clericalism of the English people. In 1286, Edward I passed the first anti-clerical statute, *Circumspecte Agatis*, which began the erosion of the powers of the Church courts. Thereafter the position of the papacy in relation to the English State was in steady, and irresistible, decline. Already, in 1318, Pope John XXII wrote sadly that 'the status and, what is more, the liberty of the ecclesiastical dignity is more depressed and trampled on in [England] than in all other parts of the world'.

This is scarcely surprising, in view of the claims of clergymen to a separate caste status, their enjoyment of between a quarter and a fifth of the wealth of the country, and their lack of a recognisable role in society: they were parasites and were seen to be parasites, and public opinion at all levels of society could be easily marshalled – indeed would marshal itself – against them. As Boniface VIII's bull, *Clericos Laicos*, admitted in its opening words, 'laymen are notoriously hostile to clerks'. The trouble with the clergy was that there were too many of them, and most of them were in the wrong places. In the late thirteenth century about 50,000 clergy were serving an English population of three million. Nearly half of them were in some 780 religious houses, fulfilling no obvious social need. There were about 9,000 parishes, but their distribution was grossly uneven. Far too many of them were concentrated in the towns – over 100 in London alone. There was an enormous bias in favour of the south-east. Clergymen did not want to serve in the wilder and poorer districts of the north and west. This certainly helps to explain the difference in the regional attitudes towards the Church. In the south-east anti-clericalism was sharper and more general, for men had ocular evidence of a swarming, idle and grasping clergy; in the north and west there was never the same animus against Rome because the clergy were less visible, and indeed in many cases were obliged by sheer lack of numbers to work extremely hard.

What the Church lacked above all was any general sense of pastoral zeal. The best minds in the Church, from the Pope down, concentrated on the maintenance and extension of privilege and jurisdiction, to the exclusion of its real spiritual purpose; the medieval Church was ruined by legalism. Men like Bishop Grosseteste, the pious and active apostle of Lincoln in the thirteenth century, were rare birds. He rightly said that clergy should not take secular offices, but should devote themselves to ministering to their flocks, and raising moral standards. But his warnings were ignored; the Crown found it cheaper to employ clerical servants because they could be paid in benefices, instead of from the Exchequer. It is true they were more difficult to punish if they proved corrupt; but on the whole it was judged the lesser of two evils. As a result the overwhelming majority of bishops were appointed from secular motives. Many of them never engaged in pastoral work, or even visited their sees. The popes could not insist on active pastoralism, for they were themselves the main beneficiaries of the absentee and pluralist system. With no pressure from the Pope, and little supervision from bishops, with absenteeism at all levels, especially in the richest (and key) posts, the ordinary clergy were naturally lax. Most of them had wives, or mistresses, and raised families. Few knew their duties. Many were illiterate, and not just at the lowest levels: the Black Prince, for instance, succeeded in getting an illiterate friend made Bishop of Lincoln. The Church did not know what was going on in the parishes; it did not even know how many there were. In 1371 it was thought there were 40,000 parishes in England; investigation (for tax, not spiritual, purposes) showed there were, in fact, less than 9,000. The truth is that the clergy, like, for instance, qualified doctors in underdeveloped countries today, were distributed according to the availability of pickings, and not according to actual need.

Moreover half of them, the regulars, had no obvious public function at all. In the Dark Ages the monasteries had served an important economic and social purpose, as well as a cultural one: they forced the pace of technological change in agriculture. Even up to the beginning of the fourteenth century the monks were very active, and usually efficient, farmers. But the collapse of high farming, and the spread of leasehold, turned the monks into a rentier class, without any role in society other than as conspicuous consumers, living on the labour of others. In the second half of the fourteenth century we get the first demands, in Parliament and outside – and often from hard-working parish clergymen – for the general confiscation of clerical estates, especially of the regular clergy. In 1385 some of the Commons wanted all the temporalities of the Church to be seized, and they were echoed by Langland in *Piers Ploughman*: 'Taketh here londes, ye lords, and let hem lyve by dymes.'

The senior Church authorities played into the hands of the confiscators. Some of the priories were offshoots of foreign mother-houses, and their profits went abroad. There was no protest from the English hierarchy when in 1295 Edward I, inspired by the xenophobia arising from the war with France, made the first seizures of alien priories. He was followed in 1324 by his son, and in 1337–60 and again in 1369 by his grandson. These foreign religious properties were wiped out and engulfed by the State. The Commons petitioned for the monks to be expelled, on the grounds that they were spies. A few bought charters of denization, and survived. The English Church got a share of the spoils, and the Pope got his cut too. In any case, the papacy was not in a position to protest: the crushing of the Templars had set a dreadful precedent, as Langland shrewdly noted:

> For coveityse of that crosse · men of holy kirke
> Shul tourne as Templeres did · the tyme approacheth faste.

The means employed against the Templars – confessions of moral turpitude, extracted under torture, to justify the seizures – were themselves consciously echoed by Parliament when, in 1536, it appropriated the lesser monasteries:

> For as much as manifest sin, vicious, carnal, and abominable living, is daily used and committed among the little and small abbeys, . . . whereby the governors of such religious houses spoil, destroy, consume and utterly waste . . . to the high displeasure of Almighty God, slander of good religion, and to the great infamy of the King's Highness and the realm . . .

The Church, indeed, was in part the architect of its own destruction. Powerful prelates had never hesitated to misuse Church property, and even to grab it, with the barest show of legality, for their own purposes. Cardinal Wolsey was merely the last of a long line of ecclesiastical confiscators when he suppressed a group of small religious houses to found his Cardinal College (now Christ Church) at Oxford.* There was nothing new about the dissolution of the monasteries: it was the culmination of a long English tradition, inaugurated with the approval of the Pope. The Church was self-devouring, and riven by bitter animosities. An archbishop of Canterbury called the Cistercians 'the worst possible neighbours' because of their greed and love of litigation. The Black Monks hated the Austin canons, and both hated and feared the Franciscans. There was constant litigation between regular and secular clergy, between diocesans, who engaged in mutual excommunications, and between bishops and chapters. A case concerning the tithes of the Priory of Lenton, begun under King John, was still being conducted in an animated manner at the Dissolution, over 300 years later. Another, involving the rights of the Dean and Chapter of Durham to administer the spiritualities of the see during an episcopal interregnum, first came before the courts in 1283, survived the Reformation, and was last argued about in 1939; it is still unresolved, though dormant. At no stage was the English Church able to present a united front against its critics; and this is one chief reason why only a tiny minority of the clergy opposed either the Henrician reformation or the Elizabethan settlement. The

* It was done with the approval of such leading papalists as Bishop Fisher of Rochester. Wolsey's agent in the business was Thomas Cromwell, who thereby acquired a closer acquaintance with conditions in the monasteries, and with the social and legal technicalities of dissolution, than any other man in England. The case against the regular clergy was not so much that they were corrupt (though some were) as that they were idle: about 8,000 men and women sitting on one-eighth of the country's wealth. The 357 lesser monasteries averaged less than four religious each. Butley Priory, in Suffolk, had only 12 canons; but it maintained two chaplains, 11 valets, a barber, three cooks, a slaughterman, a sacristan, a cooper, three bakers and brewers, two grooms, two maltsters, a porter, a gardener, six laundresses, an under-steward, a surveyor and 36 estate-workers.

English clergy nearly always sided with the authorities, even when their brethren were being persecuted. The only occasions in the whole of English Catholic history when a majority of the bishops opposed the State were during the desperate crisis years of King John's reign – and then only for a very short time – and in the first year of Elizabeth; and on this second occasion the bishops had been hand-picked by Mary for their ultramontane views. Henry VIII had to execute only one bishop, Fisher of Rochester, who was a notorious opponent of reform in any shape (he hotly defended pluralism and absentee clergy), and who was certainly guilty of treason, in 1533, when he invited the Emperor to invade England. Among the ordinary clergy, acquiescence in the changes in religion was the prevailing pattern. Less than 1 per cent of the regulars defied Henry VIII when he seized their property; less than half of 1 per cent of the seculars rejected the Elizabethan settlement. The Church Militant may not have been dead by the time the Reformation came; but it certainly put up very little resistance.

Of course the Church of England had been conditioned to lay supremacy long before Henry VIII made it formal. What killed papalism in England were the French wars of the fourteenth century. Edward III's statutes of *Provisors* and *Praemunire*, which in effect made it a capital offence to obey the Pope as opposed to the King, were the direct product of English xenophobia, generated by hatred of the French and the new consciousness of English nationalism. *Praemunire* in particular was an omnibus statute which could be used against anyone and everyone who defied the King on spiritual matters. As one papalist at Henry VIII's court remarked, no one really understood the statute, or could construct a defence against a charge under it, because it meant whatever the King wanted it to mean. Yet many, perhaps most, Englishmen would argue that it went no further than what had always been the accepted position of the English Crown in its relationships with the Pope: it was part of the great continuity of English history. William I had demanded, and got, no less. The consequences of John's surrender of the realm to the Pope was a temporary aberration, and the surrender itself illegal and *ultra vires*. The popes could do nothing against the solidity of the English State. Martin V, early in the fifteenth century, conceded: 'It is not the Pope but the King of England who governs the Church in his dominions.' In 1486 many precedents were quoted, in Parliament, by the Lord Chief Justice and the Bishop of London for the proposition that 'the Pope could not lawfully act in derogation of the King and his Crown'. Several popes tried to persuade the English monarchy to repeal *Praemunire*, but were brushed aside. The relationships between individual clergy and the papal see were rigidly and ruthlessly controlled. Papal powers were often conceded in theory but denied in practice. The appointment of bishops was typical of how the system worked. In 1446, for instance, the Pope was permitted to nominate a new Bishop of Norwich; but the bishop had formally to renounce all the provisions of the bull appointing him before being allowed to take up the see. English bishops and abbots travelling to Rome had to promise before their departure to sue or procure there nothing that was prejudicial to the King, the rights of the Crown, or the rights of his subjects. Papal envoys were liable to arrest: there was a regular form of writ for this purpose. The attitude of the English to Rome was notorious. In 1468 the envoy of the Duke of Milan wailed: 'In the morning the English are as devout as angels, but after dinner they are like devils, seeking to throw the Pope's messenger into the sea.'

The English, indeed, were perfectly capable of combining doctrinal orthodoxy with rabid anti-clericalism, though they were equally capable of favouring heresy if they thought it would suit their purposes. On the whole an orthodox king, who took an active interest in religion, was the most dangerous opponent the Pope faced. No English monarch treated papal claims more harshly than Henry v. But he was a pious, high-minded and fanatical Catholic. He personally supervised the burning of heretics; on one occasion he had a Lollard blacksmith taken out from the flames when he was already half-dead, exhorted him to recant, and when he refused thrust him back on the pyre. He evidently considered himself the effective head of the English Church, for he personally carried out a visitation of the English monasteries, examining the monks, correcting and punishing abuses, and laying down standards of conduct. From the Pope's point of view this was a most sinister precedent, a direct adumbration of the events of 1536. Henry v was regarded as a true son of the Church because he never talked during mass and had monks castrated for sexual incontinence. But it was precisely his religious zeal which made him a menace to Rome. This was Rome's fault. The claims of the papacy to divine authority as Christ's vicar led to a corresponding exaltation of kingship; the divine right of the Monarch was the secular mirror of ecclesiastical self-glorification. And the King had physical force where the Pope had only threats and curses. Henry v thought he acted on the direct orders of the Deity; God won him his battles; God told him to uphold orthodoxy, and reform abuses. He might one day issue instructions to reform abuse at the source, in Rome. By the fifteenth century there was universal agreement among the high-minded that the Church was in need of reformation. Ecumenical councils were held for this purpose; the best of the clergy preached and agitated for reform. But the papacy was incapable of reforming itself.

After a century of talk about reform, it ended up with a Borgia pope, a man of the quality of Alexander VI; a reforming monk like Savonarola ended up at the stake. The papacy did not begin to take religion seriously until the 1550s, and by then it was too late.

But perhaps it was fortunate for the popes, in the long run, that the English Reformation was delayed. Henry V would have made a much more formidable opponent than Henry VIII. He had no respect for the papacy: when Bishop Beaufort, the richest man in the kingdom, sought a cardinal's hat without the royal permission, Henry immediately placed him in peril of his life, and mulcted him of the enormous sum of £26,000. Had Henry lived a few more months he would have become King of France and a European tyrant. He was the only English king who could win and hold provinces as well as mere battles. Had he survived to middle age, the probability is that this zealous, God-inspired man would have superintended a Reformation of an altogether different kind, and have created a new Church – all in the name of orthodoxy – of European extent and Caesaropapalist flavour. In such a Church the Pope would have been a mere subordinate and functionary: the Hildebrandine programme in reverse.

How far the English would have relished such a scheme is difficult to judge. Most of them, in so far as they took any interest in religion, were Anglicans, as they always had been. They wanted an English Church, run by Englishmen. They did not object to a link with Rome provided the Pope did not interfere, especially in appointments and finance. They thought there were too many idle, dissolute and criminal clergymen, and objected strongly to the fact that some of them were foreigners. As a matter of fact the foreigners were not to blame; their numbers fell sharply as a result of *Provisors*, while the clerical crime-rate went up. The public took a prejudiced view of clerical behaviour. In 1515 the Bishop of London complained to Wolsey that any jury of twelve men in London would convict any clergyman whatsoever, 'though he were as innocent as Abel'; he spoke with feeling, for his Chancellor had just been accused of murdering a tailor.* London juries hated clergymen even

* The tailor was called Richard Hunne. He declined on principle to pay mortuary dues after the burial of his child, and was successfully prosecuted in the Bishop of London's court. He replied by serving a writ of *Praemunire*, whereupon Bishop Fitzjames, a notorious reactionary, accused him of heresy and committed him to the Lollards' Tower at St Paul's. Two days later he was found with his neck broken, and a London jury brought in a verdict of wilful murder against the Bishop's Chancellor, Dr Horsey. Horsey was almost certainly guilty; but the Bishop ignored the verdict, pronounced Hunne a heretic, had his body burnt at Smithfield and confiscated his property, making his widow and family paupers. The case aroused the fury of the Londoners, and was one reason why the Reformation was so popular in the capital. See A. Ogle: *The Tragedy of the Lollards' Tower* (Oxford, 1949).

more than Welshmen. But we should not confuse anti-clericalism with a mass movement against orthodoxy. Clerical recruitment was increasing right up to the breach with Rome. As we know from a sharp account written by Erasmus, who visited Canterbury about 1512 with a friend (probably Dean Colet), the shrine of St Thomas was still doing a roaring trade, both national and international, on the eve of the Reformation. It was pulling in £8,000 a year (half the cost of maintaining the Calais garrison), much of it in foreign currency.

On the other hand there was, and had been for nearly two centuries, an important and active minority working for radical reforms of doctrine and organisation in the Church. They represented a streak of heterodoxy in England which went back right to the earliest days of Christianity. It is very significant that William of Ockham, the fourteenth-century scholar who conducted a frontal assault on the prevailing orthodoxy of the Schoolmen, was accused of Pelagianism, and did in fact uphold the individualist tradition of free will which, as we have seen, the Briton Pelagius had founded in the early years of the fifth century. At any rate, Ockham's teaching directly inspired Wyclif; and Wyclif, writing and preaching in the 1370s and 1380s, adumbrated virtually all of the Reformation programme: consubstantiation, the English Bible and the use of the vernacular instead of Latin, the end of idolatry, the royal supremacy, the breach with Rome, and the confiscation of clerical property. He hated the Romanist bishops, and they hated him. He was reputed to be the best English scholar of the day, and the University of Oxford closed ranks behind him until its resistance was smashed by the brute power of the hierarchy. Wyclif's teachings appealed strongly to the anti-clericalism of the House of Commons; and he had an influential, if select, following among the very rich. But after the Peasants' Revolt struck terror into the possessing classes, the organs of the State were turned against his movement. In 1401 Parliament passed the statute *De Haeretico Comburendo*, empowering the secular authorities to underwrite decisions in the clerical courts by burning the heterodox. Between then and the Reformation some 100 Lollards were executed, some as late as the reign of Henry VIII. But Lollardy was only driven underground; it had both a popular following and support from individual members of the gentry; it survived as a distinct minority movement, until in the 1520s it merged with Lutheranism.* Thus, when the breach with Rome came, a very ancient English tradition, maintained admittedly only by a minority, was available to supply doctrinal nourishment.

* Lollardy was particularly popular among weavers, clothworkers, wheelwrights, smiths, carpenters, shoemakers and tailors. See A.G. Dickens: 'Heresy and the Origins of English Protestantism' in *Britain and the Netherlands II* (Groningen, 1964).

One of the reasons why the Reformation was successful in England was that there was absolutely nothing new about it. All its elements – anti-clericalism, anti-papalism, the exaltation of the Crown in spiritual matters, the envy of clerical property, even the yearning for doctrinal reform – were deeply rooted in the English past.

The breach with Rome, like the 1914 War, could have come at almost any time. The elements had been there for decades; only a spark was needed. There were sinister portents that English xenophobia was on the boil again. On May Day 1517 the London mob carried out an anti-foreign pogrom; two years later some of Henry VIII's younger friends were expelled from court on the grounds that they had 'French manners'. Once Henry had decided to divorce Catherine, it became obvious that the breach would come unless the Pope did what Henry wanted; this was Wolsey's view from the start, and he warned Pope Clement VII repeatedly that Rome's future in England hinged on the divorce. Yet oddly enough the divorce was the one issue on which Henry did not have public opinion behind him. It is a curious fact that English kings who quarrel with their wives always forfeit the general sympathy. There was no reason why Catherine should be popular; but she was. Both Houses of Parliament disliked the divorce, and the prospective marriage into the Boleyn family still more; when steering the Reformation legislation through Parliament, Thomas Cromwell was always careful to divert attention from the personal issue to the safe ground of clerical abuses.

Yet Henry was undoubtedly right to seek a divorce. As he saw it, in the light of recent English history, the provision of a male heir who would have communal backing for his title to the throne was essential to stable government, and was thus a necessity of State. It was intolerable that this vital national interest should be jeopardised by the actions of a foreign power, the papacy, motivated not primarily by spiritual considerations but by the needs of its own foreign policy. Any self-confident English king would have taken the same line. Moreover Henry believed, and may have been right to believe, that his marriage to Catherine was genuinely invalid. There had been a technical impediment of public honesty, as Wolsey pointed out; unfortunately Henry ignored this point, and concentrated his case on the more complex and intellectually fascinating grounds of affinity, where the consensus of European canonical opinion went against him.* The trouble with Henry

* Catherine had earlier been married to Henry's elder brother Arthur, who had died; when Henry married her, in turn, a papal dispensation was required to remove the obstacle of affinity. Henry now claimed this bull was invalid, on the grounds that the Pope had exceeded his powers. In 1527 Catherine, seeking to defend her marriage to Henry, declared that she had never slept with Arthur. But she should have said so in 1503, and the bull would then have been issued on a different basis. Hence the

was that he was a clever, shallow-brained pseudo-scholar, prone to sudden enthusiasms – to go on crusade, to get elected Emperor, to make Wolsey Pope, to become an author. He now embarked on a career as professional theologian, but was much too mentally indolent to get to the roots of the matter.*

Happily he was saved by the frivolity and deceitfulness of the Pope. Clement evidently did not take the breach seriously, or at least never imagined it would be permanent. At the height of the divorce dispute he wrote a letter to Henry asking for facilities to be given to a friend of his who wanted to examine some English libraries; he did not seem to grasp that Henry, and indeed the English, were playing for keeps. (In 1536, after Anne had been beheaded, Clement's successor assumed the slate was wiped clean and everyone could be reconciled; the fact that the English Reformation had taken place appears to have escaped him.) Most of all, Clement was dishonest in his actual handling of the case; it was this aspect which swung the English ruling class, not initially in favour of the royal divorce, behind Henry. A significant episode took place in London when Cardinal Campeggio, the legate, acting on secret instructions from Clement, adjourned the ecclesiastical court set up to settle Catherine's divorce. The evidence of Clement's duplicity then became manifest even to the far-from-active brain of the Duke of Suffolk. He crashed his fist on the table and said: 'By the mass, now I see that the old said saw is true, that there was never legate nor cardinal that did good in England.' He was consciously echoing the words spoken by Henry II during the Becket crisis 370 years before: 'I hope I will never set eyes on another cardinal.'

Most Englishmen understood the international implications of the Reformation no more clearly than the Duke of Suffolk; though, like him, they sensed them instinctively. But the two cleverest men in England, More and Cromwell, got the point. They saw it as a historic choice in foreign policy, no less than in religion. Both were reformers, in

impedimentum publicae honestatis, as Wolsey pointed out. Henry's argument from affinity was never strong; and it was weakened still further by the fact that he proposed to marry the sister, Anne Boleyn, of a woman with whom, on his own admission, he had had sexual relations: this also constituted a barrier on grounds of affinity: if his marriage to Catherine was invalid then so, for the same reason, would be his marriage to Anne. Whether he would have won his case if he had followed Wolsey's line of attack is, however, doubtful. Clement could not afford to grant the divorce because he could not risk offending Catherine's nephew, the Emperor Charles V. In 1530 he even proposed that Henry should stay married to Catherine, marry Anne too, and thus commit bigamy. For a recent and illuminating analysis of the canon law of the divorce, see J.J. Scarisbrick: *Henry VIII* (1968), Chapter VII.

* For a more appreciative estimate of Henry VIII's capacities, and an ingenious analysis of his methods of government, backed by copious references to the sources, see Lacey Baldwin Smith: *Henry VIII, the Mask of Royalty* (London, 1971).

that they wanted a spiritual regeneration of the Church; in all else they differed fundamentally. More was a European, Cromwell an English nationalist; they symbolised the division into the two categories which Henry VIII himself called 'Englishmen papistical' and 'entire Englishmen'. More represented the ancient minority tradition of the imperial party in 410, of Gildas in 550, of Wilfred in the seventh century, of Becket in the twelfth century, of Stephen Langton in the thirteenth. To him, England was not an island but part of a great Continental community; it could not cut itself adrift by a unilateral act; it was bound to European Christendom by an indissoluble spiritual treaty, which it might attempt to reform from within but which it could not renounce without defying God. It was nothing to him that a majority of the English people, a majority even of the English Church, accepted separation; this was something no one nation could determine for itself. The supranational authority of the community overrode national self-interest. As he told his judge in Westminster Hall:

> I am not bounded, my Lord, to conform my conscience to the Council of one realm against the General Council of Christendom. For of the aforesaid holy bishops I have, for every bishop of yours, above one hundred; and for one Council of Parliament of yours (God knoweth what manner of one), I have all the Councils made these thousand years. And for this one kingdom, I have all other Christian realms.

More, in fact, explicitly denied English sovereignty:

> This realm, being but one member and small part of the Church, might not make a particular law dischargeable with the general law of Christ's holy Catholic Church, no more than the City of London, being but one poor member in respect of the whole realm, might make a law against an act of parliament.

More can thus be presented as adumbrating modern internationalist doctrine, in which nations voluntarily relinquish portions of their sovereignty to provide a common fund of authority for such organs as the United Nations or the European Economic Community. But equally he can be seen as upholding an ancient and ramshackle structure, whose reality had never corresponded to its ideals, and which was now breaking up under the stress of nationalism: the Catholic Church, in its capacity as residuary legatee of the Roman Empire. The current of the times was against More. His European Christendom was a mirage. Continental Catholicism was not an international community, operating by consensus or majority vote, but the helpless prize in a power-struggle between emergent nations. In 1534 orthodox Christendom was coterminous with the interests of the House of Habsburg, whose head was identified with Spanish imperialism. When, that year, the Pope finally pronounced in favour of Catherine's marriage, the Roman mob screamed out in triumph: 'Empire and Spain!'

Cromwell saw this well enough. He lacked More's academic background, but he knew far more about what was going on in Europe. He had been, so he told Archbishop Cranmer, 'a ruffian in my younger days', and had made his way abroad. He is believed to have fought in the French army at the Battle of Garigliano in 1503; he had worked as a banker in Florence and Venice, and as a business consultant in Antwerp. He had negotiated with courts and popes. He knew Europe from the inside, and he knew it to be the world not of Christian unity, but of Machiavelli. In 1523, as a young Member of Parliament, he had made a very significant speech, highly critical of the Continental foreign policy waged by Wolsey and the King. These European entanglements, he said, were misjudged and likely to prove ruinous; England should look to her own national interest, and in the first place the unity of the British Isles. Cromwell never wavered in this view. England must come first. Her Church must reflect her needs, not those of some Continental despot. The King in Parliament was supreme, the ultimate arbiter of the national destinies. There could be no abridgement of sovereignty. As he put it in the statutes he drafted: 'This realm is an empire' – that is, it acknowledged no superior but God. It was no coincidence that Henry was having the archives ransacked to produce evidence that Arthur was an emperor and had renounced allegiance to Rome. Cromwell, as well as More, stood in a great English tradition; and his was the majority one. But most Englishmen lacked his clarity. They simply felt in their bones, like the Duke of Suffolk, that foreign prelates had no business interfering in English affairs. The crash of the Duke's traditionalist fist was thus the real beginning of the English Reformation.

The Reformation, indeed, was a typical piece of English conservatism, conducted with the familiar mixture of muddle, deviousness, hypocrisy, and *ex post facto* rationalisation. Henry was never quite clear in his own mind whether he wanted an actual change in religion, though there is evidence that in his last years he was moving in that direction; it is significant that he excluded Bishop Gardiner, the leading Romanist, from the council he appointed to manage his young heir. He had no plan of action, moving from one expedient to another. When he slapped a writ of *Praemunire* on the entire English clergy in 1530 he really had all the instruments he needed to control the Church. By agreeing to submit all decisions of Convocation to him, the Church in effect recognised his supremacy. Some of the subsequent acts, and in particular the Act of

Supremacy itself in 1534, were unnecessary.* The progressive seizure of monastic property was very much an *ad hoc* business, and its subsequent disposal was conducted on no apparent principles of equity, public finance, economic reason or elementary common sense.

On the other hand, with the benefit of hindsight, we can recognise two important elements of long-term policy in this apparent confusion. It may be that Cromwell, one of the ablest men who ever served the Crown, was far more deliberate and systematic in his methods than his master, or than appears at first sight. Cromwell was a parliamentary manager; it was on his advice that the Reformation was carried through by Parliament, in the most punctilious and thorough constitutional manner, providing the Crown with a massive overkill of statutory weapons for present and future use against Romanism. Why should the King place himself so completely in the hands of Parliament, beyond, as it were, the call of duty? Earlier in his reign the House of Commons could, and did, 'dash' government bills (that is, reject them); it had always to be coaxed; it was not always united, and from this reign we get the first known instance of a formal division of the House. The explanation, or rather Cromwell's explanation, was no doubt that Parliament, itself the national repository of anti-clericalism, was the best guarantor of the permanency of the breach with Rome. And so it proved. After the Reformation Parliament, it became impossible for the Monarch (irrespective of personal religious views) to decide such matters except in a parliamentary context. It was very significant that Queen Mary had to go to Parliament to get Henry's laws reversed; and on certain matters it declined to do so. Mary could not really put the clock back without destroying Parliament and operating a personal tyranny. And although her parliament restored the link with Rome, it was axiomatic that the decision could be easily reversed. Parliament's sovereignty in spiritual matters – the superiority of its statutes to any natural law, or canon law, or anything the Pope might enact – had thus been formally acknowledged even by a fanatically Catholic queen. After that the Elizabethan settlement was simple and obvious.

* Unless of course we assume that Henry wished to terrorise his enemies into abject compliance by acquiring the legal right to execute them. Thomas More, for instance, had resigned as Chancellor in 1532, in protest against the enforced surrender of Convocation. By declining to take the oath of succession (of Anne's children), he was guilty of misprision of treason, which made him liable to imprisonment during the King's pleasure, but not death. The Act of Supremacy was accompanied by a new Treason Act, which prescribed the death penalty for 'malicious' denial of the King's title. It was on this basis that More, and others, were executed. More had to wait 400 years for canonisation, a scandalous example of the way Rome treats English saints. In the next century John Aubrey noted: 'Methinks 'tis strange that all this time he is not canonised for he merited highly of the church.'

This political underpinning of the Reformation was reinforced by the creation of a huge vested interest in its permanency. By the end of Henry's reign, the bulk of the monastic lands had passed into the hands of private individuals. Their total annual value was about £200,000 at 1536 prices; the actual capital receipts by the Crown for their sale was not much more than £1·5 million; if the Crown had kept all the lands its income from them by 1547 would have exceeded the sale price. Such a policy, therefore, did not make any financial sense. But the King's Government was not composed of fools. We must do them the elementary justice of assuming they knew what they were about. The probability is that the policy of selling land in a flooded buyer's market was deliberately inspired by political motives: to give the propertied classes of England a direct, financial interest in the dissolution. After 1545, there were very few wealthy or influential Englishmen who did not have a personal stake in the Reformation.

The success of this policy became apparent when Mary set about reversing her father's work. Despite her efforts to rig elections, and to have returned to Parliament 'men of the wise, grave and Catholic sort' (i.e. the older generation), the Commons not only flatly refused to restore the monastic lands but insisted on passing a statute to safeguard their present owners. Even Mary's hands were not entirely clean. She said she would give back the lands still held by the Crown. But there was the important matter, for instance, of the 'Regale of France'. This enormous and valuable jewel, probably a ruby, had been presented to the shrine of St Thomas by Louis VII of France. It was the glory of the shrine, and had been promptly pocketed by Henry when the tomb was demolished. He had it made into a ring, which he wore on his thumb. Now Mary should have given the jewel back; but she did no such thing. She had it made into the centrepiece of a brilliant collar, which she constantly wore in public. So all the world could see that she herself, no less, was a beneficiary of the Reformation.

As it happens, it was Mary's own actions which killed Roman Catholicism as the majority English religion. She had all the murderous instincts of her father and grandfather.* The English were accustomed to seeing people burnt for their religious views; about 60 had been thus

* Physically she took after her father, being fair-haired. She had that ferocity in virtue characteristic of a certain type of Englishwoman, to be seen today at Tory Party Conferences when hanging and flogging are on the agenda. Mary must take prime responsibility for the burnings. Her husband, Philip II, was against the policy; so was his ambassador in London, Simon Reynard, who said that at least the executions should be carried out secretly. But the English, including Mary, felt that to hold executions in public was a guarantee of liberty. As late as the 1860s, public executions were defended (e.g. by Palmerston) on the grounds that to give the executive the right to put people to death in secret would open the door to tyranny.

disposed of during the first 20 years of the Reformation. Some had been Catholics and some Protestants. The English were not fanatical about religion, and regarded execution as a fair professional hazard for those who were. But what struck contemporaries was the sheer scale of the Marian persecution. There had been nothing like it seen in England before; it had the flavour of Continental excess. Over three years, Mary burnt just under 300 people, including 60 women. Moreover, these public killings were concentrated heavily in the opinion-forming areas; London and the Home Counties provided two-thirds of the total; there was only one killing in the north and one south-west of Salisbury. There was, too, an unpleasant class flavour about it all. There was no man of breeding among the lay martyrs. Mary showed a craven clemency towards the well born, even if they were traitors. This idea of one law for the rich and another for the poor again had the smell of Continental tyranny about it and was deeply resented – not just among the lower classes. The killing sickened even some of Mary's strongest clerical supporters, and long before her death it was evident to all that her policy had not only failed but had inflicted grievous damage on her cause.* The hatred her persecutions aroused became an important fact of English history for a very long time. They confirmed to most English people that their anti-foreign, anti-papal views were not just prejudices but rooted in a sound instinct for self-preservation. Foxe's *Book of Martyrs* sold more copies than any other publication after the English Bible; it was placed in churches, and kept in the homes of all classes. Every literate person read it, and it was recited to those who could not read. It was the first history of England to reach the masses, and for many it embodied everything they knew about their country. Until Mary's reign there was a real prospect of a multi-religious community emerging in England. By her death this was no longer possible.

The problem which faced Elizabeth on her accession was how to bring to an end the violent oscillations in the State religion, to de-escalate the rising frenzy of doctrinal killings, and, if possible, to take religion out of politics. By temperament she was an agnostic. To her, religious belief must be subordinate to the needs of public order and social decorum. She would have agreed with John Knox's view, indeed taken it as a compliment, that she 'was neither good protestant nor yet resolute papist'. She also agreed with the Duke of Norfolk when he told her:

* Even Bishop Bonner of London, the arch-villain of Protestant hagiography, was officially reproved for his slackness in punishing heretics. When upbraided for his severity in having an elderly man whipped, he replied: 'If thou hadst been in his case, thou wouldst have thought it a good commutation of penance to have thy bum beaten to save thy body from burning.' The real instigators may have been Mary's Spanish confessors. See A. G. Dickens: *The English Reformation* (London 1967), Chapter 11.

'England can bear no more changes in religion. It hath been bowed so often that if it should bend again it will break.' Elizabeth undoubtedly prayed to a very royalist Deity in moments of crisis and anxiety, but she took the view, shared by the overwhelming majority of her subjects, that doctrine was not a thing that any sensible person would kill or be killed for. She hated capital punishment by instinct and reason. It seemed to her monstrous to kill a man for his beliefs alone; only four people were executed for heresy in her reign, none of them Catholics, and all against her will. People should even be allowed to state their views, within reason. As she put it to the Commons: 'God forbid that any man should be restrained or afraid to answer according to his best liking, with some short declaration of his reason therein.' As for private views: 'I seek not to carve windows into men's souls.' What she was looking for was a lowest common denominator of agreement on religious matters, underwritten by statute, upheld by the State, and accepted by the public as reasonable. What she would not tolerate was anyone who strove to upset such a settlement by force; that was treason, because it was aimed at the tranquillity of the realm, and was certain to lead to bloodshed.

Thus Elizabeth was forced, with the greatest reluctance, to turn first against the Catholics and then against the Puritans. She did not want to persecute anyone; but both groups, in the end, left her with no alternative. Within five months of her accession, she had passed Acts of Uniformity and Supremacy; the clergy were required to take the oath, but the few who refused were merely deprived of their benefices. No Catholics were executed then, or for many years afterwards. But Elizabeth's difficulty was that, by the time she came to the throne, the papacy was beginning to take religion seriously, and, worse, do something about it. As late as 1541 Cardinal Contarini had advocated a *rapprochement* with the heretics; but by his death the next year he was regarded as a heretic himself. His opponent, Cardinal Caraffa, set up the inquisition in Rome in 1542, and in 1555 was elected Pope as Paul IV. He did, in fact, what men had asked for during the last 150 years: he reformed the Church of Rome, but on the basis of doctrinal fanaticism and the ruthless enforcement of central authority. He invented the Index, forced Jews to wear yellow caps and live in ghettoes; killed off what was left of the Italian Renaissance. In 1570 the equally fanatical Pius V began to take steps to bring England back into the fold by force. This was by now a forlorn venture. It was not to be expected that many English people, whatever their religious views, would wish to replace Elizabeth, a ruler with a reputation for prudence and virtue, with a sound hereditary and parliamentary title, by Mary Queen of Scots, a foreigner tainted with the double disadvantage of Scottish and French descent, a notorious adulteress, a probable murderess, and with an infinite capacity for causing trouble wherever she went. There is no evidence that the English Catholics, as a group, wanted to expel their queen. Most of them did not care a damn for the Pope; they never had done. What they did care about was the mass, and certain other spiritual comforts of the old religion. They were, like the genuine Protestants, a minority group, and Elizabeth would have been prepared to give them minority rights. But the papacy, by excommunicating Elizabeth, and by instructing English Catholics to depose her, branded them with treason. Her penal legislation was a response to papal aggression, as William I and Henry II had responded. The belated reformation of Rome brought not reconciliation but war, and in the course of it the destruction of English Catholicism. The Catholics could not logically plead that they still served the Queen without renouncing the Pope; they were either bad Catholics, by papal definition, or bad Englishmen. Campion, in his famous *Brag*, put the best construction he could on this double loyalty; it was not convincing then, and it is not convincing today. If Rome triumphed again in England – as Campion by his own admission wished – he would have had to obey orders like any other loyal Catholic, and to take part in whatever acts of treason and persecution the Pope thought fit. The English are not particularly logical; but they saw the logic of this problem quite clearly. Moreover Campion was not typical of the cohorts Rome sent to England; a more representative figure was the sinister Father Parsons, a professional international conspirator. The Elizabethan persecution of Catholics was thus justified by the needs of State and public security, and on the whole it was carried out in a reticent manner. Elizabeth throughout preferred fines and imprisonment to execution; she killed on average no more than eight a year; and nearly all of them got a fair trial. But she could not save the English Catholic community; at her death only about 10,000 were still prepared to declare themselves publicly Romanists.

The threat to Elizabeth from the Puritans was far greater, and she was in the end obliged to take it seriously. English Puritanism was born among the Marian exiles of the 1550s; it was thus an alien import. It had a consistency wholly foreign to the English. The exclusive authority of scripture, for instance, though favoured by Wyclif, appeared to most Englishmen to make no more sense than the *magisterium* of Rome. The doctrine of predestination was ludicrous. The Puritan argument with the authorities began over vestments but quickly spread to include almost everything, from the royal supremacy downwards. The Puritans, like the Roman Catholic extremists, believed that religion was the only important thing in life, whereas most Englishmen thought it was something you did on Sundays. They were influential out of all proportion to their numbers because, like the Communists in our own age, they were highly organised, disciplined and adept at getting each other into positions of power. They were strong in the universities, at a time when a growing proportion of university figures were being elected to Parliament. They oozed hypocrisy. Peter Wentworth and his brother took their stands in Parliament on the right of free speech. But they did not believe in free speech. They believed in a doctrinaire religion, imposed by force and maintained by persecution. Wentworth was a fanatical proponent of alien ideas who wanted to turn England into a Geneva run by Calvinists.

Fortunately Elizabeth was quite capable of dealing with such men. She suspended one Archbishop of Canterbury, Grindal, for being too soft with the Puritans and, in 1583, appointed another, Whitgift, for his known anti-Puritan views. She killed only four of them; but a good number were gaoled, and on the whole she held the movement in check. We need spare no great sympathy for these Puritan gentlemen. They were, in a sense, the mirror-image of the Counter-Reformation, for Ignatius Loyola was a Puritan too, though a Puritan of the Right. The privileges the Puritans claimed for themselves they would certainly have denied to others. One of the best-argued defences of persecution to come from any sect during the whole period of religious controversy was *A Free Disputation Against Pretended Liberty of Conscience* (1649), written by Samuel Rutherford, a Presbyterian. Another Puritan, Dr Reynolds of Oxford, was the scourge of the Elizabethan theatre which, as he made clear in *The Overthrow of Stage Plays* (1599), he wished to ban completely. The Puritans forced the theatres to move from the City of London to Southwark. If they had triumphed nationally, many of the greatest works of English literature would never have been written, for they could not have been produced. Shakespeare, Jonson and Webster, no doubt, would have turned to other professions. The Puritans did not believe in reason, but in the Bible. The early Protestants had rightly denounced the gross superstitions of the popish church; such instruments of the old regime as the notorious Boxley Rood had been exhibited in London to the jeers of the mob. But among many Protestant sects new and more virulent forms of superstition soon appeared, springing in many cases from the literal interpretation of the Bible. Luther himself, the castigator of indulgences, believed that the Devil deliberately created flies to distract him when he was writing works of edification. In Geneva men and women were sent to the stake for allegedly spreading the plague. Above all, Puritanism was the dynamic behind the increase in witch-hunting. Despite the efforts of the Crown and the episcopal bench, vast numbers of innocent women were put to death. It had been suggested that the hunts had an economic purpose: to rid society of pauper women who would otherwise have to be fed from public funds. But this seems too cruel and cynical even for the English. Doubtless the motives were mixed. Witch-hunting was an old English tradition. In the fifteenth century even the wife of Humphrey, Duke of Gloucester, barely escaped with her life on trumped-up charges. But the Puritan concept of evil led to an appalling escalation in the scale of persecution.*

It is against this background of murderous zeal that we must place the achievement of Elizabeth in stabilising the religious system of England on a basis of moderation, common sense and tolerance. It was a personal achievement, for most of her advisers were not noted for any of these characteristics. It was an enduring achievement, too, for the Elizabethan religious settlement survived all the shocks of the next century, and emerged into modern times roughly the same article. Elizabeth would have recognised, and approved, the services, doctrines, customs, attitudes and organisation of Anglicanism as they exist today. To be sure, she had no great respect for the Anglican Church, and still less for most of its dignitaries. But this down-grading of the priestly class was central to her attitude, and reflected the general desire of the great mass of the population. Elizabeth felt that religion was too dangerous an element in the body politic to be safely left to clergymen. It should be the servant of the public, not its master. It should provide comfort in a harsh and painful world, not add to the troubles of society by provoking controversy and division. She wholeheartedly echoed the cry of the moderate Protestant Sebastian Castellio, who expressed, in *Whether Heretics Are to Be Prosecuted?*, published in the 1560s, the view of all sensible Christians who peered through the mists of conflicting dogmas to the heart of their faith:

> O Christ, creator and king of the world, dost thou see? Art thou become quite other than thyself, so cruel, so contrary to thyself? When thou didst live upon earth, none was more gentle, more merciful, more patient of wrong. . . . Men scourged thee, spat upon thee, mocked thee, crowned thee with thorns, crucified thee among thieves and thou didst pray for them who did this wrong. Art thou now so changed? . . . If thou, O Christ, hast commanded these executions and tortures, what hast thou left for the Devil to do?

Elizabeth felt that the religion of the people must be safeguarded by the moderate intervention of the State, acting in the public interest; it was

* The exact number killed in England is still in dispute: probably about a thousand over two centuries. For a recent, and fair-minded, discussion of the subject, see A.L. Rowse: *The Elizabethan Renaissance: The Life of the Society* (London 1971), Chapter 9.

not the business of the clergy to determine religion, merely to administer it. This was a thoroughly English approach. A man's religion was a matter between himself and his God; its outward forms and organisation were a matter for the due constitutional process of law. The English have the great merit of recognising that bishops should be appointed not by those who care deeply about religion, but by those whose duty it is to preserve public order and decorum. Equally, in doctrine, the object was not to thrash out the minutiae of belief, but to draw up a code sufficiently vague, ambiguous and ramshackle to persuade the maximum number of people to accept it without too much strain on their consciences. The Thirty-Nine Articles admirably fulfilled this aim. No one man has ever been able to agree with every single element in them, or even to understand precisely what they mean; but over the centuries vast numbers of clergymen have happily sworn to uphold them because their spirit is obvious, and sufficiently enveloping to cover a wide range of belief. The English have never made the mistake of saddling themselves with a written constitution. In the mid-sixteenth century the pressure of the times left them no alternative but to adopt a religious constitution. They solved the problem by producing a document which made nonsense in detail but admirable sense taken as a whole.

The truth is that the English are not, and never have been, a religious people. That is why toleration first took root in our country. There were, to be sure, plenty of religious zealots in England; not just Protestants and Catholics, but Anabaptists and Huttites, Mennonites, Waterlanders, Socinians and men of Rakow. But all together they never made up more than a minority. It is a matter for argument whether England has even been a Christian country. The English like to be baptised, to get married in church, to be buried in consecrated ground; they pray in times of peril; they take a mild interest in religious controversy, and like to clothe the State in religious forms. But they are not truly interested in the spiritual life. We must not think of the Middle Ages in England as a religious era. It was a time when the priestly caste occupied a major role in society and the economy. But the universal levity with which the moral law was broken, and ecclesiastical sanctions defied, suggests that most healthy men and women did not take hell fire seriously. The Church was a profession. It was not, on the whole, interested in pastoral and parochial work. Religion in the towns was weak. The inhabitants of many country districts were served, if at all, by very humble clergy indeed, usually half-educated and often wildly unorthodox. Most Englishmen did not even know the principal articles of their faith. Anglicanism did something to improve the situation in the wealthier country districts, but it was never more than a middle-class affair in the towns. Protestantism was a more meaningful faith than Catholicism for the English, but only for a minority, and perhaps a small one. For about a century (1750–1850) nonconformity occupied an important place in English urban life, but again only for a minority. When the Irish immigration to England took place in the nineteenth century, Catholic priests were able to secure a comparatively high rate of church attendance, though they have always exaggerated it. But on the whole it is doubtful whether, at any time in history, more than 50 per cent of the English people have attended Sunday services regularly, or paid more than lip-service to their church. This is not true of many other countries. In the United States, even today, well over 50 per cent regularly go to services on sabbatical days. In Scotland, Ireland and Wales it is likely that, until recent decades, observance was the custom of the great majority, and religion played a meaningful role in their lives. But for the English the Deity is a social instrument, a mere part of the constitution, which has other (and more important) elements. Elizabeth, who was, as she never tired of pointing out, 'mere English', had the merit to perceive this fact, and act upon it.

We owe a great deal to this remarkable woman. To be sure, she presided over a dazzling galaxy of talent, political, commercial, military, naval and artistic. But she herself took all the really important decisions – and non-decisions – of her reign, often against the advice of her ablest counsellors. It is impossible to read the letters and documents of this period, to examine the domestic and foreign political strategy, culminating in the defeat of the Armada, to analyse the solutions to the problem of the succession, or the religious settlement, without concluding that her hand and brain were firmly in charge of the national destiny. She was a political genius of a very rare kind, for her inspiration was a sense of tolerance, springing from a warm heart and a cool intellect. She inherited all her father's will-power, but none of his murderous instincts. She loathed killing and cruelty. Her tutor, Roger Ascham, had taught her to hate war and violence; but it was a lesson she did not really need. As a young woman she had been in that horrible place, the Tower of London, in fear for her life. As a result, she determined to make England a country in which moderate, reasonable people could feel safe – even engage in controversy, provided their only weapons were words. For two centuries the public life of England had been engulfed in a rising tide of political and religious murder. The judicial killings had struck at kings and archbishops, noblemen and great lawyers, to say nothing of a mass of humbler people. Many of the country's greatest talents had been destroyed in senseless ignominy on the scaffold. If the fabric of English society was to survive, the process had to be stopped; and Elizabeth stopped it. What had become a bloody English tradition was firmly extinguished; and it was never really resurrected.

Elizabeth's personal tragedy was that she, who hated killing anyone, was nevertheless obliged by overwhelming pressures and circumstances to kill a few. No ruler ever went through greater agonies in signing a death-warrant. She fought desperately to spare Mary of Scotland. Her contemporaries thought she was mad to be so lenient; worse, criminal. 'The Queen's Majesty,' wrote Burghley, 'hath been always a merciful lady, and by mercy she hath taken more harm than by justice.' Lord Hunsdon put the point more strongly, when she delayed signing Norfolk's death warrant:

> The world knows her to be wise, and surely there cannot be a greater point of wisdom than for any to be careful of their own estate, and especially the preservation of her own life. How much more needful it is for her Majesty to take heed, upon whose life depends a whole commonwealth, the utter ruin of the whole country and the utter subversion of religion. And if by negligence of womanish pity these things happen, what she hath to answer for to God, she herself knows.

But it was not just womanish pity, though that played a part. Elizabeth had not much religion, but she had a very strong conscience. She thought it wrong to kill. She also thought it impolitic, harmful to her own reputation as sovereign, and that of the country she ruled. She would not, she told Parliament, execute Mary:

> Full grievous is the way that I, who have in my time pardoned so many rebels, winked at so many treasons . . . should now be forced to this proceeding against such a person. What will my enemies not say when it shall be spread, that for the safety of herself a maiden Queen could be content to spill the blood, even of her own kinswoman?

Elizabeth did not kill Mary; on the contrary, she preserved her life for nearly two decades, against the will of her subjects.

Tolerance and a hatred of violence were modern virtues in Elizabeth's age; if they have become English characteristics, some of the credit must go to her. She was a kind person. Though she never slept with a man, there was plenty of love in her heart. Her formal letters to her ministers and commanders are often embellished by touching and affectionate footnotes, written in her fine, firm hand. Though a lot of the romantic mystique of her court was deliberately contrived to suit her public purposes, there can be no doubt that the warmth which existed between her and her greatest servants was absolutely genuine. On his death-bed, Burghley asked his son to thank the Queen for her kindness:

Though she will not be a mother yet she showeth herself by feeding me with her own princely hand, as a careful nurse; and if I may be weaned to feed myself, I shall be more ready to serve her on the earth; if not I hope to be in heaven a servitor for her and God's church.

The Queen loved, and understood, children. To her young godson she sent a copy of her speech to the 1576 Parliament, with these words:

Boy Jack,
I have made a clerk write fair my poor words for thine use, as it cannot be such striplings have entrance into parliament assembly as yet. Ponder them in thy hours of leisure, and play with them till they enter thy understanding; so shalt thou hereafter, perchance, find some good fruits hereof when thy godmother is out of remembrance; and I do this because thy father was ready to serve and love us in trouble and thrall.

Elizabeth visited the sick; she attended her friends on their deathbeds, sometimes staying in their houses and ministering to their wants herself. To the bereaved she sent little notes of condolence. 'My own Crow,' she wrote to Lady Norris, whose son had been killed in Ireland, 'harm not yourself for bootless help, but show a good example to comfort your dolourous yoke-fellow.' She even sent a message of sympathy and reassurance to the wife of a man who had deliberately defied her, and was held in the Tower. When she said she loved the people of England – and they are not a people whom anyone can easily love – she meant it. The real measure of her achievement is that she was able to express this love in concrete terms, and impart to her people a taste for the new and unfashionable virtues she possessed. So long as the English exist, she will not be 'out of remembrance'.

PART FOUR

The Chosen Race

[1603–1780]

IT is a curious fact that the most important debate in English political history took place not in the House of Commons but in the fifteenth-century parish church of St Mary in Putney. There, on 28 October 1647, and for the next two weeks, a group of about forty men met in informal conclave, and proceeded to invent modern politics – to invent, in fact, the public framework of the world in which nearly 3,000 million people now live. There was no significance in the choice of the church; it was simply convenient. The men sat or stood around the bare communion table and kept their hats on, as Englishmen had learnt to do in the Commons House. The meeting was officially styled the General Council of the New Model Army, the force which had recently annihilated the armies of King Charles and was now the effective master of the entire country. Some of those present were distinguished generals: Oliver Cromwell, second-in-command of the army, and its real creator and ruler, and Commissary-General Henry Ireton, his brilliant son-in-law. Some were gallant regimental commanders, such as Lieutenant-Colonel Goffe and Colonel Rainborough, men of humble birth who had risen to field-rank in battle. Some were junior officers. Some were ordinary soldiers, like Edward Sexby; two are described in the record merely as 'Buffe-coate' and 'Bedfordshire Man'. There were three civilians, political radicals, or Levellers, who had come to help the soldiers put their case. It was a very representative gathering of Englishmen, covering all classes, save the highest, and a wide variety of peacetime trades and callings. The verbatim record, kept by the Secretary to the General Council, William Clarke, is occasionally garbled (he was unused to taking shorthand) and, alas, incomplete; it remained unread, buried in the archives of Worcester College, Oxford, for more than 250 years, until it was examined at the end of the nineteenth century, edited and published.* But the ideas flung across that communion table – then in all the exciting novelty of their pristine conception – had in the meantime travelled round the world, hurled down thrones and subverted empires, and had become the common, everyday currency of political exchange. They are still with us. Every major political concept known

* *The Clarke Papers*, edited by C.H.Firth, the Camden Society, 1891, Vol. 1. Next to Rushworth's *Historical Collections* (8 vols, 1659–1701), *The Clarke Papers* form the most valuable authority for this period, and it is a pity they are not available in a cheap paperback. Putney Church, incidentally, is now overshadowed by a huge and hideous office block; across the road is a pub which advertises 'drag' shows.

to us today, all the assumptions which underlie the thoughts of men in the White House, or the Kremlin, or Downing Street, or in presidential mansions or senates or parliaments through five continents, were expressed or adumbrated in the little church of St Mary.

Before we examine the debates, it is important to understand why they took place in England and why they could only have taken place in England. They might never have occurred at all, and if so the world would now be a radically different, and much more primitive, place than we find it. But certain peculiar developments in English history – developments rooted many centuries back, and ultimately resting on the geography of England, and the composition of its people – allowed this thing to happen; and so the world is as it is. Let us then trace the genesis of the Putney debates.

The ancient Greeks had begun to explore certain entirely new political and scientific concepts when their cities and culture were absorbed in the imperialism of Rome. Rome provided order of a sort, but it killed creativity. Its empire lay for hundreds of years like a vast and motionless log across the stream of human progress. The Romans were compilators and codifiers, but they could not invent new thoughts, and they successfully inhibited others from doing so. They were lawyers by temperament and their language was legalistic. They could explore backwards into the origins and precise meanings of existing concepts, but the very orderliness of their verbal apparatus – the skill with which it played endlessly on the known and finite – locked the doors firmly against the unknown and the infinite. Their mental world, like their language, was static and in the end degenerate. Their vigorous lawyers' republic became a soldiers' empire and, in turn, an oriental despotism. As it shrank and disintegrated, it embraced a religion from another race of lawgivers, the Jews; and Roman Christianity, drawing its intellectual concepts exclusively from a static body of sacred and immutable texts, its forms, organisation and discipline from a military empire in decline, became the residual legatee of Rome. For a thousand years it lay across Europe like a winding-sheet, monopolising education, culture, science, and technology, interposing a hieratic class of interpreters between the people and such learned texts as it possessed, banning any form of empirical inquiry which did not square with its fixed, received notions, and limiting its intellectual activities to formal theological exercises, which merely played on words and were wholly barren of discoveries. Its grip on the world was underpinned by secular societies whose power-structure reflected its hierarchy and which had a shared interest in preserving a comatose and unequal world. Europe was internationalist in that Church and State cooperated across frontiers in the extirpation of novelty; and it had a common language, Latin, to control knowledge and preserve it for the *élite*.

It required an extraordinary conjunction of destructive forces to shatter this adamantine mould. The ancient Greek knowledge had been filtering into western Europe since contact with the transmitting Arabs had been established in the eleventh century; and the extinction of Byzantium brought volumes of hitherto unknown texts to the West. But this was not enough. The use and development of such knowledge required a political society in which the free spirit of inquiry could act. Roman Christian Europe was a mutual protection system, organised on a supranational basis to safeguard the property of the possessing classes, those who owned both knowledge, such as it was, and land. It had successfully aborted the intellectual revolution of the twelfth century, and reimposed its negative philosophy of learning for two long centuries. It might have done so again. The Renaissance presented it with a challenge, on which the crude, practical genius of Luther seized and, backed by the new power of German nationalism, thrust brutally through the enveloping mould. He caught both Church and Empire off-balance, in disarray; they took too long to perceive the fundamental nature of the threat, and acted too late. But the breach could have been sealed: the impressive power with which the Counter-Reformation eventually organised itself, the ruthlessness with which it acted, leaves little doubt that Continental Protestantism could eventually have been extinguished, and the mould universally reimposed. But there was the little matter of England, and the English Channel.

In England the conjunction of forces operating against Roman civilisation was unique. As we have seen, it was from this country that Pelagius had first developed the dynamic, anti-defeatist philosophy of free will, and in so doing created a heterodox, nationalist tradition which had never been entirely lost. The offshore islanders were the only colony which had thrown out their Roman governors. They had received back Roman Christianity, but transmuted it into an insular form. They occupied a unitary and centralised kingdom, which meant that their religion must be identified with the national spirit. Their relations with the Continent had always been uneasy and suspicious; they rejected its norms, and the Channel allowed them to do so with relative impunity. The mutual-protection system of Roman Christian Europe stopped short at the walls of Calais. The Reformation in England thus made explicit a declaration of independence from the Continent which was rooted in a thousand years of political and intellectual development. It was carried through, ironically, by the last of the medieval kings, a man whose motives and objects were never clearly formulated even to himself, but who possessed extraordinary reserves of courage and will-power springing from his brutish nature; he had a manager of genius, Thomas Cromwell, who flawlessly exploited the resources of an ancient institution, Parliament, to anchor the changes firmly in English law and tradition; and the break with the Continent was confirmed and made permanent by the old King's matchless daughter, Elizabeth, who inherited all her father's courage but who possessed, too, a wisdom, gentleness and a sense of balance and tolerance which were completely alien to him. All these factors might be called accidental. But there were others which made a clash between the English and the Continental system inescapable: the vigorous development of the English language, which made the cultural monopoly exercised by Latin increasingly intolerable; a xenophobic hatred of priests and priestcraft, which merely waited an opportunity to vent itself; and, above all, a rising consciousness among the English that they were a people somehow different to all others, called to a special destiny.

The last factor was decisive – the keystone in the Reformation arch. It takes enormous energy to change the entire course of world history, and such energy cannot be drawn exclusively from physical forces; something metaphysical is required too. What sustained the English during the Reformation and Counter-Reformation years, what enabled them to preserve heterodoxy in England and uphold it on the Continent, to defeat the Armada and rip open the world empire of Spain – in short to thrust aside the inert log of the Roman heritage and allow the stream of progress to flow again – was not just patriotism, or nationalism, but racism, the most powerful of all human impulses. The English had come to believe they were the chosen people. They could thus answer the Continental armoury of faith and superstition with the vehement conviction of divinely inspired rectitude.

How did the English reach the audacious conclusion that God, having found the Jews inadequate for His great purposes, had entrusted the island race with the unique role of completing His kingdom on earth? They were not particularly devout. They disliked clergymen, except in a purely sacerdotal role. They built splendid churches and cathedrals, but did not frequent them except in a spirit of social decorum. On the other hand, their island situation had made them natural racists, overbearing and aggressive towards strangers, holding their own superiority to the rest of mankind to be self-evident. This was fertile soil on which to sow the seed of a national mission to reform, or indeed conquer, the world. The English would have received such a mandate as willingly from Jupiter, or Allah, or even Buddha, as they did from Jehovah. But the manner in which inspiration came was characteristic. It arose from the devotion of the English to their history, their misunderstanding of certain salient facts in it, and their breathtaking ability to rewrite it to suit their inclinations and convenience.

Throughout the Middle Ages they had delighted in manufacturing world chronicles in which the English played a prominent role. In the fourteenth and fifteenth centuries, armed with ancient claims and grudges, they had inflicted their historic visions and myths on the hapless French. Quite when they first took note of the fact that they were the successor-race to the Jews is impossible to determine. It must have occurred, in a significant sense, early in the sixteenth century. It was a period prolific in historical writing, much of it highly imaginative; and the trickle of printed books was fast becoming a torrent, spreading this knowledge, or half-knowledge, of the past among an ever-growing circle of men in positions of authority and influence. Henry VIII's controversy with Rome gave an enormous impulse to these probings into the past. Suddenly, history became politics; records and libraries were closely scrutinised for immediate public objects. King Arthur made his formidable appearance in the debate with Pope Clement. Still more shadowy figures were resurrected or invented to prove the unique relationship of England to the Christian community. If the English had read Bede they would have found the disappointing truth about themselves. But they did not read Bede; they read Gildas and Nennius, Geoffrey of Monmouth, and the successive generations of historians who had built on their fantasies. Thus a myth was publicly accepted as fact. It took various forms. Some believed Christianity had been brought to Britain by Joseph of Arimathaea, on the express instructions of the Apostles; some thought the agent was St Paul; others believed Christ himself had paid a special visit. But all the versions had one thing in common: Britain had got the faith directly from the apostolic succession – hot, as it were, from the Holy Land – without the intermediary of Rome. The popes had had nothing to do with it. As Queen Elizabeth herself put it: 'When Austin came from Rome, this our realm had bishops and priests therein.' What is more, it was through Britain that the Roman Empire had embraced the faith. Constantine had been British; his mother Helena was the daughter of the British King Coilus. So, wrote Foxe, 'by the help of the British army', Constantine 'obtained . . . peace and tranquillity to the whole universal Church of Christ'. This being so, what special authority – indeed what authority at all – could the Pope or any Continental sovereign, spiritual or secular, claim over the English?

It is, however, important to grasp that this myth was not the property of any religious sect; it was racial rather than theological. Just as the

Pilgrims of Grace, as well as the Puritans, appealed to Magna Carta, so even the staunchest Catholics were confident of England's special role. King Philip II, who had certain myths of his own, must have been outraged to hear, at his first court sermon preached in England, by no less a papist than Cardinal Pole, that England was '*prima provinciarum quae amplexa est fidem Christi*' – the first country to receive the faith. Moreover, went on Pole, 'the greatest part of the world fetched the light of religion from England'. Mary nodded her head vigorously: she believed it too. All the English sects, however they might differ on any other matter, were united in assigning a unique and Godly destiny to the English: even Laud, anxiously putting back the Reformation clock, was to teach that the *ecclesia anglicana* was the true Church of Peter, pure, solitary and undefiled.

However, it was obvious that this dynamic myth came handiest to those who wished to break away from Continental religion, especially those who wished to base English Christianity on the broadest possible national consensus. Not only did the myth identify the race with the national religion, but it enabled God's purpose in choosing the English as his race to become perfectly clear: the destruction of Rome and the renovation of the entire Christian world. Think of what an aggressive and fanatical war politician like Henry V would have done with such a commandment! Equally, the purpose allowed the creation of the most unifying force of all, a common enemy: the Papacy, huge and hideous, the terrestrial instrument of the Devil, whom God had told the English to root out and destroy, together with such secular lieutenants as King Philip, and so forth.

Thus the myth of the chosen race underlay the Elizabethan religious settlement and the extraordinary national unity she contrived to maintain. There was not much piety about it. It merely provided a purpose and ideological framework for the rank but aimless racism which had been growing in England throughout the Middle Ages. In the second year of the Queen's reign, John Aylmer, a friend of Ascham, wrote in his *An Harborow for faithfull and true subjects* that England was the virgin mother to the second birth of Christ. The English should thank God that they were not born French, Germans or Italians (they did not need any encouragement). England abounded in good things, and God and his angels fought on her side against all her enemies:

> God is English. For you fight not only in the quarrel of your country, but also and chiefly in defence of His true religion and of His dear son Christ. [England says to her children:] 'God hath brought forth in me the greatest and excellentest treasure that He hath for your comfort and all the worlds.

He would that out of my womb should come that servant of Christ John Wyclif, who begat Huss, who begat Luther, who begat the truth.

The myth crystallised in the huge volumes of Foxe's book which, despite its expense and size, had sold 10,000 copies in England before the turn of the century, more than enough for every parish church in the country. As we have seen, it made a Catholic restoration on Marian lines impossible. Even more important, it gave a complete rationale for all the characteristic features of Elizabethan England: the Queen herself, a national Church based on a degree of tolerance, the government's foreign policy, the spread of printing, education, and the use of the vernacular as the language of culture and science.*

Foxe and many other writers stressed the unique role of Elizabeth in the national mission: she was Deborah, a virtuous and virginal creature, the special spiritual servant of God divinely appointed to safeguard true religion and lead the English in victory over God's enemies. But such warfare, said Foxe, was waged not by rulers alone but by all classes of the chosen race. He proved from English history that one essential test of a people's fidelity to God was their willingness to rebel when rulers were misled by corrupt advisers. In his tales of the Marian years he exalted especially the working-class martyrs, including women: they, as well as the rich and educated, had a part to play. He related the case of Alice Driver, who told her persecutors:

> I was an honest poor man's daughter, never brought up in the university as you have been, but I have driven the plough before my father many a time, I thank God. Yet notwithstanding, in the defence of God's truth and in the cause of my master Christ, by his grace I will set my foot against the foot of any of you all in the defence and maintainence of the same . . .

Religion was thus a leveller of the classes, indeed of the sexes; all should be united in the national work of God. Foxe underlined the importance of the English standing together: his final words, in the last edition he prepared for the press before his death, was an eloquent plea for mutual tolerance in line with the Queen's religious policy:

> And if there cannot be an end of our disputing and contending one against an other, yet let there be a moderation in our affections . . . because God hath so placed us Englishmen here in one commonwealth, also in one Church, as in one ship together, let us not mangle and divide the ship, which being divided perisheth, but every man serve in his order with diligence, wherein he is called.

* The best and fullest analysis of the influence of the historic myth on English religion and politics is William Haller: *Foxe's Book of Martyrs and the Elect Nation* (London, 1963).

The mission, needless to say, presupposed an active and aggressive foreign policy, conducted in strict accordance with Protestant (i.e. English) interests. Foxe told his Good Friday congregation at St Paul's in 1570 that, though England might be weak and her enemies powerful, they would collapse like the walls of Jericho: let the rich and the mighty beware, the great Turk, the great Caliph of Damascus, the great Caliph of Old Rome, 'and all other cruel tyrants and potentates of this world which have abused their sword to the destruction of Christ's saints'. The English pirates and adventurers who, in Elizabeth's reign, were beginning to carry out the national mission all over the globe accepted Foxe's words as the literal truth. Drake took a copy of the great work with him when he set off to circumnavigate the world in 1577: he read the more sententious passages to his sullen Spanish prisoners, and, on rest days, coloured the pictures with his own hand. After his victory at Cadiz in 1587, almost his first act was to write to Foxe to thank him for his prayers.

Yet the mission was cultural as well as military. John Jewel had pointed out that English was the special language of Godliness; to which Foxe added that the invention of printing was a miracle, expressly performed by the Lord to complete the reformation of his Church:

> How many printing-presses there be in the world, so many block-houses there be against the high castle of St Angelo, so that either the Pope must abolish knowledge and printing or printing at length will root him out.

This stress on the divine value of the printed word, the imperative command to disseminate the truth as rapidly and widely as possible, brought the medieval values and defences tumbling down. Religion was the Word – the Bible – and the Word was English. The national language swept away Latin as the vernacular of doctrine and piety, and rapidly began to invade other spheres hitherto protected from public intrusion by the dead culture. Extraordinary national energies were thus unleashed, most strikingly in the theatre, but in every other branch of literature and knowledge. At a humble political level, the Government poured forth or inspired innumerable pamphlets defending its actions and lambasting its domestic and foreign enemies, many of them from the busy pen of Sir William Cecil himself. Puttenham's *Arte of English Poesie* and Sidney's *Defence of Poesie* justified the abandonment of Latin as the prime vehicle for poetic expression. The glorification of England, her countryside, her people and her history, was the central theme of an enormous literary output. A third of Shakespeare's plays concentrated on historical themes: some on Roman history reconstructed for English purposes, ten on English history alone. But there were also Sir Thomas

Smith's *Commonwealth of England*, Hakluyt's *Principal Navigations* of the English mariners and explorers, Camden's *Remains . . . Concerning Britaine* and his *Britannia*, Stow's *Survay of London*, Daniel's *Historie of England*, and, in the next reign, to cap them all, Ralegh's gigantic *History of the World*. The breaking of the Latin stranglehold brought into play whole new classes and categories of men, most notably the humble London craftsmen who were creating the precise instruments of navigation, on which the scientific revolution, and ultimately the industrial revolution, would be based.

It was, indeed, the navigators – men whose lives and fortunes depended absolutely on the accuracy of their instrumentation and maps, and therefore on the free flow of knowledge and ideas, and the progress of experimental philosophy – who were most humbly grateful for the opportunity Elizabethan England gave to the new culture, and most strident in proclaiming the doctrine of the chosen race. One of them, John Davys, put the English ideology in its extreme form:

> There is no doubt but that we of England are this saved people, by the eternal and infallible presence of the Lord predestined to be sent into these Gentiles in the sea, to those Isles and famous Kingdoms, there to preach the peace of the Lord: for are not we only set upon Mount Zion to give light to all the rest of the world? Have we not the true handmaid of the Lord to rule us . . . ? It is only we, therefore, that must be these shining messengers of the Lord, and none but we.

Expressed thus crudely, the doctrine cannot have found universal acceptance in a society which, especially towards the end of the reign, was rapidly acquiring an astonishing degree of intellectual sophistication. It ill accords, for instance, with much of Shakespeare's writing, especially his subtle and emancipated view of the national character. But if many, including no doubt the Queen, declined to swallow the myth entire, all absorbed a portion of it. There were a number of great minds whose Christianity was heavily qualified, who were unavowed Deists, agnostics, even suspected to be atheists: men like Ralegh and Francis Bacon. But each found a facet of the myth to suit his tastes and convictions. It was, in one manifestation or another, irresistible. Moreover, distasteful though it appears in retrospect, it had a kind of historical necessity. It acted like a great engine, which lifted the nation up and beyond the gravitational pull of the dead medieval world, and placed it safely in free orbit. Once embraced by the nation, the myth ensured that there could be no return to the two stagnant millennia Rome had inaugurated. Mankind had achieved a kind of liberation, and was being carried forward on a self-sustaining current of progress, which events might decelerate but could not halt, let alone reverse. The current is still driving us along, ever faster.

The management of such a kingdom and people, pullulating with newly released energies, anxious to embark on a grandiose, almost manic, world mission, posed extraordinary problems; and it is a tribute to Elizabeth's unique qualities as a stateswoman that she at least contained, if she could not solve them. Such a people threatened always to break through the normal bonds of society. There were other strains, too. Prices had been rising consistently since the 1530s, but many forms of income, including the Crown's, had failed to keep pace; sadly, and despite the most stringent economy, the Queen was forced into regular sales of Crown land; she lived heavily on capital towards the end of her reign, and at her death the monarchy was much weaker financially than at her accession. Her dependence on provision by Parliament correspondingly increased, and the House of Commons required a growing degree of conciliation on the part of Government. After Cecil went to the Lords, the quality of Commons management declined; his son was by no means as astute. The Queen's majestic personality right to the end filled many yawning gaps in the Government's armoury; she personally upheld the consensus, and her last domestic speech to Parliament, the 'Golden Speech', was venerated (as we shall see) by old MPs a generation after her death. But the country suffered grievously in her final decade: appalling weather brought bad harvests, trade was in recession, the war with Spain dragged on at mounting cost. The huge transfers in the ownership of property over the last 70 years had altered the structure of landed society at a speed unusual even in England. Elizabeth, continuing Tudor policies of holding the nobility in check, had deliberately kept the Lords small; the bulk of the landed wealth therefore passed into the Commons. The country gentry had begun to invade the borough seats early in the fifteenth century: they paid their own parliamentary expenses, for one thing, and for another their power in the neighbourhood was usually so great that the boroughs had no alternative but to elect them.* By the end of Elizabeth's reign the Commons was a gentry

* See J.S. Roskell: *The Commons in the Parliament of 1422* (Manchester, 1954), and *English Historical Documents*, Vol. IV, 1327–1485, edited by A.R. Myers (London, 1969), pp. 475–6. In the sixteenth century outsiders who sat for boroughs normally paid their own expenses; and by 1515 even the knights of the shires had begun to 'entertain' (i.e. bribe) electors. In 1601 James Harrington, MP for Rutland, calculated that membership would cost him £200. The subject is discussed in detail in J.E. Neale: *The Elizabethan House of Commons* (London, 1963), Chapter XVI. By the second half of the seventeenth century it was clearly most unusual for any member to be paid, for Aubrey notes that Andrew Marvell's 'native towne of Hull loved him so well that they elected him for their

preserve, and the gentry were the richest and most influential class in the country.

This was the royal estate James Stuart of Scotland inherited, though he was devastatingly unaware of its drawbacks. He was received with some enthusiasm, as a Protestant from birth, as a male, who had already guaranteed his own male succession with two sons, and as an experienced ruler. The enthusiasm was reciprocated. James had been King almost since birth, but had led a miserable existence buffeted by rival factions of the Scots maffia-nobility, his life frequently in danger, his purse usually empty, hectored by intolerant Calvinist clergymen, and with the meagre satisfaction of presiding over a semi-barbarous and bankrupt pocket-state on the outer fringes of civilisation. Now he was to take over an august, ancient and secure throne, a dignified and hierarchical Church, a treasury bursting with gold, a cultured and splendid nobility, a brilliant court, a country where agriculture and the arts, learning, science and trade flourished as never before; or so he thought. He even inherited the magical gift of touching for the King's Evil, which Elizabeth had treated as a traditional joke, but which delighted his superstitious mind.

Those who decry the influence of personality on history find it hard to argue away the speed, the perverse skill, and the absolute decisiveness with which the Stuarts demolished their English heritage. The English had always been devoted to the monarchy, they revered strong government, they were profoundly attached to the law, tradition and established usages, they loathed abrupt change, they had willingly surrendered to the Crown a monopoly of violence – no monarch could conceivably have asked for more. Moreover, early in his reign James was presented with an astonishing stroke of luck, the only one the Stuarts ever had. The discovery that Catholic conspirators, master-minded by Jesuits, were planning to detonate King, Lords and Commons in one gigantic explosion was a patriotic scenario which would have made even the fertile Burghley gasp in admiration; moreover it was true, and could be proved. Nothing could be more calculated to bind the King to the country's affections, and emphasise the common humanity, peril, and solidarity of all estates of the realm. It was a gift beyond computation, an event which could be, indeed was, celebrated annually to refresh the minds of all.

Yet in less than forty years the nation had been driven to armed

---

representative in Parliament, and gave him an honourable pension to maintaine him'. In 1677, in an attempt to resist court patronage, MPs tried unsuccessfully to revive the ancient statutes for the payment of Members, the last time the topic was debated for 200 years.

rebellion. And the only problem which confronts the historian is why it did not occur sooner. Everything James did, and everything he omitted to do, was certain to evoke protest. He was not, to begin with, the kind of man whom even the most infatuated English royalist could respect. Here is his portrait by one of his courtiers, Sir Anthony Weldon:

He was of a middle stature, more corpulent though in his clothes than in his body, yet fat enough, his cloathes being ever made large and easie, the doublets quilted for stiletto proofe, his breeches in great pleites and full stuffed. He was naturally of a timorous disposition, which was the reason of his quilted doublets: his eyes large were rowling after any stranger come into his presence. His beard was very thin. His tongue too large for his mouth, which ever made him speak full in the mouth, and made him drink very uncomely, as if eating his drink, which came out into the cup at each side of his mouth.

His skin was as soft as Taffeta Sarsenet, which felt so, because he never washt his hands, only rubbed his fingers, and slightly with the wet end of a napkin. It is true, he drank very often, which was rather out of a custom than any delight, and his drink was of that kind of strength as Frontinack, Canary, High Country wine, Tent wine and Scottish ale, which had he not a very strong brain, might have daily been overtaken, although he seldom drank at one time, above four spoonfuls, many times not above one or two.

James's language was appalling, and his obscene jokes brought ill-concealed shudders from a court which was by no means squeamish. Elizabeth had occasionally used very direct language, to a purpose; but she was essentially a woman who valued modesty and decorum, who had a strong sense of her dignity, gracious and courtly in her manners, soft in speech, abstemious in all things, a very regal lady indeed, who expected high standards at court, especially towards her ladies – indeed ruthlessly punished those who failed to observe them. All these qualities the English have always expected and applauded in their monarchs. By contrast, James was a loutish savage. When hunting, he liked to plunge his bandy legs into the stag's bowels, so that an old Elizabethan, Sir John Harington, commented: 'The manners made me devise the beasts were pursuing the sober creation.' The French ambassador sneered: 'When he wishes to assume the language of a king his tone is that of a tyrant, and when he condescends he is vulgar.' Unlike Elizabeth, he hid himself from the public; told they merely wished to see his face, he replied: 'God's wounds! I will pull down my breeches and they shall also see my arse.' Not merely the public, but very large numbers of influential local figures found they had no access to the King. Elizabeth had taken a lot of trouble to get to know personally everyone who mattered. She scrutinised the lists of JPs throughout the kingdom, ticking off those she wanted reappointed; she claimed she knew every one of them. James knew no one outside the narrow court and government circle; and within it, instead of carefully balancing factions, as she had done, he flung himself literally into the arms of successive favourites, first the Scotsman, Carr, then Villiers, whom he made Duke of Buckingham. James loathed women. He delighted in getting the young court ladies drunk, and seeing them collapse in vomit at his feet.* He would sit there, laughing, while he fiddled with his genitals, a distasteful habit which everyone noticed. It was, indeed, impossible to ignore his homosexuality, for it was displayed in company, James planting slobbering kisses on the lips of George Villiers and fingering his body. His letters to his 'sweet child and wife' Villiers, signed 'your dear old dad and gossip', at least were private; not so the defence which the King made to the Lords of the Council of the earldom given to the youth, and in a speech where he justified homosexuality by blasphemy:

> You may be sure that I love the Earl of Buckingham more than anyone else, and more than you who are here assembled. I wish to speak on my own behalf, and not to have it thought to be a defect, for Jesus Christ did the same, and therefore I cannot be blamed. Christ had his John, and I have my George.

The English have always loathed homosexuality in public men, and punished it savagely. More remarkable, in James's case, was that he evoked the disdain of the French ambassador, who had acquired a high degree of sexual tolerance at Henri IV's court.

> The King [he wrote] . . . has made a journey to Newmarket, as a certain other sovereign once did to Capri. He takes his beloved Buckingham with him, wishes rather to be his friend than king, and to associate his name to the heroes of friendship in antiquity. Under such specious titles he endeavours to conceal scandalous doings, and because his strength deserts him for these, he feeds his eyes where he can no longer content his other sense. The end of all is ever the bottle.

Of course knowledge of these doings was confined to a comparatively close circle, though gossip inevitably spread. What could not be concealed was James's atrocious treatment of leading public figures, and the growing evidence of vice and corruption in high places. His deliberate

* Sir John Harington (Elizabeth's 'Boy Jack') described in uproarious detail the disgraceful orgy at Theobalds in 1606, in honour of James's royal brother-in-law, Christian of Denmark. 'King James' Court,' wrote Aubrey, 'was so far from being civill to woemen, that the Ladies, nay the Queen herself, could hardly pass by the King's apartment without receiving some Affront.' For a hostile portrait of James see J.P. Kenyon: *The Stuarts* (London, 1967); and for a more sympathetic one, David Mathew: *James I* (London, 1967).

destruction of Ralegh turned a highly unpopular monopolist into a national hero. It was said at the time, and believed, and we now know it to be true, that James of set purpose leaked the details of Ralegh's last expedition to the Spanish authorities, and so ensured its failure. His execution of the old gallant and scholar was thus cold-blooded murder. James even planned a public insult to the nation he ruled, for he offered to hand Ralegh over to the King of Spain so he could be hanged in a public square in Madrid; but even the Spaniards drew back at this. Ralegh ended his world history with an eloquent salute to Death the Avenger, no doubt with the hated James in mind, though in fact it made an apt comment on his son Charles:

> O eloquent, just and mighty Death! whom none could advise, thou hast persuaded; what none hath dared, thou hast done; and whom all the world hath flattered, thou only hath cast out of the world and despised!

James had been educated, after a fashion; but it was the antique Latin learning of the medieval world. Such as it was, he was proud of it, and he was bitterly and vengefully disappointed when it failed to cut any ice with the exponents of the sophisticated new learning he found in England. 'His Majesty rather asked counsel of the time past than of the time to come,' noted Bacon. Bacon offered exceptionally shrewd advice, but the vain King shrank from contact with a man so manifestly his intellectual superior; he preferred to act the role of learned father-figure to ignorant young philistines like Carr and Villiers. His only intellectual friendship was with the Spanish ambassador, Gondomar, who had a similar schoolman's background, and with whom the King muttered lengthily in Latin syllogisms. James, indeed, feared the new learning; like the popes, he thought it subversive. Elizabeth had not censored a single work of learning, education or science. Under James, and still more under his son, it became increasingly difficult to get anything new published. Some of the central works of Ralegh, Bacon and Coke had to wait until the parliamentary resurrection of 1640 to see the light. James failed to stop Ralegh's *History of the World*, which indeed became, to his fury, a best-seller; but he confiscated many of Ralegh's manuscripts—as Charles did those of Coke, his officers ransacking the old judge's house as he lay on his death-bed.* James dis-

* In 1631, when Charles heard Coke was planning to write a book about Magna Carta, he forbade publication. In the seventeenth century, Magna Carta was regarded as an anti-executive instrument. Clarendon, perhaps lying, relates that when the Commons 'with all humility, mentioned the law and Magna Charta, Cromwell told them, their *magna farta* should not control his actions'. For making the same joke in 1667 Lord Chief Justice Keeling was attacked in the Commons as 'thought to be tending to arbitrary government in the judicature'.

solved the Society of Antiquaries, for even the exploration of the past he believed fraught with peril to the static, immobile society he wished to establish.

The Stuarts thus set their faces against the whole dynamic trend of English development, and vainly sought to arrest in flight a projectile hurtling into the future. More than this, they seemed to possess an unerring instinct for wounding the deepest feelings and prejudices of the English. The ending of the war with Spain was welcome; not so the project of a Spanish marriage, which necessarily involved, as all but James recognised, fundamental concessions to the Catholic, Continental interest. The ludicrous expedition of Charles and Buckingham to court the Infanta not only humiliated the English but cost them many of the Crown Jewels: the great ruby which Henry VIII had seized from Becket's shrine, and which had once rested on the bosom of Queen Mary, vanished into the eager palm of an Escurial courtier. Relations with the Vatican were restored. The treacherous massacre of English settlers by the Dutch at Amboyna, which received feverish publicity in England, went unavenged (until Cromwell came). Ralegh's 'heroical design of invading and possessing America' was frustrated, indeed State support of all overseas adventures was withdrawn, and James even tried to wind up the Virginia Company, whose Treasurer, Sir Edwin Sandys, he hated; if private enterprise continued the Elizabethan traditions, it was in the teeth of opposition from James and his son. In 1633 Charles even went so far as to forbid English ships to enter the Mediterranean. Meanwhile, government and court society were rocked by repeated scandals. Carr's wife was convicted of murder by poison, but merely exiled to the country. Lord Audley was sentenced to death for 'sodomy, unnatural adultery and incest' (he was also a papist). The Lord Treasurer, and then the Lord Chancellor, were convicted of corruption. Judges were dismissed for refusing to give verdicts to the Government. James's relations with Parliament finally broke down in 1611; no Stuart king ever re-established them, except for a brief moment in 1660–1.

No wonder that the English, in a growing mood of national humiliation – a kind of pious agony for their reputation and past – turned to the memory of Elizabeth's time. Her accession day, 17 November, was celebrated with bonfires and pointed allusions to the present. As Bishop Goodman, a fanatical Stuart supporter (and eventually a papist), had to admit:

> . . . after a few years, when we had experience of a Scottish government . . . the Queen did seem to revive; then was her memory much magnified – such ringing of bells, such public joy and sermons in commemoration of her, the

picture of her tomb painted in many churches and in effect more solemnity and joy in memory of her coronation than was for the coming in of King James.

Her 'Golden Speech', in various texts, was read and remembered, a form of subversion the Stuarts could not very well censor; nor could they prevent the growth of a popular Elizabethan industry.* It found expression in the splendid tribute to the infant Queen in the epilogue of Shakespeare's last play, *Henry VIII*; and in a play about the Queen by Thomas Heywood, printed in 1605, and because of its immense popularity later turned first into prose and then into heroic verse. Both James and Charles must have become sick and tired of the very name of Elizabeth, who seemed to have a mortmain on the affection and loyalty of their subjects. Where they could suppress, they did so. Sir Anthony Weldon's court memoirs, with their pointed references to Elizabethan glories, could not be published; nor could Fulke Greville's *Life of Sidney*, with its references to the 'decrepit' and 'effeminate' age of the Stuarts. There were many other examples. But the Elizabethan image – powered by the dynamic force with which the English invest their past traditions as instruments of the present and future – boiled beneath the surface, and burst into dramatic life when Charles, his government in ruins, finally summoned the Long Parliament in the autumn of 1640. These Parliament men saw themselves, indeed were, reincarnated Elizabethans. It is no accident that Cromwell's mother, his wife, and his favourite daughter were all called Elizabeth; that he referred constantly to 'Elizabeth of famous memory'; that he saw himself, in power, in some humble sense as her rightful successor. Nor is it coincidence that the Long Parliament, once met, unanimously appointed her anniversary as a day of solemn fasting, humiliation and prayer. In the morning, at St Margaret's, Westminster, the preacher, Cornelius Burges, urged the Lords and Commons: 'Remember and consider that this very day . . . eighty-two years sithence began a new resurrection of this kingdom from the dead.' And in the afternoon they were told by Stephen Marshall: 'This day eighty-two years ago the Lord set up his gospel among us.' England was God's chosen people: and Parliament, on behalf of the English, should 'enter into a solemn covenant with the Lord'. It was the recovery of the patriotic English spirit.

* The speech was reprinted in January 1642, immediately after Charles's attempt to bully Parliament, and again in March 1648 on the eve of the Second Civil War. Both James and Charles made clumsy efforts to appeal to the memory of Elizabeth, but her magic did not work for them. See C.V. Wedgwood: *Oliver Cromwell and the Elizabethan Inheritance* (Neale Lecture, 1970).

English history is a continuum; it follows certain decisive and recurrent patterns with enormous tenacity, even though they may be submerged for decades. It ultimately always rejects the alien. The Stuart kingship was an aberration. Its destruction was inevitable. It operated against a national consensus so solid and powerful that, in rejecting Stuartism, the English found themselves acting against some of their deepest instincts and taking a giant leap into a wholly unknown future. The Great Rebellion, as it is so misleadingly called, was not primarily about religion, or class, or the institutions of monarchy, or the powers of Parliament, or taxation and the rights of property, or the protection of the subject against the Crown – although it was, certainly, concerned with all of these things. At heart it was a reassertion of national self-respect, of pride and patriotism; it was an expression of the love the English feel for each other and their country. The Stuarts had betrayed the national mission; the English would redeem it. We cannot read Milton's magnificent pamphlet, *Of reformation touching Church discipline*, without realising that his exalted language, breathtaking in its audacity, is inspired not by the mere details of Church management, but by an overwhelming vision of the national destiny. The Monarchy had ceased to be the focus and cynosure of the country's attention. It no longer embodied or represented anything; it had abdicated through folly, and the trust had been taken up, of necessity, by Parliament. What, then, must Parliament do? Milton had no doubts: it was to lead the chosen people in the Lord's business. The divine purpose, he wrote, was moving towards its fulfilment; the people of England, having often served as its agents before, were to serve again in the next advance. England had been appointed by God 'to blow the evangelic trumpet to the nations'. With all the world to choose from, God 'hath yet ever had this island under the special indulgent eye of his providence'. Then, in an unforgettable phrase: the English have the glory and prerogative to be 'the first asserts of every great vindication'. England must not forget 'her precedence of teaching nations how to live'. Now that God is again decreeing 'some new and great period in His Church, even to the reforming of reformation itself, what does He then but reveal Himself to His Englishmen; I say as His manner is, first to us'.

The England on which Milton looked in 1640 seemed in ruins. It was spiritually, morally and physically bankrupt. It had lost its soul, its international credit, its domestic stability, its position of paramountcy in the British Isles. All the ancient, familiar landmarks had gone. The status of the nobility had been undermined by the reckless and shameful sale of honours. Between 1603 and 1629 alone, sales of peerages, mostly to courtiers of little standing, brought in over £620,000. Elizabeth's careful husbanding of the distinction conferred to rank by its rarity had been entirely abandoned: James sold knighthoods until no man who valued his dignity wanted one, and his son fined gentlemen £173,537 in five years for declining to take them up. This, in turn, devalued the Monarchy, which was after all the mere summit of a pyramid whose lower orders were becoming meaningless. Both James and his son found themselves obliged to insist on, even to define, the nature of kingship; something Elizabeth never had to do. Definition provoked counter-definition, controversy: a mystery based on a consensus became a matter of public argument. And what argument, on the King's side! The best Charles's cloudy mind could do, in explaining his theory of government to Bishop Juxon, was:

> As for the people, truly I desire their liberty and freedom as much as anybody whatsoever; but I must tell you their liberty and freedom consists in having government, those laws by which their lives and goods may be most their own. It is not their having a share in the Government, that is nothing appertaining to them. A subject and a sovereign are clean different things.

But the whole development of English political society had shown that a subject and a sovereign were not clean different things: they were part of the same mystical body, they worked in conjunction, they were indivisible. 'I reign with your loves,' the old Queen had said. Curiously enough, Thomas Wentworth, Earl of Strafford, the King's grim instrument, grasped the point that the State could not be analysed without peril. As he told the Council of the North when he was installed as its President: 'For whatever he be which ravels forth into questions the right of a king and a people shall never be able to weave them up again into the comeliness and order he found them.' King and Parliament had been a seamless garment; now it was wrenched into hostile components.

In the Church the consensus had gone, too. Perhaps its last act was the great translation of the Bible, a work in which divines of all tendencies gloriously cooperated. Thereafter, under royal propulsion, the hierarchy moved steadily to the prelatical Right, closer and closer to Rome. Elizabeth's policy of keeping religion out of politics was reversed. The vast majority of Englishmen, clinging sensibly to the central position on which the Elizabethan Church rested, found themselves labelled Puritans or Presbyterians. They were in fact Anglicans, driven to oppose bishops simply by the Romanising of the episcopate. Archbishop Laud and Charles may have been sincere when they claimed they had no intention of submitting to Rome, but that was the logic of all their actions. Laud's creation of a judicial empire for his church was a carbon-copy of the Hildebrandine programme against which the English nation had struggled, in the end successfully, for centuries. The Reformation was being put into reverse. Charles's attempt to impose a Roman-style system on the Scots not only brought about his physical ruin; it was a clear indication to the English that he intended to put the clock back in every respect. Laud was already doing it. Even before James died, a decree of 1624 forbade the printing or importation of any book dealing with religion, Church government or affairs of State without previous approval of the authorities; in 1637 Laud got a Star Chamber decree forbidding the printing, reprinting or import of any book without his licence. Unauthorised printers were pilloried and whipped, and leading dissenters, such as Prynne, mutilated. Characteristically, Laud banned the reissue of Foxe's great work, the book Englishmen most venerated next to the Bible. And it was significant, too, that when, in 1637, a writer applied to Charles for permission to reprint a poem he had written on the Gunpowder Plot, he was refused: 'We are not so angry with the papists now as we were 20 years ago.' It was quite clear to the English that Laud was planning to negotiate new links with Rome: his policies were moving irresistibly in that direction, the Mass was being said openly in London and elsewhere, and Charles himself was becoming not so much Head of the Church, as a mere member of it, as his medieval predecessors had been, and as Continental sovereigns still were.

In government the consensus had gone completely. The State was bankrupt. Royal revenues had risen only threefold in the last century, while prices had gone up more than four times. Under Charles and his father, the sale of royal lands had accelerated. Neither of them had the remotest idea of the value of money, or how to manage what resources they possessed. They distributed gifts and favours worth £3 million at least to the peerage alone. They turned the monopoly system, already causing disquiet under Elizabeth, into a national scandal, but to little profit to themselves; Pym told Parliament, for instance, that the wine monopoly had raised £360,000 a year, but Charles had got less than £30,000 of it. Attempts to bring in outside financial advisers, like the City magnate Lionel Cranfield, merely led to fresh scandals. When Charles tried to finance himself by raising money from City syndicates, he ended by driving them to bankruptcy, as Edward III had done nearly three centuries before. He could not get money legally through parliamentary authorisation because he would not admit Parliament's role in the Government consensus; he would not admit that there had to be a consensus. He tried to raise taxes illegally but failed – after all, Parliament had come into existence precisely to ensure that taxes could, in practice, be collected; that was what the Commons House was for. Without Parliament there was no prospect of the Crown actually getting the money, whatever the courts might be persuaded to say. Ship money raised anger and opposition, but increasingly less cash and eventually less than the cost of collection. So Charles in despair turned to England's enemies abroad. He got a little money from Spain, in return for sinister services rendered. He even appealed to the Pope. How the old Pontiff must have laughed! When, he asked, had Charles been born? Didn't he know that the popes did not *give* money, they *took* it?

The judicial consensus had gone. Devoted as the English were to their ancient system, they saw it disintegrate before their eyes. The common law courts were downgraded or bypassed, and their judges dismissed or terrorised. The prerogative courts, which the Tudors had so skilfully used to secure for the Crown a monopoly of violence and emphasise national unity, were turned into divisive forces, monstrous instruments of tyranny, which had escaped wholly from statutory control. Moreover, these courts defended themselves from critics with a savagery from which no Englishman, whatever his status, was safe. In 1638, for alleged slander of the Star Chamber, a crime unknown to the common law or statute, Sir Thomas Wiseman was fined £10,000 with £7,000 damages, deprived of his baronetcy, degraded from the order of knighthood, had his ears cut off, was pilloried, and was sentenced to imprisonment during the King's pleasure. When the law not merely fails to guarantee the safety of life and property, but directly threatens both, the subject is absolved from obedience to it, and civil society collapses.

This, indeed, is what men felt by 1640, and not just in an individual but in a collective, national sense. If men could make a contract with God, their relationship with the King, with the State, was similarly contractual: if the King defaulted by failing to provide the services of government, the contract lapsed, and a new one must be made on a more stringent basis. What was the prime service of government? The defence of the realm. It was here that Charles's failure was most evident. The years 1629–40 were the last period of prerogative government in this country, and it was a total fiasco. James had lamentably failed to defend the Protestant interest in Germany; Charles abandoned it in France, after the humiliation of La Rochelle. Thereafter England had no influence on the Continent. The Elizabethan command of the seas – even the narrow seas – was abandoned. In 1631 Turkish pirates raided the Irish and Cornish coasts with impunity, carrying off many of the King's subjects into slavery. The warships Charles planned to build with his ship money were intended merely to enforce respect for the King's titles in the Channel, not to reactivate Elizabethan policies. By 1639, in return for cash, Charles was transporting the wages of Spanish troops fighting in the Netherlands across English territory, to avoid Dutch sea power – a curious role for an English sovereign. The thing came to light when a scrimmage developed between Dutch and Spanish ships in English territorial waters, and the Dutch pursued the beaten Spanish ashore, without interference or protest from the English authorities. The English were outraged, not so much at the insult as at the lack of response: it was the first point on which Parliament wished to question the King when it was at last summoned.

The ultimate humiliation was the loss of English paramountcy at home. Charles tried to enforce his will on the Scots, and not only failed but was decisively beaten; a Scottish army occupied Northumberland and many north-east towns, and Charles could bar their further progress south only by handing over large sums of money, which he did not possess. That the English should live to see 'such beggarly snakes put out their horns', should be at the mercy of such 'giddy-headed gawks' and 'brutish bedlamites', seemed intolerable. But if the English feared and hated the Scots, they feared and hated the Irish still more: and it was the Irish, by rising in revolt, who finally demolished Stuart absolutism. Reports reached London that the Irish Catholics had risen *en masse* and butchered all Protestants; that there was no secular power left in Ireland to restrain them, and that they would shortly be in England. The Irish revolt was the unexpected blow which turned an economic recession, which had been gathering force, into a catastrophic slump, the worst the oldest inhabitant could remember. It struck the most advanced part of England, the south-east, and especially the cloth trade, with bewildering severity. Enormous deputations of people, many thousand strong, led by municipal officials and local gentry, marched on London with petitions and demands for redress, hysterical calls for action. The slump must have led to a change of government in any event; in conjunction with other factors it brought a change of regime.

What men found increasingly difficult to believe was that the calamities affecting England were a pure conjunction of chance: no conceivable degree of ineptitude on the part of Charles and his ministers, they felt, could have brought about such national ruin in every department of State. It must be a conspiracy. The impression was formed – and on the face of it there was plenty of evidence – that Charles, Strafford and Laud were engaged in a deliberate operation to destroy English liberties and the Protestant religion and install instead a Catholic absolutist monarchy of a Continental type. The understanding with Spain, the negotiations with Rome, the impoverishment of the Protestant heart-land of London and the south-east, on which the power of the Tudor consensus had always rested, all pointed in one direction. All over Europe, constitutional systems with an ancient medieval basis were being overthrown in the interests of tyranny: the Cortes in Castile and Aragon had gone, Richelieu had destroyed the Estates-General in France, Gustavus Adolphus had killed the Riksdag in Sweden. Was the English Parliament the next to go? One of Charles's own courtier-MPs had issued a direct warning in 1628:

> To move not His Majesty with trenching on his prerogatives, lest you bring him out of love with parliaments ... In all Christian kingdoms you know that parliaments were in use anciently, until the monarchs began to know their own strength; and seeing the turbulent spirits of their parliaments at length they, by little and little, began to stand upon their prerogatives and at last overthrow the parliaments throughout Christendom, except only here with us....

But to the men of the Long Parliament, it did not appear to be a prospect of 'little and little', but of sudden, absolute and imminent destruction of the constitution. Strafford would be the agent, and his instrument would be a Scottish army, or an Irish army, or a Continental army (there were wild and atavistic rumours that the Danes were landing on the East Coast), or a combination of all three. Men would arise to do the King's evil bidding from what Henry VIII had called 'the most brute and beastly shires of the realm', or, in the words of a Gloucester MP 'come out of blind Wales and other dark corners of the land'. The King had not merely betrayed the English mission: he was threatening the racial impulse on which it was based. It was not just a class or a religion that was menaced: it was English civilisation.

Seen against this background, the events of 1640–60 no longer appear to be an aberration but a reassertion of some of the central currents of English historical development. As is characteristic of the English, they are dominated by three important paradoxes. First, it is not correct that Parliament overthrew the Government of Charles. On the contrary, it collapsed of its own accord. By the time the Long Parliament was summoned, many functions of the State were grinding to a halt. Charles's only effective servant was Strafford, as Parliament recognised: once he had been attainted, and executed, 'struck on the head', as MPs said, 'like a wild beast', the Government disintegrated. Many ministers fled abroad. The rest presented themselves to Parliament and asked for orders and authority.* Charles was without a treasury, an army, a judi-

* See Perez Zagorin: *The Court and the Country: The Beginnings of the English Revolution* (London, 1969), p. 210. For the behaviour of civil servants during the Civil War, see G.E. Aylmer: *The King's Servants: the Civil Service of Charles I, 1625–42* (London, 1961), pp. 337–417. The best modern comprehensive account of the period is Ivan Roots: *The Great Rebellion, 1642–60* (London, 1966).

cature or a civil service. Angry, bewildered and almost alone, he wandered aimlessly to York, where he summoned what he intended to be (one imagines) a grand council, of a type which was already obsolescent in the twelfth century; but nobody came to it. He did not know what to do. (His only decisive act, in a desperate attempt to curry favour and prove his suspect allegiance to the State, was to seize and hang two Catholic priests, one a harmless old man of 90: to my mind this cruel and meaningless murder absolves the English, then and now, of any moral duty to pity this doomed sovereign.) Charles, in effect, abdicated; and Parliament necessarily moved in to fill the vacuum of government, and to guarantee the safety of an imperilled nation, which had been abandoned to its enemies. It formally invited the Monarch to consult with it about the management of the country; it summoned the King, as the Crown had been accustomed to summon Parliament; but there was no response.

Indeed, the King's response could in legal terms be construed as an act of rebellion: herein lies the second paradox. By setting up his standard at Nottingham, Charles made a gesture traditionally associated with rebellion (though the point was, and is, arguable). Wyatt had 'set up his standard' in 1554, Northumberland in 1569; planning rebellion in 1601, Essex had been urged by his supporters to set up his standard in Wales. Such an act usually figured in formal charges of treason. Hence the strict logic of Parliament's indictment that Charles Stuart was in rebellion not merely against the nation but against the Crown itself.

This, of course, explains the poor response Charles's action evoked. But here we come to the third paradox: the civil war came about precisely because Charles was virtually alone. Parliament raised an army merely as a bluff; none believed it would be needed. The universal assumption was that the King would be obliged to submit to the nation and accept its terms. But this assumption itself brought a shift in attitudes. The King had endeavoured to destroy the constitution, and had been frustrated. Now his very weakness and isolation threatened to produce a radical imbalance in the constitution towards the opposite end of the political spectrum. At least this was what a significant section of the propertied classes came to believe: popular insurgency in the south-east, attacks on gentlemen's houses, a feature of 1641–2, seemed to confirm such fears. There was a palpable shift of opinion towards the King, motivated not by a desire to assist his armed rebellion – no one thought it would come to that – but to widen the social base of his support sufficiently to create a constitutional bargaining-position, and so restore the traditional balance of forces as they had existed under

Elizabeth. Both sides, indeed, wanted to bring back Elizabeth's day: herein lies the tragedy. Only Charles, his family and his immediate associates preferred war to compromise. But this politically motivated access of support he turned into a military instrument: it was just sufficient, with his allies from the Continent, from the Catholic interest, and from the economically backward parts of the British Isles, to produce civil war. It was all a fearful miscalculation, a muddle from which, for once, the English genius for constructive hypocrisy could not extricate itself, until the consensus had been restored by force.

Thus the English were confronted with something they had never faced before, and from which their instincts made them recoil in dismay – a wholly unprecedented crisis, for which their historical memories laid down neither guidelines nor obvious solution. The revolutionaries in America, in France and in Tsarist Russia were to inherit a distinguished revolutionary corpus of theory and experience, ultimately derived from England. The English themselves had nothing which seemed remotely relevant. The revolt of the Netherlands had been the overthrow of a foreign despotism; the history of Venice was instructive, but inapposite. Both cases were eagerly studied, and from first-hand experience: over 100 Englishmen a year went to study at Leyden, and a number (including Milton) had been to Padua which, under the protection of Venice and outside the supervision of the Inquisition, was the freest university in Catholic Europe. But in all essentials the English were thrown back on their own history and their own intellectual tradition.

This latter was not as meagre as one might suppose. The Reformation had done its work. The ending of the old clerical censorship under Edward VI, the growth of Anglican schools and colleges, above all the invasion of learning by the vernacular, had created a robust body of independent thought, strongly tinged by nationalism but imbued also with scientific and empirical principles which rejected the closed-circuit learning of the Roman–medieval world. In the century following the 1560s England had advanced from scientific backwardness through a technological revolution – based chiefly on instruments of measurement – and at the outset of the Civil War was technically the most advanced country in the world. True, the reimposition of a strict censorship, and Laud's strenuous efforts to get a clericalist grip on the country's intellectual life, had impeded the progress and dissemination of learning. Laud silenced some scientists and drove others into exile. Exploiting the monopolistic position of Oxford and Cambridge – then, as later, the chief obstacles to the spread of higher learning in England – he tried to conjure back the medieval world. He had the enthusiastic backing of the older generation of dons, who feared the threat of the new learning to their status and incomes. He was also supported by the senior arm of the medical profession, the College of Physicians, who clung blindly to the antique pseudo-science of Galen and Hippocrates. In 1635 Laud, who with other bishops issued licences to doctors, surgeons and midwives, restricted medical practice to those with degrees: empiricist medical men were thus excluded. None of the leading scientists and mathematicians got university jobs. Bishop Williams lamented: 'Alas, what a sad case it is that in this great and opulent kingdom there is no public encouragement for the excelling in any profession but that of law and divinity.'

Nevertheless, in the face of all the difficulties, progress was maintained. John Barclay, a Scot, writing in 1614, noted: 'In philosophy and the mathematics, in geography and astronomie, there is no opinion so prodigious and strange, but in that island was invented, or has found followers and subtile instancers.' Gresham's College, which at the behest of its founder gave lectures in English as well as Latin, was a citadel of advanced learning for the London mercantile and scientific community: it had the kind of reputation which the LSE acquired under Laski, adventurous, nonconformist, dangerous to Church and State. Despite the censors, a great deal got through the net, including the immensely influential *History* of Ralegh. Subversive manuscripts passed from hand to hand, as in the Soviet Union today. Both Pym and Hampden, for instance, possessed manuscripts of banned works by Ralegh. There was a powerful intellectual underworld, which sprang into the light the moment the Long Parliament met.

The new learning was subversive by effect rather than intention. Though King James found Ralegh 'too saucy in censuring princes', the latter by no means advocated rebellion. But he pointed out that, as a matter of historical record, people did in fact overthrow tyrants; and this, combined with Foxe's doctrine of the spiritual duty to resist, was enough. Moreover, Ralegh made it clear that the nature of English society suggested that such resistance was likely to be successful. 'The husbandmen and the yeomen of England are the freest of all the world ... it is the freeman and not the slave, that hath courage and the sense of shame deserved by cowardice.' As opposed to France, where the people have 'no courage or arms', 'the strength of England doth consist of the people and yeomanry'. He emphasised, too, the new importance of the gentry, 'the garrisons of good order throughout the realm'. 'The people therefore,' he concluded, 'in these latter ages are no less to be pleased than the peers.'

Bacon, too, was an acute observer of the changes which had taken place in society, though he saw them chiefly in economic terms. He admired the Netherlands because there 'wealth was dispersed in many hands', and he foresaw an England where it 'resteth in the hands of the merchants, burghers, tradesmen, freeholders, farmers in the country and the like'. He was not a parliamentarian but a monarchist, in a sense an absolutist: but his politics were belied by his intellectual empiricism, which was what struck men as new and important in his work. What held medieval society in subjection was the intellectual consensus that no lasting improvement was possible in the material world: mankind lived in a vale of tears, from which only death and salvation would bring release. Any intellectual advance could be achieved only by obtaining a more precise definition of the received corpus of knowledge, by purely verbal methods of disputation. Bacon rejected both these propositions with scorn. 'The understanding having been emancipated – having come, so to speak, of age,' he wrote, '... there must necessarily ensure an improvement of man's estate, and an increase of his power over nature.' Verbal disputations were useless, indeed counter-productive: 'controversies of religion... must hinder the advancement of science.' They brought what Milton called 'this impertinent yoke of prelaty, under whose inquisitorious and tyrannical duncery no free and splendid wit can flourish'. The only road to advance was through action, by empirical experiment: 'The industry of artificers maketh some small improvement of things invented; and chance sometimes in experimenting maketh us to stumble upon somewhat which is new; but all the disputation of learned men never brought to light one effect of nature before unknown.' The world had been changed by printing, gunpowder and the mariner's compass, all of them 'lighted upon by chance'. If accident could produce such wonders, how much more rapid would progress be if experiments were planned, coordinated, backed by the full resources of the State. 'Nature cannot be conquered but by obeying her . . . human knowledge and human power come in the end to one. To be ignorant of causes is to be frustrated in action.' Bacon was Englishman enough to suppose that this triumph of man over nature would be a recovery of past felicity – the scientific revolution would not so much project mankind into the future, as abolish the consequences of the Fall, and restore to the children of Adam (who were, of course, essentially English) their lost birthright. But he can hardly have supposed that such progress into the past would be welcome to a monarch of James's type, who wished his subjects to be alike, 'ignorant of causes' and 'frustrated in action'.

Oddly enough, Bacon's mortal rival and enemy, Coke, supplied the practical ideology of resistance which was lacking in this broad background of intellectual subversion. One might say that the lawyers, of whom Coke was the quintessential spirit, both sired the English revolution and then aborted it. Coke had a much more direct appeal to the English, and above all to MPs of gentry origin, because he taught that Stuart monarchy was essentially an aberration from the English tradition, and its actions unjustified by ancient precedent: this was precisely what they wanted to hear. This unpleasant old judge, who tied his daughter to a bedpost and beat her mercilessly until she agreed to marry a man who would advance Coke's fortunes at court, knew more about the law than any man in the kingdom. What he did not know he invented, to suit his convenience; and no one dared contradict him. By reducing the common law to a vast series of commentaries, he in effect gave England a kind of jumbled written constitution, based on the traditional and statute law (as opposed to King's courts) and parliamentary sovereignty (as opposed to the prerogative). Much of what he believed was utilitarian myth: he taught, for instance, that Parliament went back to before the days of Arthur; and it was not until the 1660s that royalist scholars, delving into the records kept in the Tower, found that it was a mere Plantagenet institution. On the other hand it was the kind of creative history the English had always employed for political purposes – the secular equivalent to Foxe's theory of the chosen race – and provided chapter and verse for a new parliamentary consensus. As in the past, revolutionary conservatism was summoned to repel unwelcome Stuart innovation from the ship of State. The common law, Coke said, 'is the absolute perfection of reason ... refined and perfected by all the wisest men in former successions of ages ... [it] cannot without great hazard and danger be altered or changed.' How could an ignorant Scots King be expected to answer these majestic and lapidary repetitions? Thus Coke used the law to destroy the prerogative structure of the Tudor State, which Charles had inherited. He had, too, a gift for the sharp phrase which less learned MPs could repeat with mindless and impressive dogmatism. How could the King's officers dare to search a man's property, for 'the house of an Englishman is to him as his castle'. Monopolies were plainly against Clauses 29–30 of Magna Carta, and 'Magna Carta is such a fellow as will have no sovereign'. Every Englishman, said Coke, was born with a priceless inheritance denied to lesser breeds: 'The ancient and excellent laws of England are the birth-right and the most ancient and best inheritance that the subjects of this realm have, for by them he enjoyeth not only his inheritance and goods in peace and quietness, but his life and his most dear country in safety.' If such enjoyment conflicted with the King's policies, then the King was acting outside the law, and the subject was not only entitled but bound to oppose him. Thus we see how Henry VIII, in ransacking the archives to prove his power against the Pope, set a pattern of historical research which was turned against his successors, and deprived them of everything he took for granted.

Accordingly, under the banners of Foxe and Coke, the parliamentary nation went to war against a rebellious king, confident in its historic rectitude and sense of mission. But the trumpet blew with an uncertain note, and the walls of Jericho did not immediately fall. Pym was a brilliant parliamentary manager, who used the conveniently ill-defined powers and procedure of the Commons to beat Charles to every political trick; he had the true revolutionary's instinct to prefer narrow-based activism to an irresolute broad-based consensus of MPs. The royalist Members and Peers departed, and Pym was left in possession of the parliamentary and constitutional field. But he could not win battles; and nor could the Earl of Essex. So the closed circle of lawyer and gentry MPs who directed the first phase of the Civil War, and whose goal was an adumbration of the eventual Whig settlement of 1688, found themselves obliged, *in extremis*, to summon the assistance of a submerged section of the people, and bring them into the political nation in the form of constitutional warriors. Parliament won the war, in the end without difficulty, because in the New Model Army it had enfranchised the people. Its officers and men had, for the most part, been excluded hitherto from the political and religious consensus. They were tradesmen, artisans, farmers, even labourers, or men of estates so small they scarcely qualified to vote at elections. Even the regimental commanders came from a humble social class. Colonel Ewer had been a serving-man, Harrison the son of a butcher, Pride a brewer's drayman, Okey a tallow-chandler, Hewson a shoemaker, Goffe a salter, Barkstead a goldsmith, Berry a clerk, Kelsey a button-maker. Cromwell did not want fancy officers, like the Earl of Manchester, for he thought it 'would not be well until Manchester was but Mr Montague'. 'Better plain men than none.' As for troops, he demanded men with a small stake in the country and the desire and ability to increase it, 'being well armed within by the satisfaction of their conscience and without by good iron arms, they would as one man stand firmly and charge desperately'. Such were, indeed, the men he got, as Richard Naxter noted: 'These men were of greater understanding than common soldiers . . . and making not money but that which they took to be public felicity their end, they were the more engaged to be valiant.' Many of these officers and soldiers (to

Cromwell's mind) entertained weird and wonderful opinions in religion, but he insisted this was secondary to their determination and loyalty: 'The State in choosing men to serve it takes no notice of their opinion: if they be willing faithfully to serve it, that satisfies.' In fact the intransigence of their views was the best safeguard that the struggle would be pressed home to absolute victory. Cromwell had no patience with men who sought accommodation before all physical power was in safe hands. When Manchester whined, in November 1644: 'If we beat the king nine and ninety times, yet he is king still. But if the king beats us once, we shall be hanged,' Cromwell replied angrily: 'My Lord, if this be so, why did we take up arms at first? This is against fighting ever hereafter. If so, let us make peace be it never so base.' Now Cromwell knew that his men did not fear to be hanged, because they knew it was not in God's providence; they, like their general, were His Englishmen.

Thus Parliament won the war by bringing into play a new class of humble folk who, for the first time in English history – for the first time in world history – were called upon by the State to serve not just with their bodies, but with their mental and spiritual energies, not as cannon-fodder, but as sentient and thinking individuals. Naturally, they were more than a match for Charles's foreign mercenaries, his vicious courtiers and their rapacious retinues, his dragooned rabbles of Irish, Scottish and Welsh peasants. Once the New Model was formed, the King's game was up. But its chief importance was not military, but political: it was a giant step forward in the liberation of mankind from darkness.

For Denzil Holles and the other parliamentarians of the centre, the end of the war was the mere prelude to negotiations with the King on a settlement acceptable to the landed classes. These talks could be safely left to the gentry. As for the New Model, it was a genie to be replaced promptly in its bottle and firmly corked up; except, that is, for those regiments required for a campaign of attrition against the 'Irish savages'. This programme was presented to the army with a brutal arrogance which even Charles might have envied. Pym, to be sure, would have found a basis for agreement; but Pym was dead. His successors refused to pay the troops their back-wages before disbandment, declined pensions for the families of the fallen and, most sinister of all, any indemnity for acts committed by troops in the discharge of their duty. As for Cromwell and his sort, he had already expressed a view that he might, lacking a further parliamentary commission, enlist with the Protestants in Germany; let him go there. So began Cromwell and Ireton's three-way negotiations with the King and the parliamentary majority; and so, in parallel, the formation of a political movement among the troops, with the election of adjutators, or agitators, from each regiment to represent rank-and-file opinion. The King was seized; the army moved closer to London. The genie would not go back into the bottle; the question was, could it be dispersed in Putney Church?

At this point in time, October 1647, the political spectrum of England, both inside and outside Parliament, could be represented as follows, reading from Left to Right. On the extreme Left were the Clubmen (violent revolutionary anarchists); then the Diggers (communist pacifists, led by Gerald Winstanley); the Levellers (a social democrat group, the civilian wing led by Lilburne, the military wing by Sexby and the other agitators); the Independents (radical gentry officers, led by Cromwell, with a parliamentary wing of about 60); the Presbyterian centre party (the majority of MPs, led by the Denzil Holles group, who wanted a Whiggish constitutional compromise with the King); crypto- or outright royalists (mainly in custody or in exile). All these were minority groups, though the Presbyterians, followed by the Independents, were by far the largest. Cromwell's aim, towards which he worked by instinct more than by any clear and preconceived plan, was to create from these splintered groups a national consensus, on a broadly based programme of toleration and reform.

But first he had to secure unity in the army. In June, his son-in-law, Ireton, an accomplished lawyer and draftsman, had produced a declaration for the union of the army and Parliament, setting out broad, philosophic principles:

> ... that we are not a meer mercenary Army hired to serve any Arbitrary power of a State, but called forth and conjured by the severall Declarations of Parliament to the defence of our owne and the people's just Rights and Liberties; and so we took up Armes in judgment and conscience to those ends, and have so continued in them, and are resolved according to your first just desires in your Declaration . . . and our own common sense concerning those our fundamental rights and liberties, to assert and vindicate the just power and rights of this Kingdome in Parliament for those common ends promised against all arbitrary power, violence, and oppression, and against all particular parties or interests whatsoever.

The army's right of resistance was based upon 'the Law of Nature and of Nations' and 'the proceedings of our ancestors of famous memory to the purchasing of such Rights and Liberties, as they have enjoyed through the price of their bloud, and we (both by that and the later bloud of our deare friends and fellow soldiers) with the hazard of our own, do not lay claim on to'. In other words, the army's programme was modest. So long as parliaments were 'rightly constituted, that is, freely, equally, and successively chosen', with a legal, fixed duration, and summoned at definite intervals, the army would willingly submit to their authority:

> Thus a firm foundation being laid in the authority and constitution of Parliaments for the hopes, at least, of common and equall right and freedom to ourselves and to all the freeborn people of this land; we shall for our parts freely and cheerfully commit our stock or share of interest in this kingdome into this commom bottome of Parliaments, and though it may (for our particulars) go ill with us in one Voyage, yet we shall thus hope (if right be with us) to fare better in another.

At Putney Church, however, it was soon apparent that, in the eyes of the Left, this programme begged all the crucial questions. It merely guaranteed the powers and regular sittings of Parliament: it left untouched its present composition. Yet this was based on usurpation, the imposition of the 'Norman yoke' on what had once been a free Saxon society, guaranteeing the fundamental rights of all. Under the original contract of society, said Wildman, 'the true and ancient fundamental constitution', the King and Lords had no special position. Why should what they had stolen now be legally endorsed? 'The difference is whether we should alter the old foundations of our Government soe as to give to Kinge and Lords that which they could never claime before.' Moreover, even if it were conceded that the real power should lie with the Commons, the suffrage on which it was elected was so narrow as to place authority, in perpetuity, in the hands of the landed classes. So the agitators produced a programme in which the vote was given to all adult males, except those in receipt of wages and poor relief.

Cromwell was appalled. The changes the agitators proposed alarmed him not so much because they were wrong as because they were so big. 'Truly,' he said, 'this paper does containe in itt very great alterations of the very Governement of the Kingedome, alterations from that Governement that itt hath bin under, I believe I may also say since itt was a Nation.' Supposing it were carried out: might not another group of opinionated men come along and propose a further set of changes? Where would it all end? England would become another Switzerland, with a different constitution for each canton. 'Would itt nott bee utter confusion?' Yes: it would produce 'an absolute desolation to the Nation'. The English were conservative, adverse to change; a written constitution was worthless unless 'the spiritts and temper of the people of this Nation are prepared to receive and to goe alonge with itt'. They needed coaxing: there 'bee very great mountaines in the way of this'.

Moreover, said Ireton, like Cromwell a man of substance, the fundamental axiom of the English constitution was that power should go with property, 'which if you take away, you take away all by that'. A man had the right to vote by virtue of his fixed interest in the State, which took the form of land or membership of a trading corporation: he thereby invested in society, and had a legal claim to a voice in its government. 'Now I wish wee may all consider of what right you will challenge, that all the people should have right to Elections. Is itt by the right of nature? If you will hold forth that as your ground, then I thinke you must deny all property too.' The vote must be confined to men with a 'permanent interest in the State'. Give it to a man who has no more fixed property than 'hee may carry about with him', a man who 'is heere to day and gone to morrow', and there will be nothing to prevent him from stealing by confiscatory laws. If the vote is his by law of nature, then so is equal division of property by law of nature. The result would be chaos and violence. To which Cromwell chorused that, if power in the State was given, through the vote, to 'men that have noe interest butt the interest of breathing', then 'the consequences of this rule tends to anarchy, must end in anarchy'.*

This argument brought an explosion from the radical Colonel Rainborough. Why should they presume that all men were evil? Why should giving men a vote lead to the destruction of property, let alone anarchy? God had laid down a commandment: 'Thou shalt not steal.' This was the true law of property. Did they have such little faith in the people of England that they would not trust them, even with power in their hands, to obey such a fundamental command? The question to be determined was by what right, and with what means, had property in the past been acquired. Besides, added Maximilian Pettus, why should a 'fixed interest' be defined as 40s freehold and above? Why should a vote be denied to a man with a leasehold worth £100 a year? Ireton replied hastily that he was not defending the *existing* property qualification, which might well be anomalous: he simply maintained that there must *be* one.

This the radicals would not admit. As one pointed out, the concept of the 'freeborn' was more important than the concept of the 'freehold': people were more important than things. 'The chief end of this Government is to preserve persons as well as estates, and if any law shall take

* The Cromwell–Ireton theory remained standard English constitutional doctrine until the Reform Bill, 1832. The reactionary Scottish judge, Lord Braxfield, at the trial of Thomas Muir for sedition, 1793, put it almost in Ireton's words: 'A government in every country should be just like a corporation; and, in this country, it is made up of the landed interest, which alone has a right to be represented; as for the rabble, who have nothing but personal property, what hold has the nation of them? What security for the payment of their taxes? They may pack up all their property on their backs, and leave the country in a twinkling of an eye, but landed property cannot be removed.'

hold of my person itt is more deare than my estate.' Property qualifications, added Rainborough, meant that power would be confined to the rich – the top fifth of the nation. The result? 'I say the one parte shall make hewers of wood and drawers of water of the other five, and soe the greatest parte of the Nation bee enslav'd.' The defenders of the *status quo* had no argument 'butt only that itt is the present law of the Kingedome'. Then what had the war been about? 'What shall become of those many men that have laid themselves out for the Parliament of England in this present warre, that have ruined themselves by fighting, by hazarding all they had? They are Englishmen.' But now they were to be denied the vote, to have 'nothing to say for themselves'.

Rainborough returned again and again to one question, to which Cromwell and Ireton could return no answer (no one can): 'How is it that some men have property and others do not?' On what principles of law or justice is this determined? No such principles existed: accident, chance, events of the past, often bloody and unjust events, had created the present disposition of property. Was this therefore to be frozen in perpetuity by vesting power exclusively in those who already had everything else? To which Ireton replied, a touch complacently, that until God gave a manifest sign that property should be redistributed, the present arrangements must endure: 'The law of God doth nott give mee propertie, nor the law of nature, butt propertie is of humane Constitution. I have a propertie, and this I shall enjoy.' To destroy the principle of property is 'a thinge evill in ittself and scandalous to the world'.

This absolutist insistence on the rights of private property brought from the lowly Edward Sexby perhaps the most brilliant and bitter intervention in the debate. 'Wee have engaged in this Kingdome and ventur'd our lives, and itt was all for this: to recover our birthrights and priviledges as Englishmen, *and by the arguments urged* [by you] *there is none.* There are many thousands of us souldiers that have ventur'd our lives; wee have had little propriety in the Kingedome as to our estates, yett we have had a birthright. Butt it seems now except a man hath a fix'd estate in this Kingedome, hee hath noe right in this Kingdome. I wonder wee were so much deceived. If wee had nott a right to the Kingedome, wee were meere mercenarie souldiers.' However, he had news for the conservatives (eyeing Ireton and Cromwell): the 'poor and meaner of this Kingdome' had given their all, 'to their utmost possibility' for 'purchasing the good of the Kingdome'; they would not now be deprived of their rights; it was a lie to say they stood for 'anarchy and confusion' – they have 'the law of God and the law of their conscience' on their side. 'I shall tell you in a worde my resolution. I am resolved to give my birthright to none.'

It was at this point, it seems to me in studying the record, that Cromwell decided that the debate could not be pressed to its logical conclusions without shattering the revolution in fragments, and in consequence bringing the blackest reaction, and indeed death, upon them all. In logic, he knew, he and Ireton did not have a case: Sexby and Rainborough were right. But he did not believe that logic was very important. What was important was to achieve a consensus, a compromise upon a lowest common denominator of agreement. This was the only way in which an immensely lethargic, conservative and traditionalist nation like the English could be brought to move into the future. Someone had earlier used the phrase 'half a loaf is better than none'; it was not the first time it had cropped up in English political debates; and it is still in constant use today wherever Englishmen gather to discuss anything contentious. It sums up the mindless common sense with which the English approach political problems. The question was, as Cromwell saw it, at what point was the loaf to be sliced? And how could agreement be reached on where the knife should fall? 'Let us be doing,' he said, 'but let us be united in doing.' It was no use trying to force universal suffrage on Parliament, even were it desirable in the abstract; such could be achieved only by the sword, 'and what we need is a treatie'. In any case, a comatose nation would not stand such an upheaval: the minority status of the political activists would be exposed, there would be a flood of revulsion, and the King would win back all just at the moment when he had lost it.

So Cromwell moved towards a compromise, ably assisted by Ireton. The latter pointed out that he was really just as radical as anyone: did he not believe in fixed parliaments, in equal constituencies? As for a wider suffrage, he would not oppose it if it were truly wanted by 'the generalitie of those whome I have reason to thinke honest men and conscientious men'. It was agreed that those who had helped Parliament should have the vote as of right; it was agreed that further consideration should be given to the whole question of property and political power. Early in November the agitators were persuaded to disperse quietly to their regiments. On the fifteenth Cromwell had a radical soldier executed by drumhead court martial. In the next two years he de-politicised the army. In the meantime, he gave the radicals much of what they wanted. He executed Charles Stuart. He abolished the House of Lords. But he did not entrust the people with the power to dispose of property. He did not trust the people with anything. Did he save the revolution? Or did he betray it? It is not in the nature of English history to provide unequivocal answers to such precise questions.

Whether or not Cromwell betrayed the revolution, the England he created, or rather unleashed from the bonds of the past, was something entirely new, strange and wonderful. The year 1640 is the great watershed in the history of the offshore islanders: during the next 20 years many things were done in the island which all the forces of conservatism were later unable to reverse, and though the experiment as a whole was aborted, the England of 1660 contained the chromosomes of a modern country, the first in the history of the world. The image of Cromwellian England handed down to us –puritanical, intolerant, hating pleasure and the arts, restrictive of liberty, a military dictatorship informed by a narrowly religious view of life – is not only false: it is the exact reverse of the truth. The movement Cromwell led unleashed the latent energies of the English people: he himself was, as Milton later put it in *Samson Agonistes*,

> a person raised
> With strength sufficient and command from heaven
> To free my country.

All great revolutions – the American in 1776, the French in 1789, the Russian in 1917, the Chinese in 1949 – are in one vital sense patriotic, springing from a sense of national frustration, a conviction that the existing structure of society imposes intolerable restraints on the genius and capacity of the people. In this, as in other respects, the English revolution which began in 1640 set the pattern. Suddenly, as the presses poured forth a torrent of forbidden books, and new ideas, openly expressed, fought for survival in a thousand excited conversations, the English experienced the joyful intoxication which Stendhal's *La Chartreuse de Parme* portrayed so brilliantly in Milan, and which I witnessed in the early days of Castro's Cuba. For decades great talents and endeavours had boiled beneath the restrictive carapace of early Stuart England. A conservative like Donne had fearfully sniffed the winds of change:

> And new Philosophy calls all in doubt
> The element of fire is quite put out;
> . . . Prince, Subject, Father, Sonne, are things forgot,
> For every man alone thinkes he hath got
> To be a Phoenix.

It was Cromwell who allowed the phoenix to rise. 'Men,' wrote Bacon, 'have been kept back as by a kind of enchantment from progress in the sciences by a reverence for antiquity.' Many years before the Revolution, he had advised Buckingham to do 'that which I think was never done since I was born ... which is that you countenance and encourage and advance able men, in all kinds, degrees and professions'. This was precisely the formula on which the New Model had been built, and not just in a strict military sense. The antique doctors of the College of Physicians had mostly joined the royalists, whose wounded they dispatched by the thousand; the New Model had an organised medical service, staffed by the despised Surgeons, and the humble Apothecaries, a service later extended to the Commonwealth navy, and which allowed Robert Blake to give England world sea-supremacy: the physical power of Cromwellian England was based essentially on the new learning. This was a time for talent to manifest itself.* To Milton, London was 'a city of refuge, the mansion house of liberty', where men were 'reading, trying all things, assenting to the force of reason and convincement'; or, as John Hall put it in 1649, England was imbued with 'the highest spirit, pregnant with great matters ... attempting the discovery of a new world of knowledge'. Sir Arthur Haselrig, MP, summed up the whole experiment in a phrase: the country was 'living long in a little time'.

Cromwell was conscious of this sense of adventure. He had read Ralegh's *History*. He was a child of the Elizabethan age; his aim was to recreate its glories in a more sober, scientific spirit. He had lived through the national humiliations of Charles's personal government when, in the words of the Venetian ambassador, 'England had become a nation useless to the rest of the world, and consequently of no consideration'. He had thought of emigrating to get away from the pain of it all; indeed, he swore that he would unless the Grand Remonstrance was passed. He had the self-confidence in England's destiny which came from membership of an immense and ramifying family of squires. When he was first elected in 1628, nine of his cousins were MPs; 17 of his cousins and nine other relatives served, at one time or another, in the Long Parliament. He was one of the few people, even then, who could trace their origins to pre-Conquest times. His family were active in the fight for

* One of Charles's judges, John Cooke, advocated free medical treatment for the poor; Samuel Herring proposed free medical services, run by men paid by the State, something for which the English had to wait until 1948, and which the Americans lack even now. The Leveller newspaper, the *Moderate*, forecast the invention of flying-machines: 'Experience daily shows us that nothing is impossible unto man.' For Cromwell's role in these stirring times, see Christopher Hill's brilliant biography, *God's Englishman* (London, 1970). Hill, in his *Intellectual Origins of the English Revolution* (Oxford, 1965), makes the ingenious point that there was a logical connection between new scientific theory and radical politics: the parliamentarians were heliocentrists, the royalists Ptolemaics, and, as can be seen from popular almanacs, Ptolemaic theory perished with Charles I; Copernicus 'democratised the universe' by breaking the hierarchical structure of the heavens, Harvey 'democratised' the human body by dethroning the heart. (But Harvey was a royalist: he took charge of the King's sons during the Battle of Edgehill, according to Aubrey.)

liberty. Six cousins were imprisoned for refusing the forced loan of 1627; Hampden was his cousin, and so was the man who undertook his defence over ship-money. Cromwell was an integrated Englishman. To him, Charles was an incompetent, alien adventurer, a man whose pride and deceit had cost oceans of decent, innocent English blood on both sides; a man to be punished like a felon. His son was feckless, self-indulgent and unworthy: 'he will be the undoing of us all'; all he wanted was 'a shoulder of mutton and a whore'; it was intolerable that such men should sit on Elizabeth's throne and be honoured by the charge of England's destinies.

There is no doubt that Cromwell, in his slow, cautious but deeply passionate manner, was convinced of the divine mission. As Lord Protector, he declared in May 1654:

> Ask all the nations of this matter, and they will testify, and indeed the dispensations of the Lord have been as if he had said, England, thou are my first-born, my delight amongst the nations, under the whole heavens the Lord hath not dealt so with any of the people round about us.

But there was nothing messianic about Cromwell. His victories were a providential dispensation, ending in the 'crowning mercy' of Worcester in 1651, the last time he took the field. But God merely ordained; it was for man in righteous energy and prudence to fulfil the ordinance. He accepted Ralegh's view that history was like a clock, unwinding itself according to God's design: but its shape was determined by secondary causes brought about by man's efforts. Like Henry V, whom he resembled in so many ways, he knew that God was on his side in battle; but, unlike Henry, he took no pleasure in numerical inferiority; except at Dunbar, he always ensured he had a preponderance of men and cannon. He rejected the *rentier* religion of the Catholics, in which the believer drew on the accumulated virtues of the saints; his faith was active, entreprenurial, dynamic: he 'wrestled with God', as he put it.

Under Cromwell, England emerged as a world power, not through faith but through works. The New Model Army exorcised the nightmares of generations of English governments: the open 'back door' in Ireland, and the threat of invasion from the north. Problems which had baffled Edward I and Elizabeth were resolved in two swift campaigns, and Cromwell was able to summon the first imperial parliament, with the British Isles resting in perfect tranquillity (not, alas, justice) under an English paramountcy. He was so much the master of England that a royalist uprising could be treated with a leniency springing from absolute confidence, almost contempt. The hostility with which the English revolution was regarded throughout royalist Europe, even in

Russia, soon turned to nervous respect, and then fear, as the English fleets and armies began to operate. In 1649–51 alone, 40 great warships, equipped with new heavy guns, were built and sent to sea. A new generation of humbly born generals, captains and admirals, of enthusiastic civil servants, of military scientists, of stern-faced professional diplomats, backed by the latest equipment, financed by the wholly unprecedented sums that Parliament raised in taxation, carried the revolutionary enterprise abroad. Wherever the Stuarts got aid and comfort, Blake's ships and Cromwell's redcoats were liable without warning to make a devastating appearance. The decline of English power was abruptly reversed, and the awe-inspiring aggression of English racialism unleashed. In 1651 Parliament had passed the Navigation Act, imposing a strict mercantilism on England's export and import trade, leading to a rapid growth of the English civil marine, and opening a vast programme of expansion to English commerce. Under Cromwell, Portugal was reduced to an English political satellite, and Brazil entered; the Netherlands were battered and allotted a subordinate role in the international economic system. Mazarin had rightly feared that revolution would make England formidable, for he predicted that a free Parliament would willingly finance a forward policy. France suffered from Cromwell's fist, and so, in turn, did Spain, her West Indian empire ripped open, Jamaica seized, one treasure-fleet captured, another destroyed under the guns of Spanish forts, her coasts blockaded for the first time throughout the winter season by Blake's all-weather fleet. England locked up the Channel and entered the Baltic. Across the Atlantic, the English colonies at last got the economic backing and the military protection of the State. In the Mediterranean, the navy asserted an easy supremacy, putting down piracy, avenging insults to English merchants and nationals, enforcing tolerance and a respect for commerce on the rulers of Tetuan and Tunis. It was the beginning of gunboat diplomacy, as of much else. Cromwell laid down the matrix of three centuries of Empire (in the name of God, of course).

He might have gone very much further; no one who studies the records of these astonishing years can escape a feeling of relief that, in fact, he stayed his hand. A man of resolve, presiding over a revolution which has brought whole new classes – an entirely fresh range of national genius and talent – into the enthusiastic service of the State, possesses a terrible power for good or evil in the world. Napoleonic France made itself the master of Europe through the energy of such a revolution; it was only deprived of global power, and eventually extinguished, because it was faced by the countervailing power of English nationalism, itself the product of revolution. But in seventeenth-century Europe no such balancing factor existed: apart from Holland, the debilitating restraints of the Roman–medieval world still paralysed national energies everywhere. For the English it was a moment of terrible temptation, of the type to which Henry V would surely have succumbed if he had lived. In the 1650s, said Thurloe, Cromwell 'carried the keys of the Continent at his girdle, and was able to make invasions thereupon, and let in armies and forces upon it at his pleasure'. Moreover, Cromwell not only had the opportunity: he and the articulate nation had, or believed they had, a mandate from no less a person than the Deity. God's Englishmen had been told to complete His reformation; and how was this to be done except by the destruction of Continental Catholic power? All successful revolutions tend to export themselves, and thereby create tyranny not only abroad but in their own heartlands. If Cromwell had entered Europe, there can be little doubt that the Continental monarchies would have collapsed like a pack of cards, and that the Catholicism of southern and central Europe would have been torn from its secular foundations.

Certainly there were many voices urging Cromwell on. George Fox, not yet by any means a pacifist, chided the army for its failure to carry Protestantism to Spain and Italy. Andrew Marvell wrote in anticipation:

> And to all states not free
> Shall climacteric be.

Cromwell's own thoughts strayed occasionally in this direction. He is reported to have said to General Lambert: 'Were I as young as you, I should not doubt, ere I died, to knock at the gates of Rome.' Happily, perhaps, age was a barrier to adventure. Cromwell was over 40 when the Long Parliament met; he was in his mid-fifties when power came to him in its plenitude, already tired by battles and arguments, sick of the bloodshed he had witnessed, determined to hold together 'this poor tottered realm' as long as he lived, but disinclined to accept fresh and unknown responsibilities. And he had to take account, too, of the instinct of the offshore islanders – in this as in so many other ways – that the divine mission lay on the oceans; that the prudent attitude towards the Continent was to turn towards it a heavily armoured back; that in Europe the redcoats were ill-advised to venture beyond sight of their ships. Thus a sinister chapter in English history was not written.

Instead, the English concentrated on achieving a cultural, scientific and technological supremacy, on the fulfilment of the Baconian programme. Machines which were banned under Charles were licensed and brought into use. The mineral monopolies were broken. Oxford was

'Greshamised', and invaded by scientists, who, after the Restoration, regrouped themselves in London and founded the Royal Society. Revolution and education go hand in hand.* Even Clarendon, who hated Cromwell, admitted that under the Protectorate Oxford 'yielded a harvest of extraordinary good and sound knowledge in all parts of learning'. As early as 1641 the people of Manchester petitioned for a university, 'many ripe and hopeful wits being utterly lost for want of education'; York followed suit; the Commons set up a Committee for the Advancement of Learning, and universities were proposed for Wales, Norwich and Durham. Charleton, in *The Immortality of the Human Soul* (1657), rejoiced 'that Britain, which was but yesterday the theatre of war and desolation, should today be the school of arts and court of all the Muses . . .'

It is facile to present the Civil War as a conflict between the Two Cultures, parliamentary science versus royalist arts. True, some MPs disliked Charles's art collection, mainly because it had not been paid for: they resented 'squandering away millions of pounds upon rotten old pictures and broken-nosed marbles'. But the early intolerance was swept away as the Cromwellian vision unfolded. Attempts to legislate against the theatre were unenforced and unenforceable; more plays were in fact published at this time than in any previous period, and great poets were honoured and set to splendid public employments. Cromwell put an abrupt stop to the dispersal of national art-treasures, and his court at Hampton, where the envoys of the European kings came to pay him dutiful homage, was a model of modest and seemly splendour.† Perhaps with an eye to the homosexuality so rampant under the Stuarts, he authorised women to appear on the stage for the first time. He was passionately devoted to music: he sponsored the first performance of an English opera, installed two fine organs at Hampton Court, and held music parties every evening when affairs of state permitted. Smear-stories about the goings-on at Hampton are, of course, royalist propaganda, and moreover contradictory. Cromwell was accused, on the

* A point which did not escape Charles Dickens. cf. Joe Gargery on the subject of Mrs Gargery: 'She ain't over partial to having scholars on the premises, and in partickler would not be over partial to my being a scholar, for fear as I might rise. Like a sort of rebel, don't you see?' (*Great Expectations*). Modern Conservative theory is based on the denial of educational opportunity to working-class children in the name of 'parental freedom', an up-to-date version of Filmer's paternalist theory of absolute monarchy.

† And also of strict security. Charles II had offered £500 a year for life 'to any man whosoever, within any of our three kingdoms, by pistol, sword or poison, or by any other ways or means whatsoever, to destroy the life of the said Oliver Cromwell'. Charles also discussed with his brother a plan to kill Cromwell with an infernal machine, the first recorded instance of bombing for personal assassination. For details of life at Cromwell's court, see Ernest Law: *History of Hampton Court Palace*, Vol. II (London, 1888).

one hand, of introducing 'the French cringe', and on the other of encouraging an informality unbecoming an English ruler. His wife was still more viciously attacked, both for the 'impertinent meannesses' of her simple tastes, and for giving herself airs; the royalists sneered at her 'nimble housewifery', and said she had secret passages made so she could creep up on her servants unawares; they also charged her with drunkenness and low gallantries with the soldiers. In fact Cromwell's court, far from being, as the royalists claimed, a place of 'silent mummery, of starched and hypocritical gravity', was loud with songs and laughter. The solemnity was kept up chiefly to impress foreigners. For the rest, Cromwell liked 'jocos and frisks'. At the marriage of his daughter Frances to Mr Rich, he threw about 'the sack posset amongst all the ladies to spoil their clothes, which they took as a favour, and daubed all the stools where they were to sit, with wet sweetmeats'. Bulstrode Whitelock records:

> He would sometimes be very cheerful with us, and laying aside his greatness, be exceedingly familiar with us, and, by way of diversion, would make verses with us, and everyone must try his fancy. He commonly called for tobacco pipes and a candle, and would now and then take tobacco himself. Then he would fall again to his serious and great business, and advise with us in those affairs.

As for the stories of Cromwell's destruction of church treasures and stained-glass windows, they are pure inventions, propagated by generations of high Tory clergymen; the actual evidence points all the other way.* Cromwell was devoted to the arts: his own bedroom was hung with 'five pieces of fine tapestry hangings of Vulcan and Venus'. But then all Cromwell's doings are obscured by lying myths.

We have noted before that toleration flourishes in England only under strong governments. Cromwell was as strong as he was tolerant. He proudly told Mazarin of the many Catholics he had saved from persecution. He set his face like flint against the burning of witches, and encouraged efforts (characteristic of this time of freedom) to raise the

* It is examined by G.F. Nuttall in 'Was Cromwell an Iconoclast?', *Transactions of the Congregational Historical Society*, XII. Damage attributed to Cromwell had, as a rule, occurred during the Reformation. There is no proven instance of him or his men deliberately despoiling churches; but of course he systematically 'slighted' royalist castles. For an examination of the myths surrounding Cromwell's memory, see Alan Smith: 'The Image of Cromwell in Folklore and Tradition', *Folklore*, 1968. The ill-disciplined Cavaliers left far more destruction in their wake; Anthony Wood commented angrily on their sojourn at Oxford: 'To give a further character of the court, though they were neat and gay in their apparell, yet they were very nasty and beastly, leaving at their departures their excrements in every corner, in chimneys, studies, coal-houses and cellars. Rude, rough, whoremongers; vaine, empty, careless.'

status of women.* He had learnt in battle the practical virtue of toleration. As he wrote to the Speaker after the successful siege of Bristol:

> Presbyterians, Independents, all had here the same spirit of faith and prayer . . . They agree here, know no names of difference; pity it should be otherwise anywhere. All that believe have the real unity, which is most glorious because inward and spiritual . . . As for being united in forms, commonly called uniformity, every Christian will for peace sake study and do as far as conscience will permit; and from brethren, in things of the mind, we look for no compulsion but that of light and reason.

Or again, in 1650:

> Truly, I think that he that prays best will fight best. I had rather that Mahometanism were permitted amongst us than that one of God's children should be persecuted.

In 1656, against strenuous opposition from nearly every political and religious faction, he brought the Jews back to England, from which they had been banned since the days of Edward I. Thus in England, at least, the seeds of religious toleration, first sown by his heroine Queen Elizabeth, were replanted; but to do the same in Ireland proved beyond his power, or imagination.†

Where Cromwell was less successful was in achieving a stable relationship between the executive and Parliament. His aim from first to last was a broad consensus. God, he believed, spoke through his Englishmen; but the word of God was many-sided: it was heard by men with different nuances, all true in their fashion. Cromwell thus favoured coalitions; and he had a brilliant talent for devising them and holding them together. He was a good listener, very slow to make up his mind until he had heard all points of view; he liked to dine with men of all factions, sitting silent while they harangued. He was an idealist, who wanted politics to disappear; he undertook the government merely 'until God may fit the people for such a thing'. But in the meantime he worked cheerfully along the traditional lines of English pragmatism. As Abbot said: 'Cromwell proceeds with strange dexterity towards the reconciling all kinds of persons, and chooses those of all parties whose abilities are most eminent.'

* He inherited a radical tradition aimed at improving and protecting the status of women, especially against wife-beating, which goes back to *Jane Anger her protection for women* (1589) and William Heale's *Apologie for Women* (1609). Some of the Independents allowed women to share in the governance of the Church, and at this time they preached, took part in demonstrations, and presented petitions. See Christopher Hill: *Intellectual Origins of the English Revolution* (1965), p. 275.

† See Appendix II: Cromwell and Ireland.

But it was one thing to do this in practice, quite another to create a permanent, theoretical framework for it. Parliament, from being the popular underpinning of the Throne in Plantagenet and Tudor times, had passed into opposition in 1611. On the existing suffrage it was inevitably dominated by the country gentry, who were in practice interested in nothing except the security of private property and low taxes; they opposed any kind of reform, and were really opposed to the principle of government itself. The Commons controlled its own composition by its arbitrary right to debar and expel members. Cromwell took over this right in an endeavour to produce a workable House. He experimented with the suffrage. He tried an appointed Parliament. He increased the number of county seats, then the number of borough seats. But the same MPs always turned up; and, lacking the constructive genius of a Pym to lead them, proved an irritating obstacle to the grand Cromwellian vision. They would not countenance the splendid scheme of law reform upon which Cromwell had set his heart; as he remarked, angrily, the moment you raise the topic of law reform, MPs accuse you of threatening the sacred shibboleth of property. He recognised, grimly, that Parliament is the custodian not so much of liberty, as of the existing division of property, as indeed it still is today. He could, of course, have transformed the composition of the House by a radical extension of the suffrage. But he feared that a popular vote would turn royalist: 'If the common vote of the giddy multitude must rule the whole, how quickly would their own interest, peace and safety be dashed and broken.' So he turned instead to the piecemeal system of parliamentary management which, over the next 100 years, was to produce political stability in England. Cromwell was the first Whig, the connecting link between the system of Burleigh and the system of Walpole.

But he was also, as he ardently wished to be, the reincarnation of Elizabeth, her true heir. He moved only reluctantly to the abolition of the throne, and quickly returned to the view that the government must have 'somewhat of monarchical in it'. He did not want to be a king; he agreed, for want of a better solution, to be a 'single person'. Under him, the system worked well: 'All things here are in a calme, expecting what his highness will settle, and what lawes he will make. All stand bare to him.' If Cromwell had lived another year, he would almost certainly have accepted the Crown, and set up his own dynasty; this would have been endorsed by the royalists, who were attached to a *de facto* monarchy, not the Stuarts. But Cromwell was carried off by the traditional English killer, bronchitis. His son was given the office without the magic of hereditary right, and soon resigned it. Cromwell might have chosen Lambert, the ablest of his subordinates, as his heir (Ireton and Blake were dead); but the two men had quarrelled, and Lambert therefore lacked the authority and status for the delicate job of reconciling army and parliamentary interests. Thus the Scottish command of the army was able to reimpose the Stuarts, to mixed feelings on the part of the English public. The Cromwellian vision faded, and the pace of English development slowed down. Had it persisted and, as was inevitable, accelerated still further, the industrial revolution would have occurred at the end of the seventeenth century, and we would now be living in a twenty-first century world, with all its wonders and terrors.

As it was, the Restoration put the clock half-back, in a thoroughly English spirit of muddled compromise. Some Commonwealth men were victimised; others prospered; it was all a lottery. Charles venomously settled some personal scores; but he was too weak, or too lazy, to undertake the effort and danger of a general proscription, so he merely disinterred the corpses of some great English patriots. He had his shoulder of mutton and his whore. Some of the reforms were kept; others scrapped, to reappear again in different guises. Prerogative power could not be resurrected, but the King, in some mysterious way, was still expected to rule. The election of the Cavalier Parliament marked an enormous swing to the Right; but within a year the majority melted away, and the last Stuart kings always faced opposition parliaments. In the Earl of Danby* Charles eventually found a parliamentary manager with some of the Cromwellian skills; but the amount of patronage at his disposal was too limited to maintain a working majority, so Charles dispensed with Parliament and ruled through foreign subsidies.

The return to Continentalism was bound to be fatal. The Cromwellian triumphs were too recent to allow public acquiescence in Charles's mismanagement of external affairs: the disastrous war with the Dutch, the abrupt fall in the level of protection the State afforded to English trading and overseas interests, Charles's acquiescence, indeed active support, in the rise of a menacing French power on the Continent; the smell of secret treaties, so destructive of English interests that their terms could not be disclosed even to the King's closest advisers. More-

* This man illustrates the confusion caused by the habit English politicians have of changing their names as they rise up the social scale. He began political life as Sir Thomas Osborne, Bart., then changed to Viscount Latimer (1673), Earl of Danby (1674), Marquess of Carmarthen (1689) and finally Duke of Leeds (1694). Even English *cognoscenti* sometimes get muddled. Thus, Fred Robinson, later Viscount Goderich, and later still Marquess of Ripon, foxed even his own Prime Minister, Canning. In August 1827 he wrote a paper on the situation in Portugal and told his secretary: 'Send it to Goderich and Robinson.' [W.D. Jones: *Prosperity Robinson* (London, 1967), p. 152].

over, Continentalism inevitably reopened the Catholic question – the two were inseparable. The last Stuart kings were not Catholics in a meaningful sense; it is hard to accept they even believed in God (though James was a Mariolater). But they needed Catholicism as the only dependable political underpinning of a regal State. Sooner or later they were bound to seek to restore it, to bind England to a Continental system so closely that a revolt against the throne would evoke a response from the entire European community. So the last years of Charles were a continuum with the reign of his brother. As he aged, 'Old Rowley' dropped his mask of tolerance, and became increasingly suspicious, secretive and vindictive. His court acquired the seaminess of his grandfather's dotage, as Evelyn noted just before Charles's death:

> . . . unexpressible luxury and profaneness, gaming and all dissolution, and as it were total forgetfulness of God . . . The King sitting and toying with his concubines, Portsmouth, Cleveland and Mazarin, etc.; a French boy singing love-songs in that glorious gallery, whilst about twenty of the great courtiers and other dissolute persons were at basset round a large table, a bank of at least £2,000 in gold before them . . . six days after, all was dust.

Charles prophesied that his bizarre and eccentric brother would not last as King for more than four years, and he was right. But there was no change of policy, merely an increased haste to execute it, and an aggravated indifference to public opinion. After all, it was Charles who appointed Jeffreys Lord Chief Justice, with a mandate to smash the courts; and the current against the Stuart monarchy set in years before James got to the throne. But James had his own unique contribution to make to the long catalogue of Stuart folly and stupidity. His rare unconsciousness of the minds and feelings of others was reflected even in love-making, for he sought to seduce Miss Hamilton by 'giving her accounts of broken legs and arms, dislocated shoulders, and other curious and entertaining adventures'. Where Charles had despised the English, James actively hated them: 'He knew the English people,' he said, 'and they could not be held to their duty by fair treatment.' And what was their duty? James spelt it out: 'to follow his wishes blindly, and to own an attachment to his interests that was without any qualification or reserve whatsoever.' Not surprisingly, even the English recusants, to a man, declined to join such an ill-found ship. James was above taking realistic measures to secure his throne (though by no means averse to judicial murder). An exasperated Jeffreys noted: 'The Virgin Mary is to do all.' In no time he was back in exile at the French court, where the verdict was: 'When you listen to him you realise why he is here.'

So the English wearily set about the business, as they had done in the fifteenth century, of finding a royal line which would suit them. Parliament, as it were, was an employer advertising a top vacancy; but the number of candidates was small, and the qualifications (which had nothing to do with worth, brains or talent) were in some respects strict. It was partly a question of geography. How could the focus of monarchical interest be firmly anchored, as under the Tudors, in the south-east, instead of drifting away, as under the Stuarts it invariably did, to the Celtic fringe and the 'brute and beastly' north and west? William III went a long way towards solving this problem by involving England in a series of wars against the French which could be turned to commercial advantage; and by tying up court and government, almost inextricably, with the business of the City, now in the full throes of the financial revolution which the Cromwellian tax-system had detonated. The throne acquired a stake in English commercial prosperity, thus reverting to the prudential methods of Edward IV, Henry VII, and his shrewd granddaughter. But to align the court with the Home Counties was not enough: what was also required was an absolute guarantee of a formalised Protestant succession, underwritten by statute, to rule out once and for all a drift back into the Continental system. Mary did not despise the English (like her father), though she preferred the Dutch; but she had no child. Anne rejoiced in her English mother and namesake, Anne Hyde (whose only recorded accomplishment was the ability to drain a quart-mug of beer without drawing breath), and prefaced her speeches, in a conscious echo of Elizabeth, with 'As I know myself to be entirely English...' But she, too, despite heroic efforts, could not produce a child who lived. So the prize went to the Hanoverians, who had an absolute vested interest in Protestantism, who put the pursuit and spending of money above any political principle whatsoever, who were devoted to the City, and who were, moreover, prolific. So Continentalism was scotched, to raise a feeble head only in 1715 and 1745, and English policy was able to oscillate safely between the Little Englandism of Harley, Walpole and Bute, and the Cromwellian tradition of Big Englandism upheld by Stanhope, Carteret and Chatham. Moreover, there were over 50 people with better hereditary claims to the throne than George I. It was the end of the Divine Right of Kings, and the monarchy was placed where the English wanted it; on a business-like no-nonsense basis of convenience.

It was the end of much else too. The civil wars and their aftermath, with their achievements and disappointments, produced an earthquake in English society, completed the work of the Reformation, destroyed the old certitudes and assumptions, and brought into question habits and attitudes which had been taken for granted almost since the beginning of time. In his vast compilations, Aubrey noted this great watershed: the end of the old jousting court in Whitehall, the fact that gentlemen now travelled in carriages instead of on horseback, the replacement of their armed retinues by mere footmen, the disappearance of a multitude of antique customs and beliefs:

> Civill warres comeing on have putt out all these Rites, or customs quite out of fashion. Warres do not only extinguish Religion and Lawes but Superstition; and no Suffimen is a greater fugator of Phantosmes, than Gunpowder.

Society ceased in great measure to be patriarchal. It was no accident that Filmer, the ideologist of royal absolutism, had compared the King to the father of the family. Now the children had revolted not merely against the King but against their own parents. Aubrey did not regret this aspect of the old days:

> The child perfectly loathed the sight of his parent, as the slave his Torturer. Gentlemen of 30 or 40 years old, fitt for any employment in the common wealth, were to stand like great mutes and fools bareheaded before their Parents; and the Daughters (grown woemen) were to stand at the Cupboards side during the whole time of the proud mothers visit, unless (as the fashion was) 'twas desired that leave (forsooth) should be given to them to kneele upon cushions brought them by the servingman, after they had done sufficient penance standing . . . fathers and mothers slash't their daughters in the time of that Besome discipline when they were perfect woemen.

Thus revolution introduced the concept of the generation gap, which has never since been bridged, and authority in all forms was increasingly subjected to the secular tests of reason. As Lord Halifax put it:

> The liberty of the late times gave men so much light, and diffused it so universally among the people that they are not now to be dealt with as they might have been in an age of less inquiry . . . Understandings . . . are grown less humble than they were in former times . . . the world is grown saucy, and expecteth reasons, and good ones too . . .

It had also grown more cynical and corrupt. In the hearts of the sophisticated, science-orientated revolutionaries of the Commonwealth, there was a spark of pure childlike innocence, the combination of religious faith and a naïve, secular idealism, the desire to do good on earth to their fellow men. Some Englishmen – a minority, no doubt, but a strong and purposeful one – really believed that they had a mission to accomplish, and that what they did should shine forth for all mankind to see. For once in English history, the convenient hypocrisies and fictions were cast aside. It was proudly proclaimed by the Commonwealth that the deeds of the court which tried and sentenced Charles Stuart would 'live and remain upon the record to the perpetual honour of the English state, who took no dark and doubtful way, but went in the open and plain path of Justice, Reason, Law and Religion'. Or, as Milton wrote: 'God has inspired the English to be the first of mankind who have not hesitated to judge and condemn their king.'* It took an enviable self-confidence to say such things, and in the confused decades which followed the Restoration, in the empirical search for stability at the cost of almost any principle, that self-confidence evaporated, and the English turned again to draw on their bottomless wells of hypocrisy. The settlement of 1689 knocked the heart out of religion as a thing men would die for. The history of those years was skilfully rewritten, shrouded in obscurity and double-think, made to seem inevitable, a dispensation of a Whig providence and not the violent action of real human beings. Bishop Hooper of Bath and Wells epitomised the new revisionist spirit of cynicism:

> The Revolution [of 1688] was not to be boasted of, to make a precedent, but we ought to throw a mantle over it, and rather call it a vacancy or abdication; and the Original Compact were two very dangerous words, not to be mentioned without a great deal of caution; that they who examined the Revolution too nicely were no friends to it, for at that rate the crown would roll like a ball, and never be fixed.

Thus a revolutionary State, in which the monarch was a mere convenience, was cunningly legitimised in the interests of the propertied classes who had benefited from it. James II had not been expelled; he had 'made off', as the police say; England now had a perfect constitution, and force was never again to be used to obtain redress except in circumstances so remote as to be unimaginable. The new-speak doctrine was

* The trial was fully reported in six licensed newspapers, two of which issued supplements containing the verbatim record (the three unlicensed royalist newspapers refused to report it). The only form of censorship was to cut out references to differences of opinion among the judges, which emerged at the regicide trials in 1660. Contempt of court was by no means strictly enforced. When Bradshaw said the King was charged in the name of the people of England, Lady Fairfax, sitting masked in the gallery, shouted: 'Not half, not a quarter of the people of England. Oliver Cromwell is a traitor.' When she continued her noise, the guards levelled their muskets at the gallery ('By this time,' said one sitting there, 'we were very hush'), but she was allowed to slip away unpunished. Fairfax's failure, after the defeat of the King, to play the role in the political debates (notably at Putney) which his position demanded, is attributed to his stammer; but Lady Fairfax's strident views must also have been a factor. For the trial, see C.V. Wedgwood's essay in *The English Civil War and After, 1642–58*, edited by R.H. Parry (London, 1970).

unashamedly exposed to Parliament by Walpole, the practical ideologist of constitutional stability:

> Resistance is no where enacted to be legal but subjected to all the laws now in being to the greatest penalties; tis what is not, cannot, nor ought ever to be described, or affirmed in any positive law, to be excusable. When, and upon what never to be expected occasions, it may be exercised, no man can foresee; and ought never to be thought of but when an utter subversion of the realm threaten the whole frame of a constitution and no redress can otherwise be hoped for.

But of course political stability could not be bought merely by a foggy and meaningless ideology, a kind of constitutional opium. Something more practical was required. By accident, the English managing classes found the answer: bribery. There was, of course, nothing new in the principle – there is rarely anything entirely new in English public life. English governments had always sought survival by permitting enough powerful individuals and families a hand in the till to guarantee an adequate basis of support. With ups and downs, the system had survived until the end of the sixteenth century, but even then it required a fine sense of balance, of the kind Elizabeth and the Cecils possessed. With the rise of the gentry class, and their absolute domination of the Commons, too many hands were stretched out, and the Government still was too small to satisfy enough of them to guarantee a parliamentary majority. The enormous expansion of government activities under the Protectorate pointed the way to a possible solution, as did its success in tapping for tax purposes a rapidly growing national income. The trend was reinforced by the long wars with France which created thousands of new government jobs, civil and military, not only in central government but in the new customs and excise services which financed it. Jobs directly under the control of the court fell in numbers both absolutely (from about 1,500 under Charles II to about 1,000 under George I) and relatively; but the number controlled by ministers rose many-fold. There were now, to vary the metaphor, many more, if smaller, slices of an infinitely larger cake. Or, as the Duke of Newcastle put it, 'enough pasture to feed all the beasts'. A seat in the Commons now became not merely a mark of status but a definite commercial property, whose sale could be advertised in the newspapers. By the turn of the century, it was noted: 'Nothing is now more common than for members first to buy [the electors'] voice and then sell their votes, which are grown very good merchandise at court'. In the seventeenth century, no government succeeded in winning a parliamentary election (except the illusory victory of 1661); in the eighteenth century no government lost one.

The spoils system inevitably reduced the popular element of participation in choosing parliaments. In the early seventeenth century, opposition parliaments had tended to enlarge the suffrage by tinkering with the borough corporations; after 1688 the process was reversed, except in the few popular constituencies, where the gentry had been driven from the field by urban spread. And fewer and fewer electors could in practice cast their votes freely – perhaps only 5 per cent out of 200,000 by 1750. There was too much money at stake to allow the giddy multitude to sway results. Corporations themselves aided the process by restricting membership: the fewer the voters, the greater the bribes for those who still had the vote. The cost of a borough seat rose rapidly in the eighteenth century, from an average of £1,500 to £5,000; and when the life of Parliament was extended to seven years, security of tenure thus improved, and the price rose still higher. So huge was the cost of a contested election that the number of contests dropped sharply; and one fight might settle the fate of a seat for a generation.

The elements of the new system of government were created in 1688; but it took more than 30 years, and the manipulative skill of a Walpole, before it could be made to work smoothly. By a process of trial and error, he completed the debauching of the political nation. He believed in corruption with a passion and intensity which other men brought to religion. His own depredations were vast, continuous and highly professional.* But he ensured that others got their due strictly in accordance with the influence they had to offer, beginning with the King, his two mistresses (one short and fat, the other tall and thin), his German advisers, Bothmar and Bernstorff, his Turkish servants, Mohammed and Mustapha, and working steadily downwards. Government became a joint-stock company in which men invested in the hope of dividends. Sometimes it was possible to draw up a neat balance-sheet. Thus, the Duke of Chandos spent £14,000 in four years bribing the King's German ministers and one of his mistresses; in return he got a peerage for his father, the Deanery of Carlisle for his brother, and a court position for his son. But not everyone entered the game with the same objects. Some, like Walpole, sought power for money (as well as for its own sake); others,

* In the four years 1714–17, for instance, £109,208 4s 9d passed through Walpole's hands, of which £61,778 14s 9d was invested, the rest spent. He grabbed (not necessarily illegally) much larger sums when he became the chief minister. He must have spent about £250,000 on Houghton, and his collection of paintings alone was valued at nearly £35,000 at his death. It cost him about £200 a day to entertain his friends at Houghton; in 1733, for instance, the bill from one of his wine-merchants came to £1,118 12s 10d. As the estate he inherited was worth only £2,000 a year, and encumbered, the overwhelming bulk of Walpole's vast expenditure must have come from public funds. See J.H. Plumb: *Sir Robert Walpole: The Making of a Statesman* (London, 1956), and *The King's Minister* (London, 1960).

like the Duke of Newcastle, spent money to acquire power. High office left men very much poorer, as well as very much richer.* Moreover, the system only worked smoothly when the political temperature was low; if war or principles raised their ugly heads, it tended to break down. About 150 backwoods squires stood outside the circle of corruption and retained their political freedom. Silent and acquiescent as a rule, they could become unpredictable in moments of crisis. Boswell recorded some shrewd remarks by an old parliamentary hand, almost certainly Burke:

> The House of Commons is a mixed body. It is a mass by no means pure; but neither is it wholly corrupt, though there is a large proportion of corruption in it. There are many Members who generally go with the Minister, who will not go all lengths. There are many honest well-meaning country gentlemen who are in parliament only to keep up the consequences of their families. Upon most of these a good speech will have influence.

Just so. As Walpole perceived, to manage the Commons successfully, one must remain a member of it; the temptation to take a peerage as the reward of office was great; most leading ministers, from the Cecils to Chatham, succumbed to it, to their cost; Walpole preferred to enter the Elysian Fields only on final retirement. But even he, in the end, could not avoid the terrible uncertainties of war. Moreover, the growth of political stability made opposition respectable, safe, even (in the long run) profitable. So long as it grouped itself around the heir apparent, it remained personal and factional and hence no threat to the system. But George III did not have an heir until he was already on the throne. For want of a personal focus, opposition began to move into the dangerous waters of ideology. The system thus contained the seeds of its own destruction.

Meanwhile, what had happened to the divine English mission? Were the English still the chosen race? It was beginning to look increasingly doubtful. For one thing, the growth of historical studies cast a more accurate and less sensational light on English origins. The myth did not

* In 1797, after half a century of running seven seats (at £3,000 apiece), Lord Eliot calculated he had lost by his operations. Pitt the Younger left debts of £40,000. Lord Liverpool saw his fortune shrink in 33 years of office. Canning spent £60,000, out of his wife's fortune of £100,000, on politics. On the other hand, Palmerston, who spent over 40 years in office, needed his official salary to keep solvent [see Jasper Ridley: *Palmerston*, (1970)]. The last man to make a suspect fortune from office in this country was Lloyd George. Nowadays, however, top politicians can make vast sums by selling their memoirs, a fashion set by Churchill. In recent years, figures of £100,000, £240,000 and £250,000 have been paid for world rights of prime ministerial *opera*; £50,000 went to a mere junior minister. The system, to my mind, is at least partly abusive, since it is based on the convention that politicians allow each other access to official papers while denying it to the public, thus multiplying by many times the commercial value of what they write.

survive the inspection of professional antiquaries. For another, the English were now liable to be reminded, unceremoniously, that their race, far from possessing the purity which the apostolic assignment might suppose, was in all essentials mongrel. In 1701, exasperated by the filthy manner in which the English treated foreigners, especially Dutchmen, Daniel Defoe rattled off a brilliant piece of doggerel, *The True-Born Englishman*, which achieved enormous popularity. The English, he pointed out, had nothing to be proud of in their origins. They were the 'barbarous offspring' of the 'dregs of armies', an 'amphibious ill-born mob', the progeny of repeated invasions by innumerable peoples:

> A Turkish horse can show more history
> To prove his well-descended family.
> . . . These are the heroes that despise the Dutch
> And rail at new-come foreigners so much,
> Forgetting that themselves are all derived
> From the most scoundrel race that ever lived;

Moreover, the process was still going on;

> We have been Europe's sink, the jakes where she
> Voids all her offal outcast progeny

The truth, concluded Defoe, was that a true-born Englishman was 'a man akin to all the universe'.

The success of this sally indicated that some Englishmen, at least, were learning to laugh at their racial pretensions; though there was no observable decline in their active hostility to foreigners, either at home or abroad, as the ludicrous affair of Captain Jenkins' Ear showed. Indeed the circumstances of this war reflected the degeneration of a sense of mission from one in which religious duty was paramount, and commercial advantages merely secondary, to one wholly inspired by secular and materialistic motives. The voice of the new mood was undoubtedly James Thomson's. He rejoiced not only in the beauty of Britannia ('Heavens! what a goodly prospect spreads around') but more emphatically, and repeatedly, in England's world-wide commercial mission: not only in 'Rule Britannia', the song from his masque *Alfred* (1740), which became a second national anthem, but in dozens of poems. The agency in England's role has changed from a Biblical God to a mysterious, unnamed providence:

> For Britons, chief,
> It was reserv'd, with star-directed Prow,
> To dare the middle Deep, and drive assur'd
> To distant Nations thro' the painless Main
>
> *Liberty*, 1736

England, in fact, has already become a kind of economic policeman, combining moral authority, and business wisdom:

> And as you ride sublimely round the World,
> Make every Vessel stoop, make every State
> At once their Welfare and their Duty know
> This is your Glory; this your Wisdom; this
> The native Power for which you were design'd
> By Fate

Thomson, with a City accountant's eye, stresses 'this unexpensive Power' and the fact that trade, not territory, is England's object:

> . . . unencumber'd with the Bulk immense
> Of Conquest, whence huge Empire rose, and fell.
>
> *Britannia*, 1729

This bombastic tone is somehow more offensive than the equally emphatic, but naïve, credulous and hopeful tone of Milton and Foxe.* As God faded into the background, profit eased itself on to the shoulders of English racism. During the late seventeenth century, the inspiration of English overseas ventures was transformed. Of course, trade had always been an object. The younger Hakluyt, who believed in the divine mission as strongly as any Elizabethan Protestant, argued in his *Discourse of Western Planting*, written at Ralegh's request, that colonisation of North America would solve unemployment by siphoning off surplus population, make England independent of other suppliers of raw materials, especially timber, and increase her export markets for finished goods. But it would also, and more importantly, be a means to bring the Indians 'to civility'; it would enable England to break free economically from a corrupt and incorrigible Europe; it would be 'a place of safety . . . if change of religion or civil wars should happen in this realm'. North America 'God hath reserved to be reduced unto Christian civility by the English nation'. Ralegh, who liked the Indians (it was reciprocated), thought the monstrous cruelty of the Spanish in America had earned them the vengeance of God, and that it was England's duty to take over responsibility for the entire continent; had he been permitted to carry out his great 'western design', there is at

* Horace Walpole shrewdly saw through the hypocrisy with which the English invested 'commerce' and 'trade' as a justification for aggression and war. He wrote to Sir Horace Mann (26 May 1762): 'I am a bad Englishman, because I think the advantages of commerce are dearly bought for some by the lives of many more . . . every age has some ostentatious system to excuse the havoc it commits. Conquest, honour, chivalry, religion, balance of power, commerce, no matter what, mankind must bleed, and take a term for a reason.'

least a possibility that he would have saved vast territories from senseless plunder and degradation, and that Latin America would be a much happier place today. Farther north, too, colonies were established not only to further religious freedom, as in Massachusetts, but to embody advanced political ideas. The Virginia Company was an attempt to carry out some of Ralegh's notions; Sir Edwin Sandys, its Treasurer and James I's peculiar object of hatred, believed all kings had originally been elected. Like the Pilgrims, he introduced secret balloting (which Charles I forbade in all colonies), and James correctly described the company as 'but a seminary to a seditious parliament'. Many leading parliamentary radicals were involved in the Providence Island venture, and there can be no doubt that political experiments in America stiffened the reformist spirit of the Long Parliament. Embedded in the chauvinism of Cromwell's foreign policy was a powerful streak of idealism: Milton, in drawing up England's official case for war against Spain, gives the bestial behaviour of the *conquistadores* as a prime justification.

Yet the colonies, right from the start, were a perfect mirror of English virtues and vices, an extraordinary mixture of cupidity and idealism, of legalism and glaring anomalies, devoid of any logic or system, and (most characteristically) promoting glaring innovation under the guise of tradition. Where Hakluyt had called for colonies to water the parched minds of the heathen with 'the swete and lively liquor of the Gospel', Bacon rightly pointed out that the actual object was 'but gold and silver and temporal profit'. Massachusetts persecuted Quakers and witches; but it rarely used the death penalty, it did not imprison for debt, it permitted civil marriage and it raised the legal status of women. Recent custom was force-fed to produce antiquity, so that as early as 1652 Barbados petitioned to keep its assembly, as it was 'the ancient and usual custom here'. Theoretically, Pennsylvania was a private estate, a proprietary colony: but it had a free assembly, the object of its penal code was reformation rather than retribution, and its record with the Indians was almost unsullied; its founder even proposed a league of nations and a sovereign European parliament. When the English brought Negro slavery to the Caribbean in the wake of Cromwell's annexation of Jamaica, they argued that the Indians would die if forced to do heavy plantation work, and that for the virile Africans transportation was the means of salvation in the next world, and a modest comfort in this. This welter of muddled thinking produced some curious monsters. In the Restoration period, attempts were made to set up semi-feudal societies, as the crusaders had done in twelfth-century Syria. One constitution, drawn up by Locke in 1669, provided for county divisions, each subdivided into seignories, owned by the proprietors, and baronies, owned by local nobles called caciques and landgraves. Land would also be held by freemen, who would elect members to the lower house of Parliament, while nobles and proprietors would form the other estates.

But most colonies were founded on a form of contract between rulers and ruled, modelled on the original social contract which, people believed, had been drawn up in Anglo-Saxon times or even earlier. They thus possessed written constitutions, of a sort, but drawn up in accordance with current commercial practice: hence, if England was a traditionalist agrarian society which eventually became a commercial one, America was a commercial society *ab initio* – and therein lies a very significant difference. Moreover, it was also, from the start, highly legalistic: its ideological origins date from a period when English parliamentary lawyers were using the common law to rewrite history and carry through a constitutional revolution. There was, with important differences, a parallel development on both sides of the Atlantic. The colonies welcomed the Commonwealth and Protectorate; they were suspicious of the Restoration, and became actively hostile when James II, following in the steps of Richelieu, started to annul charters, and draw the colonies into a single royal dominion. In 1688 when news of the English revolution reached America, the New Englanders arrested their royal governors, claiming the right of constitutional resistance to an illegal regime, and petitioned Parliament to legalise their acts *ex post facto*. In a curious way, their behaviour mirrored almost exactly what the Britons had done in 410, and was ominous for the future of what men were already beginning to call the British Empire.

England lost the American colonies because Englishmen had already lost their belief in the divine mission. The mission was dynamic; it demanded purposeful efforts towards definite ends; it presupposed an objective, and a programme of means to attain it; it implied a society in motion, hurtling ever faster towards a millennium; it raised huge questions and demanded clear answers. Why are we here? What task has providence give us? What, then, must we do to discharge it? God was not a policeman, as in the medieval world, the ultimate resort for the forces of terrestrial order, but an imperious and scrutable taskmaster, the master builder of a vast and urgent work of construction, issuing well-defined commands to his servants. Such a belief is incompatible with stability: and the English, after a century of unrest and experiment, wanted stability, or were presumed to do so – or, in any event, were given it. But stability has to be paid for. The price, in the first instance, was the abandonment of the mission to act as divine agents in a world reformation; or, rather, to down-grade the mission morally to the mere commercial purpose of expanding world trade, something which, by its nature, was self-generating. But, secondly, a restless nation could only be induced, in practice, to abdicate from its role as God's people, and thus forfeit its guarantee of eternal felicity, by a substantial *quid pro quo* on earth. The nation, or at least the political nation, had to be bribed into quiescence; and this was done. As Walpole put it, *quiete non movere*; sleeping dogs would safely lie, if they were well fed first.

It is a sad comment on human societies that they can usually be persuaded to accept bribery as a system of government, provided the circle of corruption is wide enough. As we have seen, this became possible in the early eighteenth century with the expansion of the State. But if the circle was large, it still had very definite limits, and excluded whole categories of people: one might argue that it broke down at the end of the eighteenth century because, with the growth of population, the area of exclusion became intolerably large. But it also excluded whole nations. Thanks to the Act of Union, Walpole found it desirable to bring Scotland into the system, for the votes it exercised in both Houses of Parliament were valuable and worth buying. Thus Scotland, or at least the lowlands, became a contented and increasingly prosperous member of the community; indeed most Englishmen argued fiercely that the Scots got far more than their fair share of the spoils. But Ireland was rigorously excluded. Its own parliament was emasculated by the provisions of Poyning's Law, which forbade Irish legislation without the permission of the English Privy Council; and, of course, it had no votes to offer at Westminster. Hence Irish patronage was reserved, very largely, for Englishmen, in both Church and State; and yet another governing class was superimposed on the geological layers of injustice which the Celtic Irish carried. Rich and poor, Catholic or Protestant, the Irish resented the unfairness of it all. But though they might cry to heaven for vengeance, they could not get it on earth, for Ireland lay under the shadows of English guns. With America it was a different matter. America, too, had no votes to deliver at Westminster; she, too, was therefore very largely excluded from the spoils system; but America was 3,000 miles, and six weeks, away from the sources of English authority. This made a crucial difference, especially when, for a brief moment, England lost absolute control of the sea.

Would America have remained loyal if enough Americans had been given a share of the spoils? This was asked (though not quite in these terms) at the time. Jefferson wrote in the first draft of the Declaration of Independence: 'We might have been a great and free people together,' but was forced by his colleagues to delete the phrase. The question has been asked many times since. In 1900, Lord Rosebery, the first man to popularise the phrase 'the British Commonwealth', told the Glasgow students in his Rectorial Address that 'but for a small incident', America would in time have become the senior partner in a vast oceanic dominion, the seat of government would have been 'moved solemnly across the Atlantic, and Britain would have become the historical shrine and the European outpost of the world empire'. The 'small incident' was Pitt's acceptance of a peerage which, said Rosebery, deprived him of 'his sanity and his authority' and thus disabled him from preventing the breach.

History is, indeed, composed of small incidents; but the difficulty was more serious than Rosebery supposed. It might, of course, have been solved if Americans had been accorded some form of imperial representation, either in America or at Westminster, for this would automatically have earned them an appropriate quota of the spoils. But in both cases there were insuperable constitutional objections. The whole theory and practice of English stability rested on the assumption that the English constitution had reached its final form, and had achieved balanced perfection. To set up an American parliament with limited powers (for instance, over taxation) would mean a division of sovereignty which would make the constitution unworkable. As Blackstone, the arbiter of constitutional theory in the eighteenth century (as Coke was in the seventeenth), insisted: 'there is and must be in all [forms of government] a supreme, irresistible, absolute, uncontrolled authority, in which the *jura summa imperii*, or the rights of sovereignty, reside'. In England this was Parliament, whose actions 'no power on earth can undoe'. Such absolute sovereignty was the only alternative to the horrors of a written constitution (which, of course, the Americans did not fear). Power to tax 'is a necessary part of every supreme legislative authority'; and, therefore, if Parliament 'have not that power over America they have none, and then America is at once a kingdom of itself'.

So much for a local parliament in America. What of representation in Westminster? Here the objections were still more weighty. America was a series of joint-stock companies; its local directors, or delegates, in its colonial assemblies were strictly mandated by those who appointed them: such control by the electorate is central to the whole theory of American politics, then as now, and helps to explain why the United States has never developed ideological parties. But English theory was entirely different; it had to be different in order to justify the 'perfect constitution', which in fact was an illogical shambles. English Members of Parliament were not delegates of their voters; they were trustees of a nation, indeed of an empire. How could an MP be delegated by a close or a pocket borough? No: each and every MP was himself a custodian of the public interest, acting from his own judgment and conscience (as Burke made laboriously clear to the electors of Bristol). The Member for Old Sarum, which had no actual voters at all, was just as capable (indeed more capable, since disinterested himself) of representing the true interests of the Americans as a man put forward by a demanding rabble in Massachusetts. When Americans argued that it was intolerable that flourishing cities like Boston and Philadelphia should have no voice in Westminster, the English establishment retorted that neither did Manchester, Birmingham or Sheffield. But this cut absolutely no ice in America. The raucous Boston demagogue James Otis simply replied: 'If those now so considerable places are not represented, they ought to be.' The truth is, the Americans could not be accorded constitutional rights without granting them to the vast, unrepresented multitudes in England itself; this would make the spoils system, and so the 'balanced constitution', unworkable, and bring about a return to anarchy. The English ruling class had to choose between stability and empire; and, much as they valued both, they chose stability, as they were again to do in the mid-twentieth century.

Thus the axis of attack deployed by the American independence movement sprang from a radical, left-wing critique of the English constitution. The Americans conceded that the concept was sound; but, as Englishmen had argued throughout history, it had somehow got perverted, and a reform – a return to its pristine and perfect origins – was urgently needed. For one thing, it was supposed to guarantee property as sacrosanct; how could it be said to do this, when the goods of Americans could be seized by King's officers over whom they had no control? As Massachusetts said to Chatham in 1768: 'That grand principle in nature, "that what a man hath honestly acquired is absolutely and uncontrollably his own", this principle is established as a fundamental rule in the British Constitution.' The Americans had a splendid precedent, writ large in history: they went back to 1640. 'What we did,' said Jefferson, 'was with the help of Rushworth, whom we rummaged over for revolutionary precedents of those days.' The United States was thus the posthumous child of the Long Parliament.*

* Non-English influences on the American rebels were of little importance. In 1774 John Adams cited Plato as an advocate of equality and self-government; but he had not then read the *Republic*; when he finally did so, he was so shocked he thought it must be a satire! Wilkes's successful skirmishings with the Commons played a notable part in educating the Americans in popular opposition; so did the writings of radicals like Priestley, and, above all, Paine. It was Paine, in *Common Sense* (1776) who finally destroyed, for the Americans, the mystique of the English constitution. The concept of balance, he said, was nonsense; what liberty existed in England was 'wholly owing to the constitution of the people and not to the constitution of the government'. Where was the

Indeed, the Americans of the 1760s and 1770s, like the English gentry of 1640, were armed with a ramifying, circumstantial and (to them at least) utterly convincing conspiracy theory. They drew heavily for inspiration from both right- and left-wing critics of the Walpolean system, as perpetuated under Bute and George III. On the Left they read and admired the *Independent Whig*, the Letters of Cato in the *London Journal*, Bishop Benjamin Hoadley's rejection of the theory of submission, Molesworth's description of how the free state of Denmark degenerated into absolutism. On the Right they rejoiced in the scathing assaults on corruption conducted by Bolingbroke's *Craftsman*. They drew heavily on the Whig theory put forward by the Huguenot exile Thoyras, in his *Histoire d'Angleterre* (translated 1725–31) which warned of the 'formed design' of the Tories to restore Stuart absolutism. After 1763, when the taxation issue became acute, the menace appeared to be taking definite shape, albeit under a Hanoverian. Sir Lewis Namier may have proved from the documents that George III was not operating a personal Tory government; contemporary Americans did not agree with him.

To be sure, critics of the English system were taken far more seriously in America than they were in their own country. Yet many well informed Englishmen also believed in the conspiracy theory. Liberty appeared to be on the retreat everywhere in the world; barbarous tyrannies were growing in strength and numbers daily, in Asia and Africa as well as Europe. Burke warned in his *Thoughts on the Present Discontents* that 'a certain set of intriguing men . . . to secure to the court the unlimited and uncontrolled use of its own vast influence under the sole direction of its own private favour [were pursuing] a scheme for undermining all the foundations of our freedom'. Carried across the Atlantic, the conspiracy theory assumed weird and wonderful forms, into which historical myth, race prejudice and current events all fitted with astonishing aptness. A relatively harmless proposal to appoint bishops in America was, said John Adams, a plan to impose 'the canon and feudal law'. Colonial officials, especially customs and excise men – hated in America and England alike – were the prime agents of the conspiracy; in the words of Otis, 'a little, dirty, drinking, drabbing, contaminated knot of thieves, beggars, and transports . . . made up of Turks, Jews, and other infidels, with a few renegade Christians and Catholics'. Behind them came the 'standing army' of redcoats, against which all good English Whigs had warned, now billeted in

King in America? 'I'll tell you, friend, he reigns above, and doth not make havoc of mankind like the Royal Brute of Great Britain.' See Bernard Bailyn's analysis, *The Ideological Origins of the American Revolution* (Harvard, 1967).

Massachusetts, and providing in the 'Boston Massacre' a clear portent of things to come.

James Otis, the most successful, rabid and hysterical of the American independence propagandists, formulated the New England theory of history. The Saxons had a Parliament universally elected by all freeholders; this was overthrown by the Normans; then, through centuries of struggle, culminating in the crisis precipitated by the 'execrable race of the Stuarts', liberty had gradually been restored in 'that happy establishment which Great Britain has since enjoyed'. But this was itself now in peril; just as the Saxons had migrated to England in search of liberty, so the Americans had crossed the ocean to create a purer and freer England. There was a great deal more of such nonsense. One of the ironies of the American struggle is that the English, for the first time, faced a people who could dish out quantities of hypocritical humbug and sanctimonious myth-making of precisely the type they themselves had invented.

The conspiracy myth took every conceivable form, and was often self-contradictory. But even the most pro-English elements in America came to feel there was something in it, as exasperated English governments resorted to coercion in the early 1770s. Some believed the conspiracy dated back to the restoration of Charles II; others to Walpole, or to 1763. Bute was often assigned the role of villain; his retirement was a subterfuge; he would return with his 'Scotch-barbarian troops'. Alternatively, or in addition, a Stuart–Tory faction was to blame, backed by the 'corrupt, Frenchified party in the nation', acting 'not improbably in the interests of the houses of Bourbon and the Pretender'. But all versions concentrated on the gross, visible and indeed acknowledged corruption of the English political system, particularly in electioneering, which was emasculating a once stern and unbending nation, as Rome had been ruined, and turning England into 'an old, wrinkled, withered, worn-out hag'. In 1770, the Boston Town Meeting summed the whole thing up:

> A series of occurrencies, many recent events . . . afford great reason to believe that a deep-laid and desperate plan of imperial despotism has been laid, and partly executed, for the extinction of all civil liberty . . . The august and once-revered fortress of English freedom – the admirable work of ages – the BRITISH CONSTITUTION seems fast tottering into fatal and inevitable ruin. The dreadful catastrophe threatens universal havoc, and presents an awful warning to hazard all unless, peradventure, we in these distant confines of the earth may prevent being totally overwhelmed and buried under the ruins of our most established rights.

However – and it is at this point that the Americans snatched the

The Insular Tradition

PELAGIUS incubated in colonial Britain the doctrines of free-will and self-determination which challenged the international absolutism of Rome and its clerical legatee, the church of St Augustine. Only in 5th and 6th century Britain did his teachings find political expression.

WYCLIF recreated English Pelagianism in the late 14th century, and adumbrated the Reformation. High-placed anti-clericals in parliament, foreshadowing the Whigs, gave him initial support, soon submerged by the wave of ruling-class panic which followed the Peasants' Revolt.

THOMAS CROMWELL finally carried through the Reformation programme by employing the characteristic weapons of English insularity: anti-clericalism, xenophobia, the parliamentary statute, the Royal Navy. 'This realm,' he wrote, 'is an empire.' His work turned 16th century England away from the Continent, towards the oceans.

The Art of Ruling

Top to bottom:

CHARTER OF THE CONFESSOR reflects both the high standards and the enervating conservatism of late Anglo-Saxon bureaucrats. Then the Normans brought flexibility and experiment to the machinery of government. By 1130, England was already the best-administered territory in Europe, with an unrivalled system of public finance.

COUNCIL MINUTE of 1475 instructs commissioners to end the Hundred Years War by arranging 'a treux and abstinence of werre with intercourse of merchaundises'. Louis XI is to pay Edward IV a lump sum of '75 thousand scutes' (crowns), and an annual pension of '50 thousand scutes during their bothe lyves'. The first two signatories are the hapless Duke of Clarence and the Duke of Gloucester, later Richard III. Edward put business before chivalry, and foreshadowed the political realism of the Tudor state.

LAWYER'S LETTER from Archbishop Cranmer to Henry VIII. His divorce is at last imminent: 'Pleace yt your Hieghnes to be advertysed that your Gracys grete matier is nowe brought to a final sentence to be geven apon Fryday now next ensueing.' The signature shows that Tudor prelates knew their place: 'Your highnes most humble bedismane and Chaplain, Thomas Cantuar.'

PEACE TALKS at Somerset House end the Anglo-Spanish war in 1604: the English team, led by Robert Cecil, on right. The gravity and high seriousness of Elizabethan government survived briefly into the Stuart epoch. By 1624 (below) the English war-council already had a raffish air: corruption, favouritism, incompetence made political revolution inevitable.

Rebellions that Failed

PEASANTS' REVOLT of 1381 had the Plantagenet state at their mercy, but failed because their aims were even more conservative than the regime in power. English popular movements always tried to put the clock back to a mythical Arcadia, and so exposed themselves to the ruthless realism of the ruling class.

LOVAT EXECUTED in 1746, after the collapse of the Young Pretender's coup. No revolt in Britain has prospered without a popular base in the south-east. A Cockney woman screamed at Lord Lovat on his way to the scaffold: 'You'll get that nasty head of yours chopped off, you ugly old Scotch dog.' He replied: 'I believe I shall, you ugly old English bitch.'

CHARTISTS attack the Westgate Inn, Newport, in 1839. They got their programme from London, their mass-support from the north, midlands and Wales, a certain recipe for failure. Mid-Victorian prosperity did the rest.

CORNWALLIS SURRENDERS, (below) crowning the American revolt. The rebels were in a minority, but they won the propaganda battle by skilfully foisting a hysterical conspiracy theory on their compatriots. British troops do not relish shooting insurgents who have white faces and speak English.

GREAT REBELLION (left) sprang from the south-east and the vast economic resources of the metropolis. Charles I relied on the backward north, 'the most brute and beastly shires of the realm', and on 'blind Wales'. The outcome was predictable: he lost his head.

ULSTER REBELS Carson and F. E. Smith arrive for a demo at Blenheim Palace in 1912. This right-wing revolt could count on a section of the Tory hierarchy, some high-placed public servants, and many cavalry officers: nothing else. The Liberal government failed to call its bluff, and the consequences are with us still.

The Chosen Race

JOSEPH OF ARIMATHEA brought Christianity to Britain straight from the Holy Land – or so Englishmen were taught in the 16th and 17th centuries. The belief that Christ selected the English to replace the Jews as the chosen race was the emotional dynamic of the English Reformation and the colonisation of North America.

BURDENS OF EMPIRE forced British governments to compromise with subjects of diverse creeds, and so undermined the theory of the chosen race. A 1774 engraving attacks the Quebec Bill, which gave toleration to Canadian Catholics. Bishops dance with their joined hands symbolising a cross; a devil hovers over Lord North's head; he is attended by a villainous Scotsman, playing a bagpipe.

THE LORD PROTECTOR grants the request of English merchant ships for a naval convoy, 1657. God's Englishman, as Milton called Oliver Cromwell, incarnated the apostolic mission the Deity had entrusted to the Offshore Islanders. He never doubted God's will, or his duty to enforce it with cannon.

To his Highnesse Oliver Lord Protector of ye
England Scotland and Ireland &c.

The humble Petition of ye Merchants
Tradeing into East India.

Sheweth

Oliver P.

Wee recommend this answeringe of this
pition to the Comrs of our Admiraltye
Desiringe them to doe herein what they
may for the incouragement of the East
India trade. Given att Whitehal this
6th of November 1657.

And yr Petrs shall
daily pray &c.

Maurice Thomson

Sam Moyer

Thomas Kendall

Thomas Andrews

John Banks

Martin Noell

John Frederick

John Bull's England

HOGARTH painted 'The Roast Beef of Old England' after he had been deported as a spy for sketching the gate at Calais: the first, and last, time he left England. A Gallic friar drools over prime English sirloin; a starving Scot and a decrepit Irishman complete this exercise in English xenophobia.

O THE ROAST BEEF OF OLD ENGLAND, &c.

SAWNEY IN THE BOGHOUSE: anti-Scottish (and anti-papal) propaganda by Gillray. Hatred of the Scots, associated both with Stuart despotism and (perversely) with Hanoverian court-rule, was a powerful engine of Whig populism.

FRANCOPHOBIA shaped English foreign policy, with brief intervals, from the 14th to the 19th century. Gillray illustrates one of the 'Consequences of a Successful French Invasion'. A Buonapartist officer slave-drives English farm-labourers: 'Me teach de English Republicans to work.'

Forces of Change

SHAFTESBURY'S FUNERAL in 1885 drew a vast gathering from the deprived classes he had befriended. He epitomised the English tradition of achieving social progress by private, non-party pressure groups. In politics he was a Tory; the workers on his own estates were grossly underpaid.

GLADSTONE in 1884 dining at Lady Aberdeen's, one of the few great houses where this paladin of propriety was still welcome. Rosebery sits on her left. Over sixty years Gladstone graduated from High Toryism to left Liberalism, and carried more Acts of Parliament than any other man. A. J. Balfour called him 'a Tory in all but essentials'.

REFORMED COMMONS, painted by Hayter in 1833, began the slow process of adjusting the legal structure to modern industrial society. 'I have never seen so many bad hats in my life', said the Duke of Wellington; but statistics show that the Reform Bill brought little change in the social composition of the House.

Forces of Inertia

THE SQUIREARCHY fought a rearguard action for 200 years to maintain the game-laws, and as magistrates judged their own causes. In the early 19th century, poaching formed the biggest single category of criminal offences. The laws, as reformers suspected, were counter-productive: their relaxation was followed by a rapid increase in game.

RARE PHOTOGRAPH of Queen Victoria smiling shows her marked resemblance to the present Queen. In the 1860s she became a virulent Tory, and in 1892, when her party was turned out, she summed up her political philosophy: 'These are trying moments & seems to me a defect in our much-famed Constitution, to have to part with an admirable Govt like Ld Salisbury's for no question of any importance, or any particular reason, merely on account of the number of votes.'

'SOAPY SAM' – Dr Wilberforce, Bishop of Winchester – led the high-minded reactionaries on the Episcopal Bench. Victorian bishops voted solidly against progress, especially in education and women's rights. Sam met a spectacular end in 1873: while riding with Lord Granville he was thrown from his horse, turned a complete somersault, and died instantly. His posture in death, wrote Granville, 'was absolutely monumental'.

CARD VOTE in progress at the Trades Union Congress. Some English unions go back to the late 17th century, and look it. In their attachment to ritual and archaic, self-defeating rules, they form the modern equivalent of the 18th-century squirearchy. Their record of achievement on behalf of British workers has been meagre, possibly negative.

Architects of Nemesis

SIR EDWARD GREY, Liberal Foreign Secretary 1905–16, was the willing puppet of the FO Francophiles, and systematically misled parliament and cabinet colleagues. He devoted his leisure to protecting birds and killing fish.

GENERAL HENRY WILSON was the centre of a web of intrigue which committed Britain, in 1914, to a vast land-war fought on behalf of French interests. In 1922 he was murdered by Irishmen on the steps of his own house, sword in hand.

ALFRED MILNER, prototype British Imperialist, directed the war policy in South Africa which first created Anglo-German antagonism. Oddly enough, during the 1930s, his followers from the 'Milner Kindergarten' pursued the policy of appeasing Germany with equally disastrous results.

racial myth, lock, stock and barrel, from the English – America was forewarned, just in time. She would save herself, and preserve the flame of liberty in 'the *country of free men*; the *asylum*, and the last, to which such may yet flee from the common deluge'. America 'may even have the great felicity and honor to . . . keep Britain herself from ruin'. America would be 'the principal seat of that glorious kingdom which Christ shall erect upon earth in the latter days' and would 'build an empire on the ruins of Great Britain'. Thus: 'The hand of God was in America now giving a new epoch to the history of the world.' From this, it was only a short step to the inscription on the Statue of Liberty inviting the 'poor huddled masses' to seek refuge from the horrors of Europe.

It was all very well. But a myth which had a certain validity, a certain honest and genuine enthusiasm behind it in seventeenth-century England, faced with a real and dangerous (if incompetent) Stuart tyranny, had an altogether more suspect and specious ring about it in the age of Enlightenment. The version of history which the new myth itself fostered – of a nation of heroes rising as one man against a ferocious and alien imperialism – cannot survive a careful reading of the documents. George III was probably nearer the truth when he maintained that certain Americans were in a conspiracy against *him*. His famous letter to Lord North, setting out the ruinous consequences to the kingdom and the Empire if the slightest concession of principle were made to the American case, was wholly and ludicrously belied by subsequent events; it can be cited as an object-lesson in folly and misapprehension to any imperial power which fears that one timely and justified withdrawal will imperil the entire structure. It provides, for instance, powerful and ironical ammunition against the American presence in south-east Asia. Nevertheless, George was right in believing that the pacesetters among the American rebels did not want a compromise settlement in any form whatsoever; they wanted, almost from the very start of the controversy, outright and absolute independence; and they were prepared to use any means to persuade the bulk of the American settlers to seize it.*

No one who studies the published correspondence, notably the careful letters of Governor Bernard of Massachusetts, can have any doubt

* George III was the first exponent of the 'Domino Theory' (letter to North, 11 June 1779); in the case of Ireland he was not so wide of the mark, for the Irish used the American crisis to rid themselves of Poyning's Law, and the 1798 rebellion was clearly related to American experience. George could also claim, like President Nixon, the support of the 'silent majority'. As Rockingham wrote despondently to Burke: 'Violent measures towards America are freely adopted and countenanced by a majority of individuals of all ranks, professions or occupations, in this country.' But military defeat inevitably brought despondency, and so a decisive change in English public opinion. On the news of Saratoga, Rockingham wrote: 'My heart is at ease.'

that the independence movement was the work of a minority, and that until the actual fighting started it was a very small minority. The real practical grievances were slight. English legislation, resented in theory, was generally evaded in practice. The customs and excise, the acts restricting American manufactures, could not be, and indeed were not, enforced; on the contrary, they created a huge vested interest in systematic smuggling and evasion, of which the 'patriots' were among the ringleaders and principal beneficiaries. When outrages were committed against the authorities, it was very rare for anyone to be punished; convictions could not be secured, because of perjured evidence; it was in fact like Ireland, in this respect; but, unlike Ireland, there was no evidence of arbitrary oppression. Force, not argument, was the chosen method of the patriots right from the start. It was, in a sense, the only method open to them, for the mass of the people were indifferent or loyalist. As Bernard put it: 'Though the driven and the led are many, the drivers and the leaders are few.' The independence movement was an unholy alliance between the great Southern landowners, the swarming legal profession, and the Boston city mob, the first two groups manipulating the third from behind the scenes. America was born in organised violence masquerading as idealism. Loyalists and officials, printers who refused to publish subversive propaganda (often barefaced lies), merchants who declined to boycott English goods, went in peril of their lives; they were often assaulted or assassinated, their families terrorised, their houses destroyed. In Boston the Lieutenant-Governor's house, designed by Inigo Jones, and containing a priceless collection of manuscripts and papers, was burnt to the ground; his family had been settled there for 130 years; but Massachusetts passed an act of indemnity for the rioters. The Boston Massacre itself was a deliberately provoked incident, as the legal depositions show; it was ruthlessly exploited as atrocity propaganda. The aim of at least some of the patriots was to goad the authorities into sending troops, and then to goad the troops into savage reprisals. Why else, in April 1775, did the insurgents scalp and mutilate bewildered British redcoats, as we know from the letters of Anne Hulton? Such tactics have become familiar to us, in the terrible guerrilla struggles of our own lifetime.

Many well-informed Americans themselves questioned the motives of the popular leaders. It was known that Otis had been bitterly against authority since his father was refused a judgeship. Others had a direct financial interest in defiance: the Tea Party was carried out by the Boston smuggling interest with the deliberate object of keeping up the price of contraband; the fact that it brought massive retaliation was a bonus. Many of the populists were teetering on the edge of bankruptcy.

Speaker Joseph Galloway of the Pennsylvania Assembly, whose compromise plan was ruthlessly scotched – indeed erased from the journals of Congress – asserted that many of the ringleaders were hopelessly in debt to British merchants, and believed independence alone could keep them solvent.* It was, in a sense, a Cataline conspiracy. As in the Russia of 1917 a small group of single-minded and ruthless men hustled along a multitude. A Maryland merchant said bitterly of the first Continental Congress: '[Sam] Adams with his crew and the haughty Sultans of the South juggled the whole conclave of delegates.'

Once warfare was engaged, the inevitable polarisation took place. Even so, it is doubtful whether, at any time, a majority of the colonists actively favoured independence. A quarter of the nation remained neutral; a quarter was loyalist – 40,000 of them later migrated to Canada, and many more wished, but could not afford or feared, to go. One of the loyalists who returned, in disgust, to England, was the Reverend Charles Woodmason, who knew from his own experiences in South Carolina that the political structure of the States already contained brazen economic, class and regional inequalities, maintained by terrorism. The Petition and Remonstrance he drafted on behalf of the wretched Carolina back-country settlers gives an alarming insight into conditions in parts of America on the eve of independence: many thousands, he said, lived as in Hungary or Germany, 'in a state of war, continually exposed to the incursions of hussars and pandours'. These men fought desperately to retain the protection, however feeble, of the imperial government against the local oligarchies. The truth is, independent America proved no more capable of giving justice to the poor than George III's England, less capable, indeed, of providing domestic tranquillity. Free Americans continued to kill each other in the lapidary shadows of the windy rhetoric from Philadelphia.

Moreover, there was the little matter of slavery. Negro slaves had been brought to Virginia as long ago as 1619, and in the eighteenth century slavery had become perhaps the biggest single item in world

* Sam Adams lost the money he had inherited in trying to run a brewery; Patrick Henry twice failed as a shopkeeper. But the connection between the 'patriots' and smuggling should be seen in its contemporary context. Nearly everyone in England and America engaged in some form of smuggling. Parson Woodforde regularly bought smuggled spirits, though he disguised the entries in his *Diary* which related to it. John Wesley, as we know from his *Journals*, found that many of his most faithful West Country supporters were smugglers. Adam Smith said that the smuggler 'would have been, in every respect, an excellent citizen, had not the laws of his country made that a crime which nature never meant to be so'. Members of the Government smuggled. When Walpole was a junior minister, he teamed up with the Secretary of the Admiralty, no less, to smuggle a large quantity of claret, burgundy and champagne, using an Admiralty launch. Walpole even smuggled when he was Chancellor of the Exchequer!

commerce, certainly the most profitable. In 1768 alone over 100,000 had been brought across from Africa; of these, English and American traders sold 53,000 in the West Indies and 6,000 in the Continental colonies. By the time of Independence slaves formed nearly one-fifth of the American population. The anomaly did not go unnoticed. If, as the patriots contended, nobody need be bound by laws they have not consented to themselves, or through their representatives, where did the slaves stand? The question was asked vociferously by many New England idealists. They recognised that Samuel Johnson was entitled to ask, in *Taxation no Tyranny*, 'How is it we hear the loudest *yelps* for liberty among the drivers of Negroes?' Early and vigorous efforts were made from New England to get the transportation trade, at least, suppressed. No attempts were made to justify slavery on grounds of morality and logic. But the arguments for the economic necessity of slave-labour were regarded as unanswerable. In any case, if the Southern oligarchs were prepared to suppress agrarian revolts of the poor whites with ruthless terrorism, no power in America could, as yet, compel them to relinquish what they believed to be the chief source of their wealth.*

So the English gave birth to a noisy, noble and flawed offspring, lavishing on it their traditional christening-gifts of idealism and hypocrisy. The taste for violence from which the English had always wished to free themselves – and were at last beginning to do – passed across the Atlantic, where it struck deep and constitutional roots. England also handed on to America the birthright of the chosen race, while she herself assumed a secular role, increasingly shaped by the necessities and moral problems of empire, the 'white man's burden'. 'God's Englishmen' became 'God's Americans', and the lingering consciousness of divine destiny, even today, still informs American attitudes, though often, alas, in a hideously debased and perverted form – as the CIA and the KGB, like God and Satan, fight Miltonic battles across five continents.

Yet not all Englishmen were prepared to surrender the badge of the elect. At the end of the century, William Blake, with a mind both anachronistic and prophetic, reaching back to the days of the Commonwealth and forward to the Welfare State, resurrected the almost forgotten legend of St Paul's conversion of England, in one of the noblest poems in our language. Most people call it 'Jerusalem', but its real title

* For the texts of the documents cited above, including the letters of Governor Bernard and Anne Hulton, Captain Thomas Preston's deposition on the 'Boston Massacre' and Woodmason's Petition and Remonstrance, see Merrill Jensen (ed.): *English Historical Documents: American Colonial Documents to 1776* (London, 1955). On slavery, American liberal opinion was reflected in the Reverend Samuel Hopkins's pamphlet, *A Dialogue Concerning the Slavery of Africans*, published in 1776. That year, in April, importation of further slaves was ended, but even in the north chattel-slavery continued.

is 'Milton', and justly so, for Blake recognised in Milton the purest voice of the celestial patriotism which the myth enjoined on the English race. If Blake posed the legend diffidently, in the form of a question, if he lacked Milton's heroic certitudes, yet he was equally resolute and sure that an earthly Jerusalem could be built, and that the English would do it. No one who has heard Blake's lines sung at great gatherings of the British working class can doubt that the myth still retains its magic, or that Buffecoate and Berkshire Man live on.

PART FIVE

Splendours and Miseries of Progress

[1780–1870]

IN July 1791 a working-class Birmingham mob, shouting loyalist and Anglican slogans, took possession of the city. They smashed the windows of a hotel, where a meeting to further the cause of parliamentary reform had been held, and began a ferocious pogrom against Dissenters, especially against those known to entertain advanced political and social views. One special object of their hatred was an elderly Calvinist minister, Dr Joseph Priestley. The doctor was a brilliant experimental scientist and polymath. He had, in effect, invented modern chemistry. Enlightened Frenchmen regarded him as the greatest living Englishman. He was revered in the new American republic and, indeed, throughout the civilised world. He was one of the chief architects of the dramatic process which men were already beginning to call the industrial revolution. But he was guilty of the crime of advocating modest changes in society and of questioning the orthodox tenets of the state religion. The mob failed to murder him. But they solemnly cut off his head in effigy. They burnt down his house and laboratory, destroyed the unique collection of precision instruments it contained, seized his papers, thought to contain treasonable matter, and handed them to the authorities (who kept them). They then burnt down two Dissenting chapels, and set about the systematic pillage of any house which did not carry the slogan 'Church and King for Ever'. Twelve of them broke into the cellars of another Dissenter, Mr Ryland, got incapably drunk on his wine, and were roasted alive when the burning roof fell in. Others killed an innocent coachman, who was attempting to defend his master's property. Many other people, chiefly humble folk, were killed or injured in the confusion. Three days later, the Warwickshire Yeomanry restored order, and two of the rioters were hanged. But the group of progressive Dissenters were scattered. Priestley, who had sadly watched his laboratory consumed by the flames, from the safety of a nearby hill, left the district, never to return. Three years later he emigrated to the United States, where he was honoured as a great philosopher, and a martyr in the cause of humanity.

This shameful episode serves as a bleak and fitting introduction to the century of change and reform which made England a modern State, and to the processes by which the English transformed the entire world beyond recognition. The English are a huge force for good and evil: producing, with relentless energy and fertility, new ideas and concepts, and men of dauntless courage to thrust them on society; rich, also, in instincts of decency, imperious in asserting the moral law, remorseless enemies of injustice, avid for philanthropy, profoundly anxious to refashion the globe on lines of purity and reason; but also, and simultaneously, blind and prejudiced, clinging desperately and often violently to the past, worshipping unreason in a thousand ways, uniquely vulnerable to the corruptions of class and snobbery and xenophobia, cruel by indifference and conservative by tradition. In this century we witness a great intestinal struggle among the English between the native forces of reform and reaction, light and darkness, a struggle which was ultimately inconclusive, because if reform emerged the victor, it did so only after the expenditure of irreplaceable energy, and after delays which were to prove disastrous. The modern history of the English is a tragic record of missed opportunities, of chances recklessly squandered or thrown to the winds, of great men dying in despair, of genius and energy poured into the sands of thoughtless indifference, of advancing reason slowed to the pace of a glacier, and of the slow, confident retrenchment of privilege, injustice and obscurantism. It is, to a great extent, a history and an explanation of everything that is wrong with the world in which we live, and a lesson to all races.

By the 1780s, the English had acquired, through the accident of geography and the merit of their own efforts, a unique conjunction of advantages: a free, though oligarchic, political constitution, and all the elements of an economic revolution. Only two other countries, the United States and the Netherlands, had a non-authoritarian system of politics; and no State whatever, except England, had the physical means to produce an unaided and self-sustaining acceleration of economic growth. England was the one dynamic element in a static universe.

For half a century, foreign observers had been conscious of the connection between political freedom and economic prosperity in English life. In the 1720s Voltaire had noted in *Letters from England:* 'Commerce, which has enriched the citizens of England, has helped to make them free, and that liberty, in turn, has expanded commerce. This is the foundation of the greatness of the State.' England was an open society. There were no barriers between the classes, at least in legal terms; Englishmen enjoyed absolute equality before the law. Peers could claim judgment in their own parliament house; but the other residual privileges of the military tenure system had been swept away in the 1640s. In 1679 the English had acquired the right of *Habeas Corpus*; in 1701 life security for judges. Juries were not accountable to the State for their verdicts, and accused men were innocent until their guilt was established to the satisfaction of courts beyond the reach of the executive. Freedom of speech, subject to closely defined laws of treason, was absolute; and freedom of publication, except in the theatre, was qualified only by the risk of subsequent prosecution, the equivalent of the presumption of innocence in legal terms. Government restricted the sales of newspapers by stamp and paper duties, but they grew steadily in circulation, numbers and influence. In the half century to 1760 some 160 provincial papers came into being, the majority critical of Government; in 1782 there were 18 daily papers in London alone, with an average readership of up to 500 a copy, for desultory efforts to suppress coffee-houses as centres of political discussion and disaffection had long been abandoned. The provincial press reflected the information and views of metropolitan journals, thus creating a national public opinion; and this was shaped in a vast and anarchic capital, twice as big and many times as rich as any other city on earth, and virtually subject to no authority other than self-restraint. In 1780 London had been abandoned for a fortnight to the rule of its own mob, which had terrorised Catholics, foreigners and both Houses of Parliament. There was no professional police force, and only a tiny army subject to annual parliamentary vote. London and other chartered cities were autonomous, and the rest of the country was governed by unpaid country gentlemen, meeting as amateur magistrates four times a year. The civil service, even including the highly-efficient postal, customs and excise system, was minute, and most of those who composed it were immune to dismissal. England was the minimal State: no such has ever existed, before or since.

Indeed, in a sense, England was a private State, in that its prime purpose was to guarantee the individual possession and enjoyment of property. Its rules had been established by the common lawyers through the ethic and mechanism of the contract. The great debates of 1640–60 had been about the source of power: was it to be monarchy, property or personality? It had been decisively resolved after 1688 in favour of property. Membership of the House of Lords was an unqualified hereditary freehold. Seats in the Commons were also, in many cases, freeholds. Some boroughs were the personal possessions of families who nominated members generation after generation. One MP, protesting against the Reform Bill in 1831, claimed in anguish that his seat in the family borough, which the bill proposed to abolish, was a hereditary possession, to be handed on to his son and grandson, and that the bill was 'robbing him of his birthright' – thus ironically echoing the angry words of Edward Sexby in the Putney debates. Where the hereditary principle was inapplicable, the lifetime freehold was paramount.

There were freehold bishops, rectors, vicars and perpetual curates An army commission was a freehold, to be bought and sold at current market prices. A judgeship was a freehold, and so were the overwhelming majority of posts in the public service. Even the humblest servants of the Crown held their jobs by secure life-tenure, which only outrageous conduct could invalidate. In June 1804, a king's messenger was found to have forged a key to the Cabinet boxes he carried, and was suspected of using his illicit knowledge to speculate in the Funds – a rewarding activity at the height of a European war. Only the discovery that there was no statute whatsoever under which he could be prosecuted was felt to justify taking the drastic step of dismissing him.

A society constructed on such a clear and consistent principle was thus highly resistant to change. That it enshrined a multitude of anomalies, that it was grotesquely inefficient, that the fundamental structure of the suffrage, laid down in the fifteenth century and since altered in detail in response to a variety of private pressures, bore no relation to the needs of a community which had changed beyond recognition – this seemed less important than the sense of overwhelming security which the absolute guarantee of existing rights provided. The political nation was a mere 400,000 and its members were unequal: but each man's title deeds were beyond challenge, and if he could get more, under the law, he could keep it. There were plenty of opportunities. England was a prosperous country, the heart of a boundless empire. A man could make money; he could buy himself the right to vote, a seat in Parliament, even the hereditary ownership of a borough. He could acquire an estate big enough to make his claims to a peerage, in time, irresistible. English society was open. The circle of power was charmed, but admission at any of its levels could be secured, at a price. Mr Robert Peel, senior, was the son of a yeoman who founded a cotton business. Peel expanded it, enormously. He bought an estate in Staffordshire, which carried with it the right to a seat in the Commons. He was useful to government, and raised and paid for a regiment of yeomanry; he was made a baronet. His son got a parliamentary seat at 21, was brought into government at 24; and in time became Prime Minister. This was an exceptional success story; but there was a multitude of others, known to all, which proved the same point. England was a stable society, and a secure one; but it was not static. The great game of success was rough, and difficult; but the rules were plain and universally understood, sanctified by tradition and the blood of political martyrs. Once men began to change the rules, where would the process end? To remove an indefensible anomaly – a parliamentary seat which had no electors at all – would prepare the way to removing one that had few.

Whose vote, whose seat, would then be secure? If you applied the principle of logic once, must you not apply it always? A man had rights because he had property; or a man had rights because he was a man. Both systems were consistent in themselves; they were fundamentally inconsistent with each other. Society was mobile; but subject that mobility to logical processes, and stability and security flew out of the window. If a political freehold was vulnerable, what other kind of freehold was safe? As the anguished MP said in 1831, 'this year you take away my seat, next year you will take away my castle'. If a man could be stripped of his vote, which after all could be valued in terms of hard cash, when might he lose his freehold tenement? The great majority of the political nation refused to admit there could be a half-way house between the existing constitution and what Cromwell had called 'a leap in the dark'.

Yet nevertheless change came, and it came in a characteristic English manner: in a welter of muddle and confusion, for a variety of reasons (most of them wrong ones) and after infinite and exasperating delays. To begin with, the English adopted their customary backwards posture, moving into the future with their eyes firmly fixed on an imaginary past. The constitution, as the political nation unanimously agreed, was perfect; but it had become corrupted and deformed by wholly unwarranted, indeed illegal, innovations. It must be restored to its pristine state. The City of London, for instance, had somehow been deprived of its time-honoured privileges: Wilkes fought for ten years to 'restore' them, and thereby drove a damaging wedge into the existing system, using popular agitation as his motive-force. One by one, ancient and important bastions fell to the blows of revolutionary traditions. General warrants were declared illegal. The Commons dropped the self-purging process of stripping validly elected MPs of their seats. It refused to give explicit authorisation to the printing of its debates, but it no longer prosecuted offenders. A popular press thus emerged, focused overwhelmingly on the activities of Parliament, and read by a multitude well beyond the confines of the political nation. And in the 1790s the judges lost to the jury the vital right to decide on the fact of libel.

Once again, too, the conspiracy theory served as an engine of change. As we have seen, its emergence in a distorted transatlantic form lost Britain the American colonies. But in England it served equally well to erode, and ultimately to destroy, the political power of the Monarchy. George III was not an innovator. The most he confessed to aim at was the removal of the worst features of corruption which disfigured a constitution which he (like everyone else) said was perfect. Of course everyone was against corruption, as they were against sin. The question was: corruption in whose interest? George inherited a constitution which was unwritten and therefore flexible; any element in it might push forward its claims without being seen to break the law, and so provoke an open crisis. But beyond a certain point, such pressures became objectionable, and provoked counter-pressures. Walpole had created a one-party State, in the Whig interest. George felt this to be an unwarranted distortion, and an infringement of his political rights. He sought to restore the balance by working towards a non-partisan State, in which the Crown would be freed from the illegal restrictions of party pressure, and govern in the general, as opposed to the factional, interest. It was unfortunate that his instrument was a Scotsman, Lord Bute; still more so that his victims felt they had lost their birthrights. They willingly subscribed to – indeed they actually believed – the theory that the King, or rather his evil advisers, were attempting to overthrow the verdict of 1688. Modern historians know that this is not true: but then they are privileged to read the King's correspondence as well as Lord Rockingham's, something denied to the Whigs. Historians see both sides of the hill, whereas the Whigs were enveloped in the smoke of battle, and felt themselves threatened by imaginary horrors beyond it. Thus myth determined events. The great Whig families had no objection to corruption as the normal method of government; it was their métier; they had invented it. They had no objection to making the King a party to the system. But that he should operate it without their participation was intolerable.

The map of English politics in the eighteenth century was like a map of the Holy Roman Empire: a multitude of small, independent states plus two big ones, the Crown and the Whigs. When the big two agreed, there was normalcy; when they disagreed there was crisis. The threat from the Crown could be met in two ways: by the political reform of changing the suffrage and the distribution of seats; or by the financial reform of removing the means of Crown corruption. The first would destroy the Crown's parliamentary freeholds, but it would destroy those of the Whigs as well. So 'economic reform' was born, and flourished. It was originally a Whig monopoly, and a crooked one: they admitted among themselves that, under the guise of saving the taxpayer's money, they aimed to strip the Crown of its influence. But the movement gained its own momentum. Ministers began to see efficiency in the public service as something desirable in itself, especially when the country was at war. The revolt of the Americans, and the abject failure of the Crown either to conciliate or to beat them, confirmed the Whigs in their belief that conspiratorial forces were at work; but it also led Lord North to a modest filching of their clothes.

Thus a tradition of economic reform grew up within government. It soon acquired an outstanding evangelist. Lord Shelburne was an intellectual; he knew, and corresponded with, the leading lights in Britain, France and America; he read Adam Smith; he took advice from such dangerous men as Priestley, and his Dissenting colleague Dr Price, and Mr Jeremy Bentham – systematic thinkers who did not share the prevailing English view that all change must be a restoration of the past, who had the temerity to advance entirely new concepts of government, which measured institutions and offices by their utility. This was a radical departure for the English, a true leap in the dark. But Shelburne rejoiced at the prospect. He believed in new systems. The Whigs, in their brief spells in office, sought to advance economic reform by parliamentary statute. Shelburne worked from within the machine. He began to disentangle the extraordinary skein of government departments, and their ramifying financial relationships. His activities set up fearsome tremors throughout the body politic, and brought on his head an avalanche of unpopularity. He saw himself as making government work; he was, but he was also dismantling the Walpoleian system of politics. Quite what he was up to the Whigs did not understand; but he was plainly a conspirator of sorts, 'the Jesuit of Berkeley Square' as they called him.* Moreover, he could not explain himself in the Commons, as he was a peer; indeed, he could not explain himself to his colleagues – for an intellectual he was curiously inarticulate, and his angry autobiography, or apologia, does not make much sense. But in falling, he handed the torch to his young Chancellor of the Exchequer, William Pitt.

Now Pitt was not an intellectual. He occasionally reread the classics he had learnt at Eton, but otherwise there is no evidence that he ever opened a book which did not relate to the work of government. On the estate he bought in Kent he cheerfully demolished the site of one of the most important Iron Age forts in the country, and laughed derisively when antiquarians protested. He gave the poet laureateship to a retired hack MP. He did not cultivate men of learning except on business. But he had an administrative brain far more powerful than Shelburne's, and shared to the full the noble lord's passion for efficiency. Adam Smith emerged from a meeting with Pitt dazzled, and confessed he now understood his own theories properly. Moreover, Pitt could work the parliamentary, as well as the government, machine. His 'blue

* After the all-powerful Portuguese Jesuit, Fr Malagrida. This occasioned one of Goldsmith's characteristic lapses in tact: 'Do you know,' he said to Shelburne, 'I never could conceive why they called you Malagrida, for Malagrida was a very good sort of man.' See John Norris: *Shelburne and Reform* (London, 1963) for an analysis of Shelburne's restructuring of central government, illustrated by an illuminating diagram.

paper' style of speaking was not to everyone's taste, but it had the enviable merits of clarity and *gravitas*. He made government, especially public finance, sound a mighty serious business, but he also made it comprehensible. For the silent knights of the shire who held the parliamentary balance, this was a new and welcome phenomenon: at last they understood how the Sinking Fund worked.*

Pitt was not just clever, he was pure. There were some famous instances in which Pitt not merely turned down time-honoured perks, but declined a permissible favour to the most important figure in his own constituency. He even allowed his own salary to get into arrears, a common fate among the humble, but not one which had yet befallen a First Lord of the Treasury. The silent knights rejoiced. After all, they largely stood outside the spoils system, and did not wish to perpetuate it if a better way of running the country could be found. Even better, there was an elevating contrast between the efficiency of Pitt's public finances and the chaos of his private affairs. As the French exile Chateaubriand commented admiringly, he was *criblé de dettes*. Quite how Pitt, whose style of life was modest, contrived to spend so much, the most recent and minute examination of his papers does not reveal; but it seems, for instance, that on a salary averaging about £10,000, he was charged £7,000 and more a year for horses and stabling, but nevertheless had to hire cabs and post-horses to get around. Obviously he was robbed by servants and tradesmen; he had no wife to supervise them, indeed used his indebtedness as an excuse to repel menacing advances from Miss Eleanor Eden; and, as the bills and household wages were rarely paid, the system had a certain equity.† What most impressed MPs, however, was that Pitt unhesitatingly rejected a handsome offer from the City merchants to pay his debts to the tune of £100,000, without strings attached. Perhaps Pitt regarded his debts as a valuable

* For a more jaundiced view of Pitt's oratory, see Sydney Smith's letter to Francis Jeffrey, 30 January 1806, commenting on Pitt's death: 'I must say he was one of the most luminous eloquent blunderers with which any people was ever afflicted. For 15 years I have found my income dwindling away under his eloquence. . . . At the close of every brilliant display an expedition failed or a kingdom fell, and by the time that his style had gained the summit of perfection Europe was degraded to the lowest abyss of Misery. God send us a stammerer, a tongueless man.'

† For Pitt's financial excuses for not marrying, see his letters to Miss Eden's father, Lord Auckland (*Journal and Correspondence of William, Lord Auckland*, (1862), iii, pp. 373–4). On Pitt's one trip to France, in 1783, Madame Necker made preliminary moves to marry him off to her daughter, later Madame de Staël; Pitt promptly returned to London and never crossed the Channel again. He probably had no sex-life at all. His bachelor status provoked tiresome English jokes (especially when he levied a tax on female servants), of exactly the same type made about Arthur Balfour in 1902–5 and Edward Heath in the 1970s. The English sense of humour does not change, or improve. Jokes made about Mrs 'Gladys' Wilson in 1964–70 were almost identical with those made about Mrs 'Joan' Cromwell in the 1650s.

token of his rectitude; this was certainly the view of many. At all events, by such means Pitt ruled the Commons. Where his predecessors had bribed, he gave peerages, garters, lord-lieutenancies: the honours system more or less as we know it today. The burden of work he assumed was enormous; for many years he had no private secretary (the post was a sinecure; the King himself did not employ a secretary until he had been on the throne over 40 years). The strain was tremendous. Pitt used to vomit painfully just outside the chamber of the House before making a speech. From the middle-1790s he took refuge in alcohol in a systematic and disturbing manner.* But he nevertheless contrived to hold supreme power, with one interval, for over 20 years, and in doing so he created the pattern of modern government: regular accountability, the systematic inspection of departmental expenditure, unity of receipt, Treasury control, and paramountcy of the annual budget. Such a system seems simple and obvious today, now that all States (in theory at least) practise it. But it had hitherto eluded mankind.

Pitt's main object was to promote efficiency. In pursuing it, he inevitably made government more honest, and the probity of the public service slowly became a feature of British life and (more quickly) was hailed as a British tradition since time immemorial. Hence, almost by accident, the direct power of the monarch was finally destroyed. In 1809 an Act was passed prohibiting the sale of Commons seats: this effectively inhibited direct cash intervention by the Treasury in elections. The rest was a matter of tidying up; and what Pitt had done by stealth, his successors continued with enthusiasm. The wars against Napoleon had brought an afflatus of Crown appointments which maintained the illusion of influence; once over, the contraction in government service revealed the reality. Throughout the 1820s jobs at the disposal of ministers were steadily reduced. Wellington, as Prime Minister, stated flatly: 'No government can go on without some means of rewarding services. I have absolutely none.' Peel, the true heir to Pitt, welcomed the change. In two years, he said, he had not had a job worth £100 a year to dispose of, and the government was the purest in any man's memory; henceforth, ministers must base themselves on public opinion. Althorp, for the Whigs, agreed: he 'thanked God the time was passed when the Government . . . could be carried on by patronage'. Thus a

* Ministerial drunkenness was aggravated by the practice of holding cabinet dinners, which persisted until the Reform Bill. In April 1828, Lord Ellenborough, Lord Privy Seal, noted in his diary: 'The Chancellor said to me: "We should have no Cabinets after dinner. We all drink too much wine and are not civil to each other."' When Lord Sidmouth or Lord Bathurst were present, or hosts, little business was done, as all tended to be drunk.

constitutional revolution occurred which no one planned, and which few noticed until it was accomplished.

If the English could transform their system of central government without fuss or argument, why did they make such a mouthful of parliamentary reform? The answer is not obvious, but there are some instructive pointers. One is the attitudes of society towards crime. Eighteenth-century crime – above all unpunished crime – was dominated by smuggling and offences against the game laws. Smuggling was a universal habit; it was also big business. By the 1780s it brought in goods worth about £3 million a year, against legal imports of about £12 million. Few thought it morally wrong; Charles Lamb put the popular view when he said smugglers 'robbed nothing but the revenue'. But it was a threat to the financial stability of the State. It almost ruined the East India Company, the world's largest trading organisation. And it led to a vast amount of violence and bloodshed. Lord Pembroke asked: 'Will Washington take America or the smugglers England first?' As many as 700 armed men guarded the smuggling trains inland; 1,000 or more supervised the beach landings. Here was something which appealed to Pitt as a challenge, because it could be solved by administration. Tea was the chief battlefield. Pitt took Adam Smith's advice and cut the tea duty, raising the window tax to balance the revenue. The smugglers were faced with ruin as the price of legal tea collapsed. They responded by trying to corner available supplies; but Pitt intervened vigorously on the London and Continental tea-exchanges, financing his operations by borrowing £300,000 from the Bank of England. Then he turned to wine and spirit smuggling, using the same techniques. He employed direct legislation, in the extension of the Hovering and Manifest Acts, because this involved no new principle, and provoked no political resistance. But in the main he simply exerted the authority of the executive. In January 1785, the news came that winter gales had forced the largest single group of smugglers to draw their boats up high on the Deal beaches. Pitt told the Secretary at War to send troops to cordon off the area while the excise smashed in the boats. The Secretary said he had no legal authority; so Pitt invoked his own, as First Lord of the Treasury, and the operation was carried through in triumph. Within a few years, the back of the problem was broken, and English smuggling entered the age of the suitcase.

If the smuggling problem could be solved, why not poaching? Because smuggling was a classless business, whereas the game laws were the spectacular underpinning of the class structure. They were what English society was about. Medieval kings had enforced the forest laws with unspeakable ferocity for the same reason: they drew a decisive line between monarch and subject. But the forest laws had finally collapsed under the Commonwealth, in a wholesale carnage which permanently changed English ecology; private hunting parks went with them. The focus shifted to game, and after the Restoration sporting guns threatened extinction. So a statute was passed in 1671 forbidding the killing of game except by owners of land worth £100 a year, or leases worth £150, the eldest sons of esquires, and the holders of franchises. A stockbroker, attorney, surgeon, or 'other inferior person', might beat while accompanying a qualified sportsman, but might not actually kill. The sale of game was prohibited, its unauthorised possession made illegal, and there was a multitude of other vexatious provisions, especially about the ownership of dogs.* The laws did not, and indeed could not, work. Their net effect was to deliver most of the game into the hands of poachers. Since game could legally come only by gift, its prestige value was high, and so, accordingly, was its black-market price. The middle classes and the new-rich got their game by the 'silver gun' in Leadenhall Market, where it arrived in excellent condition, from as far as Scotland, thanks to a nation-wide network of innkeepers and coachmen. Higglers, or travelling poultry-dealers, bought stolen game-eggs, which also ended up in London, and were bought back by the landowners, often the victims themselves. When landowners tried to sell their own game, they were undercut by the far more efficient poaching system. What is more, they had no remedy against a qualified intruder, except an action for trespass after due warning off; so the gentry could, and did, poach against each other with impunity.

The absurdities of the laws were apparent from the start. They were nevertheless maintained, with blind tenacity, for 150 years. Desperate attempts were made to strengthen them; 32 new laws were passed under George III alone. The notorious Ellenborough Law of 1803 imposed terrible penalties, including death; and it was reinforced by a still more draconian act in 1817. But poaching continued to increase. The trouble was that the farmers hated the laws because the squires insisted they keep up the hedges; so the farmers helped the poachers, and vice versa. Poachers formed professional gangs, and shooting affrays became common. Countrymen from dukes downwards bore the scars of conflict. Territorial armies were assembled on both sides. Lord Berkeley employed eight head-keepers, 20 under-keepers and 30 night

* The law of 1671 also entitled gamekeepers to search the houses of the lower orders without warrant, thus granting to the landed class the privilege enjoyed by King's officers under twelfth-century forest laws.

watchmen, plus extras when it was known gangs were about. But the worst poachers were the gamekeepers themselves, and more keepers usually meant less game. Technology was roped in to provide man-traps and spring-guns, but Parliament banned them in 1827 because they often killed and maimed the innocent, including the landowners themselves. Sometimes entire villages, led by the constables and the gamekeepers, formed poaching syndicates. By the 1820s, one-sixth of all convictions were for game offences; and since only a tiny proportion of poachers were caught, let alone convicted, the real volume of this type of crime must by far have exceeded any other. The laws made life miserable for everyone in the countryside, but most of all for the gentry: they spent a fortune in protection, and yet got very little game. But they fought to the last ditch to maintain the system, because it was a legal expression of the social structure they believed in. Needless to say, when the worst aspects of the game laws were swept away with the debris of the old regime in 1831, the immediate consequence was an enormous increase in game; and Continental mass-battues, beloved by the Germanic element at court, became possible.*

The social instinct which led the English ruling class to regard even the birds of the air as private property expressed itself throughout the criminal code, whose ferocity against the person, in theory at least, was unique in Europe. There was a certain grim logic in this. If all rights and power sprang from property, as opposed to personality, then the State correctly assumed that stolen property worth five shillings or over was of more weight in the social balance than the life of the person who stole it – which, under a statute of William and Mary, was forfeit. It was no accident that the century following the 1689 settlement, which sanctified property as the basis of political life, saw a massive expansion in the number of statutory crimes carrying the death penalty, from 50

* For the rise and fall of the game laws, see E.W. Bevill: *English Country Life 1780–1830* (Oxford, 1962). Oddly enough, the judiciary, so savage in protecting the game-preservers, showed no sympathy for foxhunters. There was a universal belief that fox-hunting could freely take place on another man's lands, springing from a judgment of 1656 that 'the fox is a noysom creature to the Commonwealth'. This was overthrown in *Essex* v. *Capel* (1809), when Lord Ellenborough's summing-up left no doubt that fox-hunters were common trespassers in law. Judges shot game but did not hunt foxes, reflecting the preferences of the more 'civilised' section of the ruling class, which regarded hunting, as opposed to shooting and fishing, as barbarous. But, with characteristic English perversity, the judgment made little difference, since farmers, who hated the game laws, on the whole favoured hunting. Moreover, the hunting fraternity took pains to conciliate the farmers: Hugo Meynell, who created modern foxhunting, would wait only 10 minutes at the covert-side for a duke, but 20 for a local farmer. Thus foxhunting entered its golden age after the law, in theory, made it impossible. For a brilliant account of the social pressures exerted on farmers who defied the hunt, see Anthony Trollope: *The American Senator* (London, 1871).

to about 200*. Nearly all the new capital offences concerned property: appropriating stolen goods, killing or wounding cattle, destroying growing trees, cutting down river-banks or fences, maliciously cutting sedges, damaging lock-gates or sluices, stealing fish from private rivers and ponds, or damaging the ponds – above all, ordinary petty theft.

Much of the medieval and renaissance apparatus of judicial savagery was still in being at the end of the eighteenth century: it was dismantled slowly but steadily, and on the whole without much argument or resistance. The burning of women went in 1790, the pillory in 1816 (except for perjury), the public whipping of women in 1817, and private whipping three years later; gibbeting was abolished in 1834, though public executions had to wait until after the death of Lord Palmerston. But the movement to restrict capital crimes to atrocious offences against the person came up against certain bedrock assumptions which proved immensely difficult to dislodge, particularly since the judges considered themselves the guardians of the property-state. It was useless to point out that only a minority of capital sentences (sometimes as little as one in 13) were actually carried out, and that juries often deliberately undervalued stolen property to avoid the mandatory sentence of death. The judges were concerned to defend the principle. Equally, appeals to consider the tender age of those sentenced fell on deaf judicial ears. What if a girl aged seven and a half was in solitary confinement, and denied even the comfort of a doll? She was already an enemy of the system, and likely to grow into a more dangerous one. In 1816 a boy aged ten lay under sentence of death in Newgate; but the recorder who sentenced him had declared: 'It was the determination of the Prince Regent, in consequence of the number of boys who have been lately detected in committing felonies, to make an example of the next offender of this description who should be convicted, in order to give an effectual check to these numerous instances of depravity.'† A substantial majority of offences, at least in London, were

* A contributory factor to the rise in capital offences was the well-founded belief of the ruling class that the concept of eternal punishment was no longer an effective deterrent to crime. As men ceased to believe in hell fire, the gallows arose from its ashes. See D.P. Walker: *The Decline of Hell* (London, 1964).

† The Home office did not begin to issue criminal statistics for the whole country until 1811, so it is difficult to compute the total of those condemned to death, or executed, until that date. For instance, in 1598 in Devon alone, 74 persons were condemned to death; but how many were actually hanged? Certainly, the ruthlessness of the Elizabethans in killing thieves impressed foreigners, including even Ivan the Terrible. In 1607–16, the yearly average of executions in London and Middlesex was 78. But in the eighteenth century the disparity between the numbers of those sentenced, and those executed, widened steadily. The largest numbers of executions, of which we have accurate figures, took place during the post-Napoleonic reaction, 1816–22, when the yearly average was over 100. By 1831, death sentences had risen to 1,549, but executions

in fact committed by those under 21; not surprisingly, since the young constituted more than half the nation. If thieves were not hanged, how could they be sufficiently punished? As Robert Peel, the Home Secretary, wrote to Sydney Smith in 1826, it was extremely difficult to make prison conditions and diet more unpleasant than anything the criminal classes experienced outside, and so maintain what he termed a 'salutary terror'.

The debate continued for half a century, and was passionately argued on both sides, for a principle of enormous importance was at stake, which went to the root of social values. The achievement of such reformers as Bentham, Brougham, Romilly and Mackintosh seems insignificant if seen in terms of statutory results. Yet in the end they forced, and Government conceded, an ideological victory of a radical kind. Were the English to be treated as property-owners, mere functions of their possessions? Or should they be seen primarily as human beings? It was not merely a battle between the trustees of the property-state and the humanitarians: the answers would determine the whole direction of future policy. If society concluded that persons were more sacred than goods, then the whole axis of its operations must eventually be swung round. Not only must personality triumph over property as the basis of politic right, but the state must actively assist the conditions in which the person could flourish: it must protect the person, by public health and factory legislation, feed and clothe him if necessary, educate him, and give him a variety of rights to protect and advance his interests. There was no logical barrier between ceasing to hang a thief and making him the beneficiary of the Welfare State.*

No logical barrier, indeed; but many English ones, of peculiar powers of resistance. Why was it, as we have seen, that a working-class mob in the Midlands could be raised to burn the homes of moderate reformers? In the 1780s, when administrative reform was getting its teeth into the whole body of government, there seemed excellent chances of political reform, too. The loss of the American colonies, which Englishmen saw

---

had dropped to 52. By 1838, reform had reduced capital crime virtually to murder: 116 death sentences were passed, but only six carried out (on the other hand, two years earlier over 52,000 were still serving terms of transportation, varying from 7 to 14 years). Incidentally, the recorder was certainly wrong about the Prince Regent: he disliked hanging intensely, and the ability of his women to secure remissions for favoured offenders was one of the scandals which hastened reform.

* Nineteenth-century judges, perceiving that the changing philosophy of the law would ultimately guide social progress, took an elevated view of their status. In 1848, while trying rioters at Liverpool Assizes, Mr Baron Alderson heard a hiss in court. He said angrily: 'Where is the man that hissed? Let me see anyone who defies the law! I sit here alone, and with the whole majesty of the Kingdom of England upon me; and let me see the man who dares to face it!'

as a catastrophe springing from the weakness of the system, created a climate favourable to change. In the closing stages of the conflict, and for some time afterwards, economic distress made it possible for reformers to put insistent pressure on Parliament through the traditional method of mass petitions. All the young men of outstanding ability in the Commons favoured reform in some shape. Moreover, there was a substantial body of propertied opinion outside Parliament which was willing, indeed eager, for change. It is true that schemes varied greatly, and were in some respects contradictory. There were those who had taken the point of the American case, and wanted the seats from the pocket boroughs to be redistributed among the new towns; wanted, too, to award the suffrage to many categories of people whose wealth, though substantial, was not in the form which qualified them under the existing system. But there were other weighty groups, especially the powerful association of gentry and yeomen in Yorkshire, who preferred a massive increase in the county seats. Such a proposal was in a sense reactionary, and deliberately so. Its object was to reinforce the essentially territorial basis of the consitution. But all schemes of reform were, characteristically, presented in the guise of the restoration of ancient perfection. There was no other way of getting the back-bench gentry to listen to, let alone vote for, any change whatsoever.* But equally, any change, even if defended on retrogressive principles, was welcome in that it served to shatter the mould which imprisoned English political development. This was Pitt's private attitude in 1785, when his reform scheme was defeated by a mere 74 votes (248 to 174). The division was regarded as encouraging, the augury of future success. It proved, in fact, the high-water mark of reform for nearly half a century; and in the meantime the forces of resistance were able to erect, with overwhelming support from the political nation, an unprecedented apparatus of violent repression. How this happened is one of the great tragedies of English history.

Some of the blame must rest on Pitt himself, and on Charles James Fox. Both possessed astonishing gifts, and were given unrivalled opportunities to exercise them from earliest manhood: they embodied such virtues as the old system possessed. Both were liberal-minded, indeed open-minded. Both were anxious to change the world for the better. Their talents were complementary. In combination, they could have carried through a peaceful revolution in that decade of missed

* Pitt introduced his reform motion by saying that its object was to erase defects from 'a beautiful frame of government ... and it would not be innovation ... but recovery of constitution, to remove them'. See John Ehrman: *The Younger Pitt – The Years of Acclaim* (Cambridge, 1969).

opportunities before 1789. In fact they became not merely rivals but mutually destructive enemies; their conflict nullified the political virtues of each, and force-fed their political vices. As public men, both degenerated, and public life with them. Their initial contacts were friendly, even warm. It is ironic that they fell out over Shelburne, the man nobody liked. When Fox told Pitt he would serve in no government of which Shelburne was the head, Pitt not only broke off negotiations but declared (and he kept his word) that he would never again hold a private conversation with Fox without the presence of a third party. He thought Fox irresponsible to allow private feelings to override the public interest; this was true. Fox thought Pitt a cold fish (in fact he was shy: 'The shyest man I ever met,' said his friend Wilberforce). But such progress as this nation has attained springs from the combined efforts of the irresponsible and the cold, or those who appear so. When Aneurin Bevan called Hugh Gaitskell a 'desiccated calculating machine', he echoed, unconsciously, the contempt Fox hurled at Pitt; when Gaitskell pointed to Bevan's lack of realism, he reasserted Pitt's principle that civil government imposed restraints and limitations which all politicians, however brilliant, must accept. 'The trouble with him,' I heard Gaitskell say, 'is that he never does his homework.' This was the voice of Pitt, for whom a blue book was bedside reading, and the national accounts the delight of his few idle hours. Now Fox never did his homework either. 'Though I like the House of Commons itself,' he told his friend Fitzpatrick, 'I hate the preparatory business of looking at accounts, drawing motions, etc.' The comparison can be taken further. Neither Gaitskell nor Pitt were heartless men, as their enemies supposed. Gaitskell was devoted to his friendships, often nourished them without regard to the consequences. Pitt's feelings grew with the years: his connection with the worthless Dundas, once based solely on official business (for Pitt thought the Scots lawyer a social inferior, as his letters to him show), eventually generated an emotional spasm, when Pitt failed to save his friend from parliamentary censure by the mere casting vote of the Speaker; as the division figures were announced, he burst into tears, and his anguished supporters crowded round him to hide the sight from the jeering Foxites. That Fox had such a heart is doubtful; like Bevan, he accepted the offerings of the multitude of admirers his genius and charm attracted, but there was little reciprocation. George Selwyn wrote of him: 'Charles, I am persuaded, would have no consideration on earth but for what was useful to his own ends. You have heard me say, that I thought he had no malice or rancour; I think so still and am sure of it. But I think that he has no feeling neither, for anyone but himself.' Philip Francis thought much the same: 'The essential defect in his character, and the cause of all his failures, was that he had no heart.' Be that as it may, these two great men fell out; after their failure to work together, the gladiatorial principle in English politics did the rest, and what might have been a human combination of unique potential became an engine of self-destruction. The other victims were the English people.

The split between Pitt and Fox damaged the prospects of reform; the French revolution destroyed them for a lifetime. The English hatred of foreigners, and especially Frenchmen, is such that no reformer can afford to be branded with Continental associations, however far-fetched. One of the great strengths of the Cromwellians was the geographical isolation, and the racial uniqueness, of their revolution: no foreign brush could tar them, indeed they could and did savage their opponents as the puppets of Continental intervention in English affairs. The tragedy of the English reformers of the late eighteenth century is that they became the victims of guilt by association. The events of 1789 in Paris were welcomed by the English political *élite*, but in a very cautious and limited manner. What happened in France was of growing concern to the English nation. Three years earlier, taking advantage of what our ambassador in St Petersburg called 'a Phrenzy for concluding Treaties of Commerce with this Country which prevails throughout Europe', Pitt had negotiated a tariff-reduction agreement with France which was immensely to the advantage of English traders and manufacturers. He had taken this step after much anxious thought, aware of the strength of anti-Continental feelings at all levels of English opinion. His own Foreign Secretary thought France, in particular, 'our natural and inveterate rival', and felt that the suspiciously generous terms of the treaty 'revived, if not confirmed' his fears. Pitt admitted 'the great difficulty is how to lay the foundations of such Connections, keeping clear at the same time of being too soon involved in the Quarrels of any Continental power', and bearing in mind 'the necessity of avoiding, if possible, the entering into any engagements likely to embroil us in a new war'. In short, England had taken a cautious, if profitable, step away from isolation, and was correspondingly nervous.

Now the English would have been happy to see their new trade links with France strengthened by a French adoption of English political practices. 'The Constitution of Great Britain is sufficient to pervade the whole world,' said Shelburne in 1782. Even those who wanted to improve it felt it was for export. There seems to have been a common assumption, in those early days, that the French Estates General would simply take over the famous 'balanced constitution', lock, stock and barrel. But the French do not like adopting foreign ideas, and if they were in the market for them at all in 1789, they looked to Philadelphia rather than Westminster. Naturally, the English did not like the role of spurned pedagogue: their pleasure turned swiftly to concern, then to fear, and finally to outright hostility. The journals of that reasonable and open-minded man, Arthur Young, who travelled through France in that fatal year, beautifully mirror the change in English opinion, with mounting irritations at France's inexplicable refusal to adopt the English model, yielding to horror at the violence and confusion, and ending on a note of pure xenophobia.

The change came very fast, and by the end of 1789 sympathy with the French insurgents was already a political liability in England. The response evoked by Dr Price's sermon, in which he compared events in France to 1688, was generally critical: it produced, among other things, the furious lucubrations of Burke in his *Reflections,* the underlying burden of which was that the spread of French ideas would destroy that Ark of the Covenant, the English common law. It was soon almost useless for Fox to ask the English to 'be as ready to adopt the virtues, as you are steady in averting from the country the vices, of France'. There was a marked refusal to analyse what the French were doing, to differentiate between the various facets of the Revolution. Equally, prevailing public opinion insisted that Englishmen who offered modest support to the French were in fact wholeheartedly endorsing their worst excesses. The year 1789 initially brought a distinct radicalisation of the English reform movement, the entry into the arena of lower-middle-class and working-class elements, who formed information and correspondence societies, and got in touch with the National Convention in France. Such elements were small in number, and surprisingly diffident in their objectives. Though Tom Paine's works enjoyed an astonishingly wide sale, only a few thousand people actively engaged in political agitation. And even Paine, though more extreme than any British-based reformer, was a moderate by French standards. His views on private property and the virtue of self-interest were broadly those of Adam Smith. In French politics he was Girondiste, and the only member of the National Convention who fought openly against the execution of the King. The attitude of the English lower-class radicals was typified by the initiation-oath of the Sheffield Constitutional Society (December 1791):

> I solemnly declare myself an enemy of all conspiracies, tumults and riotous proceedings, or maliciously surmising any attempt that tends to overturn, or otherwise injure or disturb the peace of the people; or the laws of this realm; and that my only wish or design is, to concur in sentiment with every peaceable and good citizen of this nation, in giving my voice for application to be

made to Parliament, praying for a speedy reformation, and an equal representation in the House of Commons.

Such moderation was wasted on the English alarmists. The caterwauling of the first French refugees to arrive in England was itself drowned in the hysterical descants of English travellers, and residents on the fringes of the convulsion, most notably Gibbon. Any gesture to the spirit of reform, he wrote to Lord Sheffield, would be fatal in the light of France's terrible experiences:

... if you admit the smallest and most specious change in our parliamentary system, you are lost. You will be driven from one step to another; from principles just in theory to consequences most pernicious in practice; and your first concessions will be productive of every subsequent mischief, for which you will be answerable to your country and to posterity ... If this tremendous warning has no effect on the men of property in England; if it does not open every eye, and raise every arm, you will deserve your fate ... You may be driven step by step from the disenfranchisement of Old Sarum to the King in Newgate.

In this atmosphere, the reform movement came to a complete halt, and was soon desperately on the defensive. The vicious xenophobic obscurantism of the Birmingham 'Church-and-King' riots spread in varying degrees through the country. The Commons accepted Wyndham's mindless dismissal of any scheme to alter the suffrage, 'One does not repair one's house in a hurricane', as an unanswerable truth. Pitt grasped at the rising hostility to France as a formidable weapon to brand the opposition as unpatriotic, just as Walpole had belaboured the Tories with the treasonable Stuart court of St Germains. Many of the opposition, indeed, scuttled hastily to cover, and some joined the Government. Burke's increasingly mad voice rose to a metaphysical scream as he apostrophised the virtues of the English miracle-constitution:

... the well-compacted structure of our Church and State, the sanctuary, the holy of holies of that ancient law, defended by reverence, defended by power, a fortress at once and a temple ... this aweful structure shall oversee and guard the subjected land ... (*Letter to a Noble Lord*, 1796).

The coming of war intensified the public pressure on anyone who could be associated, however remotely, with the Continental peril. In Ireland, the officers of a yeomanry regiment had a schoolteacher flogged because he was heard to speak French, and was therefore presumed to be a rebel. Paine was elevated to the status of a monster, and reading, praising, printing and distributing his works became an absolute proof of disaffection.* In May 1794 the Committee of Secrecy of the House of Commons, reporting on seditious practices, and relying almost wholly on the evidence of paid and unscrupulous informers, accused the harmless correspondence societies of planning a *coup d'état*. The movement, it said, merely paid lip-service to parliamentary reform, and its real object was 'to supersede the House of Commons in its representative capacity, and to assume to itself all the functions and powers of a national legislature'. It was 'a traitorous conspiracy for the subversion of the established laws and Constitution, and the introduction of that system of anarchy and confusion which has fatally prevailed in France'.

There followed repressive legislation of a type very similar to the code which emasculates opposition in contemporary South Africa. Under Section II of the Treasonable and Seditious Practices Act, 1795, anyone writing or speaking words which could be construed as inciting hatred or contempt of the King, the Government or Constitution, could be transported for seven years on the second offence. The Seditious Meetings Act, 1795, forbade meetings of over 50 people, unless previously licensed; if 12 or more remained after the order to disperse, they became liable to the death penalty; even at licensed meetings, anyone advocating altering 'anything by law established except by the authority of King, Lords and Commons' could be taken into custody; magistrates and constables were indemnified if anyone were killed or maimed in the course of dispersal by force; those forcibly obstructing the arrest of offenders were to suffer death; and unlicensed houses where 'lectures . . . on . . . any supposed public grievance, or any matters relating to the laws, Constitution, Government or policy of these Kingdoms' took place were to be deemed disorderly, and their owners fined £100 for every day the act was not complied with. So draconian was this measure that exemption clauses for schools and universities had to be inserted. It was, in effect, a crime publicly to advocate reform in any place except Parliament.

These legal restrictions were reinforced by a torrent of smearing abuse

* Oddly enough, Pitt had a high opinion of Paine. Lady Hester Stanhope, his niece and housekeeper, recorded: 'Mr Pitt used to say that Tom Paine was quite in the right; but then he would add, "What am I to do? If the country is overrun with all these men, full of vice and folly, I cannot exterminate them. It would be very well, to be sure, if everybody had sense enough to act as they ought; but, as things are, if I were to encourage Tom Paine's opinions, we should have a bloody revolution; and after all, matters would return pretty much as they were."' Cromwell had taken the same view of the Levellers. Paine has never been given his due in England; and, like many other famous Englishmen, he met posthumous misfortune, too. In 1818 Cobbett dug up his bones in America and brought them back to England for public exhibition. After Cobbett died, his son was arrested for debt, and the bones passed into the hands of the receiver, where they were subjected to many indignities.

from the presses and the print shops, with the savagely effective Gillray leading the hired pack. He neatly combined Francophobia with reactionary sentiment. The 'Promised Horrors of the French Invasion' (1796) shows French troops marching up St James's. Pro-Government MPs from Whites are being hurled from the balcony or hanged on lamp-posts; sacks of stolen gold from the Treasury are being taken into the Foxite stronghold of Brookes', where a guillotine has been set up on the balcony; and in the foreground Fox himself is scourging Pitt. The same year Gillray was paid by Sir John Dalrymple, an elderly and eccentric Scotsman, who hoped for a peerage if he pleased the Ministry, to produce an even more damaging series, 'Consequences of a Successful French Invasion'. The French are seen taking over the Commons, fettering Government MPs for transportation, setting up a guillotine in the Lords, and murdering clergymen. In the countryside, 'a row of English people in Tatters, and wooden shoes, hoeing a Field of Garlic', are being lashed like Negro slaves by sneering French officers; and in a final scene in Parliament, a French lieutenant points to the Mace and says (in a stroke neatly combining traditional anti-Cromwellian sentiment with anti-French racism): 'Here, take away this bauble, but if there be any gold in it, send it to my lodgings.'

The counter-revolutionary tempest swept all before it. The 'gag' acts were passed by overwhelming majorities and, as even a radical like Francis Place admitted, with the full backing of public opinion. Whig lawyers were prepared to defend the victims of the acts, but they could do little more. In 1797 Grey bravely asserted that the French Revolution 'in the end ... will tend to the diffusion of liberty and rational knowledge all over the world', but his reform proposals, the last to be brought forward for many years, which adumbrated the suffrage-extension of 1867, mustered a mere 91 votes, the hard core of the Foxites. The French Revolution thus retarded British democracy by almost a century. The Whig leaders can scarcely be blamed for not trying harder. Some of them were already branded as unpatriotic for their support of American liberty; and as Lloyd George was to say in 1914, with his mind on the Boer War, no public man can be expected to set his face twice against overwhelming popular sentiment. The mass base of the reformist movement had vanished in the war fever. By 1798 the London Corresponding Society was even proposing to raise a 'loyal corps' to resist French invasion. John Thelwall, who had been acquitted of treason in 1794, gave up the hopeless cause; the English, he said, were 'enslaved because degenerate'. Fox despaired in 1801 at the complacency with which the mass of the English accepted Pitt's system of reaction: 'Till I see that the public has some dislike ... to absolute power, I see no use in stating in the House of Commons the principles of liberty and justice.' There was something, indeed, to be said for the Whigs declining to lend their countenances to a Parliament they were powerless to influence. As Sydney Smith wrote to Lady Grey: 'Of all ingenious instruments of despotism I must commend a popular Assembly, where the majority are paid and hired, and a few bold and able men by their brave speeches make the people believe they are free.'

The long wars against France were a disaster for the English, for the French, and indeed for the world. The English decision to assist and finance the European absolute monarchies in their attempts to suppress the popular movement in France inevitably induced in the French people the familiar psychosis of encirclement, diverted and unleashed energies of a great nation from civil construction and reform to military adventure, and helped to transform a promising experiment in mass democracy into an aggressive dictatorship. The direct cost of the wars to Britain was £831 million (not until the end of the nineteenth century was British public expenditure even to approach the level of 1810–15) and the indirect cost incalculable. The benefits of the astonishing rise in the growth rate of the British economy, which marked the first phase of the industrial revolution, were thus largely siphoned off into purely destructive channels. The exigencies of war-finance and, still more, the economic warfare waged by Britain against French-occupied Europe, combined with French efforts to retaliate, impeded the development of a world trading economy by many decades and, in Britain, produced distortions in the embryonic structure of the new industrial economy which were to have permanent and tragic consequences. Britain sacrificed the splendid isolation of the eighteenth century to no purpose, and became the paymaster in a Continental crusade without a cause. The subsidies she lavishly provided merely kept afloat bankrupt and tyrannous states who used the cash to massacre the Poles and partition their country, to preserve antique social systems plainly due for demolition, and to delay across the Continent the emergence of societies based on the rights of man. Pitt had sensibly remarked, at the time of the trade treaty with France, 'To suppose that any nation could be unalterably the enemy of another, was weak and childish. It had neither its foundation in the experience of nations nor in the history of man.' But as the war continued, such weak and childish notions took possession. By June 1808, George Canning, the Foreign Secretary, was telling the Commons: 'We shall proceed upon the principle that any nation of Europe that starts up with a determination to oppose a Power which, whether professing insidious peace or declaring open war, is the common enemy of all nations, whatever may be the existing political relations of that nation with Great Britain, becomes instantly our essential ally.' Resistance to the 'common enemy', he continued, even the interests of such allies, were to have precedence over 'peculiarly British interests'.

Thus to the flagrant disregard of the interests of the English people – often in direct opposition to them – Britain imprisoned herself in the ideological disputes of the Continent. The cautious but promising liberalism of the young Pitt was transformed into a self-perpetuating series of right-wing coalitions, geared solely to war abroad and repression at home. The destruction of the Napoleonic regime became the solitary and obsessive object of policy, and in the final years of the struggle, control of it fell largely into the hands of two Anglo-Irish adventurers, Wellington and Castlereagh, drawn from the *colon* aristocracy, the most blindly reactionary class in the British Isles. Both were Continentalists by temperament, conviction and self-interest. Their fears of the demon democracy at home mirrored those of the European autocrats, and in this alliance of privilege English interests were disregarded. In January 1814, at the Treaty of Chaumont, Castlereagh created the concept of the great powers acting in concert across national frontiers. Each was to provide 150,000 troops for this purpose, and Britain an additional £5 million. Castlereagh rejoiced at this prodigal unburdening of British blood and treasure: 'What an extraordinary display of power! This I trust will put an end to any doubts as to the claim we have to an opinion on Continental matters.' It would ensure, after the war, the maintenance of 'the order of things'. But what was this 'order'? A Bourbon in Paris, a Hapsburg in Vienna, a Romanov in St Petersburg, a Hohenzollern in Berlin. Britain gained nothing from the war, or from the peace; except putative membership of an international insurance system against revolution, of a type which the English Continentalist minority have always sought, fortunately in vain. The only real beneficiary of the war was Prussia, whom Castlereagh brought deliberately to the Rhine, thus planting the seeds of a future predominance in central Europe. England is adept at creating new monsters to crush old ones already in decline.

Continentalism abroad meant Continentalism at home. The end of wartime inflationary finance brought a collapse of wages, huge unemployment, Corn Laws to keep up the price of food and so the rents of the ruling class, industrial unrest for the first time on a nation-wide scale, repressive legislation, mass hangings and transportations, a cavalry massacre of the Manchester poor. This was the Ireland with which Castlereagh and Wellington were familiar; but it was not an England the English would tolerate. Huge and frightened majorities still endorsed Government policies in the Commons – its refusal to hold an inquiry into Peterloo was carried by 243 votes – and liberal men noted in despair no obvious signs of a crack in the united and brazen front the Tory oligarchy presented to an increasingly angry nation. 'There are our masters!', wrote Sydney Smith, '... it is always twenty to one against the people. There is nothing (if you will believe the opposition) so difficult as to bully a whole people; whereas, in fact, there is nothing so easy, as that great artist Lord Castlereagh so well knows.'

Yet in time the great reaction sickened from within. It was geared to events in France: 'Everything,' wrote Sir Alexander Cockburn afterwards, 'was connected with the Revolution in France, which for twenty years was, or was made, all in all, everything.' But the noise of the tumbrils was fading, and fear of France a wasting political asset. The Government continued to maintain the scenario of conspiracy. 'They are absolutely pining and dying for a plot,' wrote Cobbett. Wellington, a man prone to hysterical delusions, feared a mutiny in the Guards. Castlereagh thought increasingly in terms of violence and lived on the fictions of informers. On the discovery of the alleged plans of Arthur Thistlewood and his companions to murder the Cabinet during a dinner at Lord Harrowby's, Castlereagh proposed that the dinner should proceed, that Ministers should arm themselves to the teeth, and blaze away when the assassins entered. The plan was rejected with raised eyebrows. Castlereagh was already accustomed to take ether before speaking in Parliament (his boss, Liverpool, did the same). His mind was moving remorselessly towards madness, and to fears of political plots he now added a manic conviction that he was being blackmailed as a homosexual. Some of his colleagues now sought an escape from the impasse of his policies.

Continentalism abroad was the first to go. Castlereagh saw the conference system he had invented as the means to promote a united, and reactionary, Europe, an immobile confederation, sterilised of radical infection, 'a new discovery in the European Government ... giving the counsels of the Great Powers the efficiency and almost the simplicity of a single State'. It was the old Roman dream, which the offshore islanders had rejected so many times before. Only Wellington, among his colleagues, showed any enthusiasm. Most shared the view of Greville: 'The result of his policy is this, that we are mixed up in the affairs of the Continent in a manner which we have never been before, and which entails upon us endless negotiations and enormous expense.' The eccentric Russian emperor, with his childish scheme for a Holy Alliance of Catholic, Orthodox and Protestant autocrats, succeeded in caricaturing the scheme, to Castlereagh's fury, and raising atavistic English hackles, even in the Cabinet. In 1818 Cabinet pressure forced

Castlereagh to issue a protest against the Russian proposal to intervene against revolution in Latin America, and in 1820 he was forbidden to attend the conference at Troppau, in which the monarchs claimed the right to impose order on any insubordinate populace. Indeed, he was obliged to publish a Cabinet paper specifically rejecting the doctrine, except in self-defence. He told the Russian ambassador that his heart bled at having to write it: not strictly the truth, since the stronger passages were almost certainly penned by his enemy, Canning. Castlereagh's policies were dead two years before he slit his carotid artery with a penknife. Thus the English escaped from Continentalism for a hundred years, until they were swindled into the Great War.

The collapse of the cross-Channel wing of the reactionary superstructure inevitably imperilled the home base. Canning, with mounting popular approval, reverted to the Big Englandism of Cromwell and Chatham, using the Navy to hold the ring while constitutionalists toppled the orthodox and opened up the obscurantist world to English goods and ideas. Such vigorous liberalism abroad was incompatible with the maintenance of a fortress-state at home. Yet the last stage of the road to reform, though now open, was characteristically paved with English paradoxes.

The first of them was supplied by one of the great suppressed characters of English history, Henry Brougham. This Westmorland squire and Scots-trained lawyer was by far the ablest public man of his generation. His extraordinary capacity to irritate even his warmest admirers has buried his achievements under a landslide of malicious anecdotes, half-truths, slanders and destructive fictions; and the historian must struggle through the debris to discover the salient truth that he was the greatest radical of them all, and the real architect of the age of reform. Whatever aspect of the Victorian reconstruction of society and the State we examine, we find that Brougham had been there before, usually by many decades. Before Waterloo was fought, he had adumbrated the age of Gladstone. Popular education, secular universities, personal freedoms, law reform, the mass suffrage, modern electioneering, free trade – in all he was a pioneer. In 1812 he defeated the Government over the vicious Orders in Council: not soon enough to prevent war with America, for the news of the revocation reached Washington too late, but decisively enough to make an early peace possible. It was the first significant Whig victory for many years. In 1820 Brougham turned the Whigs into a popular party. But he did so by adopting the very traditional opposition device of exploiting a split in the royal family. He played the reform game according to the old English rules.

In the eighteenth century, it was impossible for opposition to defeat government at a general election. Its only hope of achieving respectability, let alone power, was to focus itself around the heir to the throne. It was the fact that there was no adult royal heir during the first two decades of George III's reign which led the Whigs to experiment with the idea of an opposition based on political principles. Once the Prince of Wales was old enough to engage in public life, he automatically took his place as patron of the opposition. But by 1810, when his father went irrecoverably mad, the Regent was too old, and too conservative, to dance the Whig minuet; his mentor, Fox, was dead, and after two years of futile negotiations, he finally decided to stick to his Tory ministers. It was the end of the line for the old Whig system. The Regent had no legitimate son to quarrel with. Fortunately for the Whigs, however, he had an estranged wife, and Brougham was her lawyer. Caroline was an unlovable woman, and beyond much doubt an adulteress. The Regent had tried for years to divorce her. But he was himself an egregious fornicator, and quite possibly a bigamist too. Moreover, his last attempt to divorce her had coincided with a brief spell of Tory opposition, and Percival and Eldon had taken his wife's part – had, indeed, concocted her statement of defence which, on returning to power, they suppressed by legal injunction. So no one had clean hands. When George at last became King, in 1820, he flatly refused to allow Caroline to be crowned Queen alongside him; moreover, against the advice of his Cabinet and his bishops, he instructed her name to be removed from the Anglican liturgy, to avoid, as he claimed, the blasphemy of asking the populace to pray for her.

This was too much even for the English, inured as they were to the traditions of State hypocrisy. Anglicans in private, Dissenters and Catholics in public, prayed for the wretched woman with an enthusiasm and energy which had little to do with religion. Worse still for the Tory ministers, they were now driven by logic and a hysterical monarch to ask the House of Lords to declare that Caroline had committed adultery with her servant Bergami, and that her marriage was null and void. For Brougham, the greatest lawyer and orator of the day, the opportunity was beyond price: a monumental state trial, fought in a blaze of publicity, with opinion overwhelmingly behind him, and ministers in complete disarray. He was privately convinced of Caroline's guilt. But, quite properly, he insisted that the case against her should be proved, knowing this was impossible. The English, as the example of Catherine of Aragon showed, always rally to an injured Queen, even when her cause runs against other instincts; and to this powerful emotional force Brougham added the formidable engine of anti-foreign sentiment. Without exception, the key government witnesses against her were Italians, presumed to be corrupt liars, and indeed known to have been bribed. Brougham's cross-examination of Theodore Majocchi, the most important of them, was a black masterpiece of forensic terrorism. Completely demoralised, the trembling creature found himself saying '*Non mi ricordo*' no less than 87 times, often to questions to which his previous evidence-in-chief had provided emphatic and confident answers. The phrase was taken up by a delighted nation, and it could still raise a laugh in London pubs and drawing-rooms 50 years later. Brougham's principal speech for the defence was declared the finest ever delivered, by men who had heard Fox, Pitt, Sheridan and Burke in their prime. With sublime and magisterial humbug, he concluded: 'The Church and the King have willed that the Queen should be deprived of its solemn service. She has instead of that solemnity the heart-felt prayers of the people.' Brougham thus aligned the populace against the ruling establishment. But if the case embodied politics it also transcended them. Here was a simple issue of right and wrong, which ordinary people could decide for themselves; and in opting for the Queen (and therefore for opposition), they could not be accused of disloyalty, or lack of patriotism, of seeking the overthrow of the law and the constitution. You could not turn out the yeomanry to scatter crowds cheering for the Queen. The repressive apparatus of government seemed suddenly irrelevant. Public opinion and the mob coalesced. At Eton, Caroline townsmen fought a pitched battle with Georgian schoolboys. Greville recorded: 'Since I have been in the world I never remember any question which so exclusively occupied everybody's attention, and so completely absorbed men's thoughts and engrossed conversation.' For the first time the victor of Waterloo was hissed and nearly dragged from his horse. A Tory MP, Edward Wilbraham, wrote nervously to Lord Colchester: 'Radicalism has taken the shape of affection for the Queen and deserted its old form.' Ministers, dismayed by sliding majorities, dropped the bill in confusion. Brougham became the most popular man in the kingdom. His chambers were crowded with gold boxes containing the freedom of towns and cities, and many scores of pubs were renamed the Brougham's Head. By teaching the Whigs a lesson in mass-politics, and by aligning popular unrest behind a constitutional cause, Brougham diverted the ruling class and the people from a collision course, and opened the way to peaceful reform. The trial cut the last links between the opposition and royalty, but it forged a more enduring one with the nation. Thus a ludicrous incident became a political watershed, and a worthless woman made a valuable contribution to English history.

The lesson was not lost among the more intelligent Tories, either. They became increasingly aware that the political power of the Crown was vanishing, and that to survive they must come to terms with public opinion. During the 1820s, the Tory monolith split down the middle. The liberal elements – Canning, Goderich, Palmerston, Grant, Huskisson – moved steadily towards the Whig camp, leaving Wellington and Peel exposed on a dwindling rump of reaction.* All the same, a final paradox was required to end the half-century of paralysis. Pitt had promised the Irish in 1800 to remove Catholic disabilities as the price of union with England; and the King had forced him to renege. This was the issue, above all, which separated the ultra Tories from the Liberals, with Peel and the Duke as the sacramental custodians of the Protestant cause. In 1829, dismayed by Daniel O'Connor's famous victory for Clare County – a seat he could not legally occupy – and terrified by the prospect of a mass uprising of the Irish (25,000 out of 30,000 troops in the United Kingdom were deployed in or against Ireland), the ultra ministers ratted, and the government levies carried Emancipation through Parliament. It was a betrayal of the Protestant back-benchers without any mitigating circumstances whatsoever, and resented accordingly. The back-benchers glimpsed a searing light of revelation: the old, corrupt system, buttressed by their silent votes for decades, had made the treachery possible: it was the Members for pocket boroughs, and the peers who owned them, who carried the Bill, and formed the rank and file of the Duke's turncoat army. Moreover, with legal restrictions removed, there was nothing now to prevent rich papists from buying their way into Parliament and overthrowing the entire Anglican settlement. At last they saw a case for reform! Thus by a supreme irony, extremists of both wings found a common cause, and a motion for reform was jointly moved by the ultra-Tory Marquess of Blandford and O'Connor himself. The death of the wretched George IV precipitated a general election; Brougham campaigned in Yorkshire, addressing meetings of 20,000, even 30,000, in a foretaste of the Midlothian campaigns half a century later; in July 1830 the French threw out the Bourbons without bloodshed, so that for once cross-Channel

* Brougham had argued for many years that the reformers could achieve power only by splitting the Tories. Thanks to him, at the election of 1826, many Whigs supported liberal Tories, and vice versa. He fought consistently against the power of the great landed families by bringing into play the opinion of rank-and-file MPs, especially during the negotiations for the forming of Canning's government in 1827; it was almost certainly he who wrote a leader in *The Times*, 16 April 1827, pointing out that the government of the country was vested, by law, not in the great families but in King, Lords and Commons. Brougham finally sacrificed his political career in 1830 by accepting the Chancellorship; he was persuaded that, if he refused, Grey would resign the task of forming a government, and reform would be delayed by another 25 years. But if Brougham had remained in the Commons, he must surely have become Prime Minister, and advanced the era of Gladstonian reform by a generation. See Chester New: *Life of Henry Brougham to 1830* (Oxford, 1961).

politics worked against English reaction; and in November the angry ultras helped to turn the Duke out. Thus the last piece of the jigsaw fell into its place: now that the Whigs had their hands on the levers of power, reform in some shape was inevitable. As Macaulay put it: 'I know of only two ways in which societies can be governed – by public opinion or by the sword.'

The Great Reform Bill, like Magna Carta, was drafted in haste and carried in confusion. Largely by accident, it turned out to be a miracle of English social engineering, a famous non-victory for the people. It doubled the electorate, redistributed a third of the seats, and rationalised the franchise. It was thus radical enough to persuade the Tories to fight it almost to the last ditch, and thereby convince the innocent populace that they were getting something significant. In fact it was timid compared to the bill its architect, Lord Grey, had sponsored nearly 40 years earlier. It skilfully postponed the advent of a mass-franchise for another half-century. While admitting a significant section of the middle class to the fortress of the property state, and so enormously strengthening the garrison, it slammed the door on the workers: none of them got the vote, and those who already possessed it were disenfranchised. By making the minimum concessions to avert revolution, it effectively denied the use of the sword both to the forces of reaction and those of democracy. Granted the instinctive conservatism of the English people, and the long experience of the gentry in exploiting it, the latter found no difficulty in making nonsense of the bill's provisions. They invaded the new boroughs just as their predecessors had taken over the old ones in the fifteenth century. In 1867, a detailed analysis of the background and connections of MPs showed that the changes in social composition, over 35 years, had been almost imperceptible: the 'aristocratic element' held 326 seats, more than half, and they were almost equally distributed between the two great parties. Indeed, the beneficiaries of the old system were even more securely in control of the new, because its indefensible anomalies had been removed, and it was far less vulnerable to frontal attack. Thus the possessing classes learned a valuable lesson in consolidation through reform, and the modern pattern of British politics took shape. Moreover, the blood transfusion, which set the constitution on its feet again, permitted, as we shall see, the systematic refurbishing of a variety of institutions, whose net effect was enormously to strengthen the resources of the privileged classes. Over the Reform Bill, reaction lost the battle, but conservatism won the war.

. . .

The fact that the English avoided a political breakdown in the early nineteenth century is all the more remarkable in that they were undergoing social and economic changes of unparalleled scope and severity. The English industrial revolution of 1780–1820 is the great watershed in the history of mankind. It liberated the body, as the Reformation had liberated the mind. Indeed the two were intimately connected. It was in the light of the escape from Rome, and the break-up of a static intellectual system, that Bacon saw the Fall reversed and forecast man's conquest of a hostile and grudging environment. He regarded the prospect as stupendous and imminent, and so it might have been, for he wrote on the eve of great events. The collapse of the English republic undoubtedly decelerated the process, but it was beyond anyone's power to halt it. Indeed, we can trace from the middle of the sixteenth century a majestic chain of events, each projecting the next, which made the outcome of the modern world inevitable.

Geography had always placed the English significantly apart from the Continental conflux of societies whose very proximity and interaction secured their conservative elements in possession. The Channel gave us a certain eclectic freedom in the reception of Continental ideas: we could take by choice; we could not be made to receive by compulsion. The act of separation might have occurred much earlier, and the film of history speeded up in consequence. At all events, the change was decisive when it came. The religious revolution made possible a revolution in education, not just in scope but in quality. The new education bred the first scientific revolution, and it was the impact of scientific rationalism on society which brought the political and constitutional revolution of the 1640s. From this convulsion we can date the agricultural revolution, which completed the break with the subsistence economy, and made possible the commercial and financial revolution of the late seventeenth century. The flow of cheap money thus secured, the stability of credit, the rapid development of world trade and, not least, the emergence of a sophisticated consumer market at home, combined, in the 1780s, to produce the revolutionary combination of capital and technology in the mass-production of goods by powered machines. This transformation, paralleled by the administrative revolution in the central organs of government, in turn projected the social revolution of the nineteenth and twentieth centuries. English religion died in the process: the Reformation God did not live to see His handiwork. Nevertheless, He was the prime mover in it all. The Gospel according to Karl Marx, or to Mao Tse-tung, or to Keynes, all spring by direct intellectual descent from the Protestant Bible. And behind it all lies the enigmatic, mocking smile of Pelagius.

By the early 1780s, when all the economic indices took a sudden, and sharply upwards, turn, England's situation was unique. Some of the elements required for an economic transformation were present in other parts of the world, but only England possessed them all in combination. The 'miracle' had been brewing for 150 years; or, to vary the metaphor, a number of conventional factors of economic growth had been drawing together, and in the late eighteenth century the resultant mass became 'critical', and the explosion took place. One of the problems which Roman, medieval and Renaissance societies had failed to solve was how to make long-term investment in agriculture both safe and rewarding. It was a legal, rather than an economic, problem, for capital could not be raised or usefully employed unless the law underwrote mortgages in the interests of both parties and guaranteed the integrity of estates on inheritance. The triumph of the common lawyers in the 1640s provided a two-fold solution. The law of strict settlement ensured that entailed estates passed intact from generation to generation; while equity of redemption made the mortgage on the one hand a secure form of investment and regular income, and on the other a respectable way of raising capital for improvement. Men need no longer bury their money in holes in the ground; nor need landowners sell land to raise working cash. These simple, but original and highly effective devices, led to a rapid and sustained rise both in agricultural production and in productivity. The scientific knowledge already existed, for the most part: from the 1640s books and pamphlets ensured its wide diffusion, and once the cash began to flow it could be widely applied and improved by experiment. By 1670 the revolution of the land was complete in all essentials, and during the next century farmers and landowners systematically exploited its techniques.* A great deal more land came into cultivation: fenland drainage schemes alone added 10 per cent to the total farming area. The enclosure of the commons, pushed forward by a gentry-dominated Parliament, completed the reorganisation of the structure of English agriculture – formed by the three tiers of landowner, tenant farmer, and landless labourer – which began in the decades following the Black Death. Enclosure involved great cruelty and injustice (as it did, on a much more ferocious scale, in the nineteenth-century Scottish highlands and

* The elements included the floating of water-meadows, stock-breeding, techniques of drainage, introduction of fallow crops, like turnips and clover-grass, and the use of natural, and even of artificial, fertilisers. Most of the enclosures had actually taken place by the end of the seventeenth century. See E. Kerridge: *The Agricultural Revolution* (London, 1967). Production for the market was assisted by the fact that, by the absence of internal tolls, England was already the largest free-trade area in Europe, an area further expanded by the union with Scotland.

Ireland), for it deprived great masses of the rural poor of marginal sources of food and income, and left them almost wholly dependent on wages. On the other hand, in many ways it merely brought rural poverty, which had always existed, into the open; forced society, indeed, to use the poor law to underwrite rural incomes, an obligation already acknowledged by 1730, and made almost universal in the south-east and midlands by the adoption of the Speenhamland system in the 1780s.

Moreover, the agricultural revolution undoubtedly prevented more misery than it caused, simply by allowing more food to be produced and marketed. Even by the end of the seventeenth century, England was exporting grain worth £250,000 every year on average. These foreign earnings were not particularly significant. More important was the fact that England achieved a surplus while managing to feed a rapidly increasing population. What exactly caused the population to rise is not clear: it seems to have been a general phenomenon, at any rate in the Eurasian land-mass. It may have been due to the exceptionally fine weather which marked the half-century 1700–50. In England a contributory factor was certainly the virtual disappearance of plague before the end of the seventeenth century. But a rise in living-standards, especially of food-consumption, cut death rates in all age-groups, and markedly among infants; and it seems, too, to have produced a marginally higher birth-rate. At all events, English population, which had been 3 million in the early sixteenth century rose to 4 million in 1600, and to an estimated 5·5 million in 1650. By 1750 it had reached and passed the 6 million mark, and 30 years later it was 7·5 million.* Such upward movements had occurred before in all parts of the world, including England; and the inability of agriculture to keep pace had produced the 'natural' adjustments of famine, plague and war. In the eighteenth century they were not required: the theories of Malthus, though generally accepted, were not so much false as obsolete.

English agriculture not merely fed the new masses, it fed them better. There were periodic famines in Continental Europe and in Ireland: indeed in Ireland an equally rapid increase of population was fed only by the universal adoption of an inferior, and desperately vulnerable, potato diet. But in England there was a steady improvement. Almost everywhere, horses, which consumed more but were four times more productive, replaced oxen as the motive-power of the fields. Even the poorest labourers switched to white bread, and there was a huge increase in meat consumption, made possible by systematic stock-

* These population estimates are still a matter of controversy. For a recent critical analysis, and a useful table of rival calculations, see L. A. Clarkson: *The Pre-Industrial Economy in England, 1500–1750* (London 1971), pp. 25–41.

breeding. Bakewell proudly, and accurately, asserted that he bred his sheep for the masses as well as the classes. Though England ceased to be able to export food by the 1780s, even in 1830 some 90 per cent of her food was still home-grown. It was an astonishing achievement on the part of English agriculture: without it there could have been no possibility of concentrating such numbers in the new industrial units. In the twentieth century, industrialisation has invariably been accompanied by an absolute decline in the agricultural sector. In eighteenth-century England, the two expanded simultaneously, the latter with a vastly reduced work-force. The rise in agricultural productivity must have been phenomenal, and in the high-yield areas, as Cobbett's *Rural Rides* makes plain, it was achieved at great human cost. But the alternative was far more terrible: it could only have been widespread starvation. In fact nobody starved, and most ate better than ever before in history.* English agriculture did not, as historians once believed, finance the industrial revolution: that, we shall see, was not necessary. But it did something more important: it enabled the new industrial proletariat to stay alive, for if home supplies had failed, there was no alternative source.

Capital for investment was no problem. The English had always been able to save, and had always done so. They had had a strong currency since the eighth century; and, despite the occasional follies of their rulers, it remained the most stable in the world until recent times. The difficulty lay in persuading the English to fork out their cash – that is, to guarantee security in return for a much lower yield. Here again, the Commonwealth years were decisive. Absolute monarchy is the enemy of safe investment. The Stuarts were opposed to a central bank in principle, as a rival power-structure, and in any case they could always use their executive power to renege on their debts. The result was they had to pay between 12 and 20 per cent for their money, and general rates were over 10 per cent as long as the Stuarts were around. Even so, money stayed underground. The Commonwealth showed that a broadly-based government, committed to the sanctity of private property, and with an open ear to the mercantile interest, could raise money at modest rates even while fighting a civil war. After the Stuarts were finally expelled, the lesson was rammed home. In 1694 the Bank of England took over the role of the City as lender to the Government in its corporate capacity; it could mobilise monetary resources from all over the country, issue paper, and lend to Government at 8 per cent

* In 1688 Gregory King estimated average English incomes at between £8 and £9 a year; the average for labourers, cottagers and paupers was just over £3. By 1780, on the eve of the industrial revolution, incomes had doubled.

with a parliamentary guarantee. It made lending to the Government safe, just as the strict settlement and equity of mortgage made lending to landowners safe. Thus, in a curious way, the Great Revolution saved both the property-state and the landed interest, just as Roosevelt was to save the capitalist system in the teeth of its main beneficiaries. The strength and possibilities of the new system soon became apparent. New techniques and international clearance of debt ended the primitive old business of shipping thousands of sacks of coins from country to country. A stock exchange developed to mobilise and distribute internal capital; and early in the eighteenth century marine insurance drastically reduced the speculative element in overseas trade.* The system survived the new and alarming experience of the South Sea Bubble: the crisis was not solved by Walpole, it cured itself; and the lesson was learnt that a country could not operate two central banks, competing with each other and backed by rival political factions. By 1727 the rate on government stock was reduced to 4 per cent, and in 1757 to 3 per cent. Thus by the mid-eighteenth century there was ample capital available for canals, roads and other improvements to the infrastructure.

Technology was no problem either. A surprising amount of industrial machinery was in use in England long before the age of steam. More inventions were knocking around, waiting for exploitation. Scattered across the country were pockets of industry, some organised in comparatively large units. Even in 1700, silk manufactories, for instance, employed up to 700 hands. The huge size of London made inevitable the creation of supply industries geared to mass-production and demanding the increasing use of machinery. London consumed vast and growing quantities of coal, chiefly from the Newcastle fields: the business of getting it up and shipping it was shaping the modern Newcastle even in the mid-seventeenth century. There was a keelman's strike for higher wages there in 1654; by 1699 they had a strike-fund; and when they struck again in 1719 it took a regiment of regulars and a man-o'-war to keep the peace. By the mid-eighteenth century total coal production was already in millions of tons, and steam power was increasingly used to mine it. The central problem in industrialisation

* Cheques for internal use came into circulation about 1675; paper settlement of international transactions followed after 1688. See J. Spelling: 'The International Payments Mechanism in the Seventeenth and Eighteenth Centuries', *Economic History Review*, 2, XIV. The process was assisted by the rapid development of government statistics, following the appointment of Charles Davenant, the economist, as Inspector-General of Imports and Exports; his office supplied information to the Treasury and the newly formed Board of Trade. Accurate statistics made various forms of cheap insurance possible; and it is significant that, about this time, the English ceased to use the abacus. For further information see P.G.M. Dickson: *The Financial Revolution in England* (1967).

has always been – is still today – an adequate supply of men skilled in the intermediate technologies of metal-work. By 1750–60 a whole range of industrial developments in England had created this supply, and there were no institutionalised and class barriers between metal-workers and engineers. It was thus comparatively simple to move into the advanced technology of the factories. Machines were improvised; one development bred another. Most of the pieces of the jigsaw lay around: it was a matter of fitting them together, and then inventing the missing bits. The critical moments were not delayed by inadequate technology, but by the absence of demand, or rather by the failure to recognise it existed.

The increased circulation of money in the seventeenth century, promoted by the high-taxation policies of the Commonwealth, created a new attitude to consumption. Shops charging fixed prices, even in villages, slowly replaced fairs. The end of haggling marked the dawn of the modern world in England: this was what was meant by the 'nation of shopkeepers', a phrase otherwise meaningless. Home demand did not create the factory revolution, but it prepared the way. It made men think in millions instead of thousands. The great leap forward in the 1780s, however, was essentially what we would call an export-led boom. English mercantilism, born in the Commonwealth, carried forward by Blake's ships, was adapted to the creation of closed foreign markets for English goods. The Navy enforced strict protection for English trade until English industry was strong enough to risk free trade, and then preached it to the world as an article of moral faith. By the 1700s the English were already maintaining a two-power naval standard.* The overseas markets were of vital importance. There were already 4 million Americans in the 1780s, worth, in consuming power, over 40 million Europeans. And the English were opening up a market of 100 million Indians by direct annexation, enforced treaty, and the suppression of native crafts. This was the background to the cotton explosion, which dominated the first phase of the industrial revolution. More than any other product, it linked universal demand with new methods of mass-production. Cotton factories grew up in the

* After the battle of La Hogue in 1692, the Navy could always deny French colonies continuous help from Europe. England was already the leading naval power. During the War of the Spanish Succession she adopted the principle that the Royal Navy should equal or surpass the naval strength of any two powers combined. In 1756, for instance, Britain had 130 ships of the line; France and Spain, despite recent big increases, only 63 and 46. This standard was maintained until 1912, when the object of naval policy was limited to maintaining decisive superiority over the German Navy in the North Sea. In 1918 Britain was still the world's greatest naval power; but in 1922 she accepted parity with the United States, and a 5 to 3 standard with Japan, at the Washington naval conference.

hinterland of the great ocean ports, eventually concentrating in the Lancashire catchment area of Liverpool, the greatest of them all. In the first half of the century, industries supplying mainly home demand increased output by 7 per cent; export industries by 76 per cent; in the years 1750–70 the figures were 7 and 80 per cent. After that point, the revolution took off, and export figures climbed astronomically, with cotton supplying the bulk. The major inventions came in the 1780s, as soon as the demand justified them; and the factories were built around them. Their size and concentration, in Manchester and elsewhere, astonished contemporaries. People knew immediately that something extraordinary and irreversible was happening, and that the world could never be the same again. Revolutions in religion and politics, in science and education, bring devastating changes in men's lives; but they do not alter the physical appearance of things. The factory revolution did: it provided ocular evidence of a monster growth, new shapes, colours, smells; it changed the very air and the rivers and the fields, abolishing the seasons and transforming the daily pattern of existence. It brought an entirely novel psychology of growth and motion, and a new relationship between man and nature.

Yet if the English recognised they had given birth to a new and sensational event, this does not mean they understood their offspring, or had the least idea of how to bring it up. Indeed, they botched the accouchement; the creature was malformed and ailing from the start, and nothing in its upbringing and education was calculated to ensure purposeful and healthy growth. This is not surprising. The English are a pragmatic people. They work through practical expedients rather than majestic conceptions. The industrial revolution was the product of a thousand empirical solutions to separately considered problems, devised over centuries, which by a process of accumulation suddenly produced a qualitative change in the way economic society operated. It was, in fact, an unplanned muddle, and it remained one. If the political events of the seventeenth century had taken a different shape, if the Commonwealth had survived, if the English had chosen to direct their social development, rather than to buy stability at the cost of surrendering to their blind, traditional instincts of evolution, the industrial age would have come sooner, and would have been subjected to the disciplines of foresight, and a goal. The English in the mid-seventeenth century had the courage and the optimism to juggle consciously with dramatic ideas about their destiny. They felt – they knew – they were radically different. God had great plans for them. They still possessed the spiritual audacity to seize on a new phenomenon like industrialism, to identify its divine purpose, and to intellectualise the part it was designed to play in their creation of God's kingdom on earth. Industry came to England not too soon, but too late. Even by 1700, property had already replaced moral purpose as the framework of English society. By 1800 God was restricted to the churches and the chapels. The rest of the patrimony was parcelled out among an individualistic, secular society, operated by secondary causes, according to rules which, if superficially rational, depended in fact on anarchic change. It was no longer possible to fit the march of events into a recognisable scheme, and to advance it accordingly. Indeed, those few who still tried to think in such terms were inclined to interpret the new phenomenon as inimical, even hostile, to God's will. Milton would have hailed the factories as divinely inspired; to Blake they were the 'dark, Satanic mills'.

The truth is, the industrial revolution caught the English mid-way between faith and reason. The laws of God were hopelessly eroded; no one even thought of applying them to the practical business of running industrial society; at the most they could be brought to bear on mitigating its effects. On the other hand, the idea that man was in sole and unrestricted charge of his destiny, and must himself write the rules in the light of reason and experience, had not yet been born. In this half-way house it was assumed, instead, that the rules sprang themselves from the operative processes of nature, were self-formulating and self-enforcing. The only sin was to attempt to interfere with them. The only duty was to discover what they were. The concept was half-scientific, in a sense Newtonian. But it was also half-obscurantist; it failed to differentiate between natural forces which were irresistible, and social forces which could be controlled or reversed. After all, the belief that the end of intellectual effort was purely interpretative, that the body of knowledge was complete and finite, and had merely to be extracted from the dross of error, was the root cause of the medieval paralysis. It was the essence of scholasticism.

Here we come to the tragedy of the industrial age, a tragedy which is still with us. The economists took over the role of the schoolmen. They forgot, if they had even grasped, Bacon's assertion that the object of analysing nature was to learn how to control it. They confined themselves, as had the schoolmen, with dogmatic theology, to elucidating the law, and terrifying the secular power into allowing it to enforce itself. The industrial revolution was born in pragmatism; it grew to twisted maturity in an intellectual climate of blind theory, which forbade in any circumstances the use of physic or the surgeon's knife. A few men who were close to the actual physical events and tried to intellectualise, as it were, from the factory floor, came to radically different conclusions. As early as 1815, Robert Owen, in his *Observations on the Effect of the Manufacturing System,* forecast that the revolution 'will produce the most lamentable and permanent evils, unless its tendency is counteracted by legislative interference and direction'. He could not have put the modern, rationalist case more clearly; but his voice was wholly ignored. The schoolmen-economists had the monopoly of the public ear. They were not men of action. Some, like Malthus, were indeed clergymen, or clerical academics. Others were financial manipulators, like the stockbroker Ricardo. They knew nothing of the new factories, except as observers. They tried to solve the problems of the world in the quiet of their studies, inside their own heads. Their systems were as majestic, as logical, as complete and perfect as the great *summae* of St Thomas Aquinas and St Bonaventure, and as irrelevant; and, like the schoolmen, though they differed in detail, and argued acrimoniously among themselves, they shared common premises, and reached a common consensus. For the laws of the Canon they substituted their own 'iron' laws of wages, and theories of value. They produced a new vocabulary of mumbo-jumbo. It was all hard-headed, scientific, relentless, immensely appealing to an intellectual *élite* brought up in the atmosphere of the minimum state. The iron laws must be allowed to operate: society would be crushed if it sought to impede their remorseless progress. It is an astonishing fact that all the ablest elements in English society, the trend-setters in opinion, were wholly taken in by this monstrous doctrine of unreason. Those who objected were successfully denounced as obscurantists, and the enemies of social progress. They could no longer be burned as heretical subverters of the new orthodoxy, but they were successfully and progressively excluded from the control of events.

Such scientific inquiry as did take place was concerned exclusively with trade and finance, with credit and the money supply, with paper currency and the bullion problem. These matters were closely examined by parliamentary committees, and hotly debated in Parliament. In a curious way, the factory revolution had grown up outside the financial system: for the most part it was locally or self-financed. It produced dramatic consequences for the trading and financial community, which was closely linked to Parliament and administration; and it was these consequences which were analysed – the basic industrial cause was ignored. No one in authority visited the new factories. They agreed, with reluctance, to have them inspected by the State: but for moral purposes, to prevent the abuse of persons, not to discover how they worked, or how they could be made to work better. Throughout the industrial revolution, the English saw themselves as a trading, not as an industrial community; indeed they still do. Peel, who had more influence than any other man on the public response to economic events for nearly 40 years, took little interest in industrial matters as such, though his father had been the greatest of the early cotton pioneers. He was the skilled chairman of the bullion committee, and later the architect of unrestricted free trade. But he visited no factories, did not seek the company of engineers and inventors. He turned his back on the machine and built up a remarkable art collection from its profits.

Thus the English botched the greatest opportunity in their history. The real creators of the revolution, the mechanics, the inventors, the chemists, often died in obscurity or poverty, even, in the case of Priestley, in exile. The big fortunes were made by the second generation of industrialists, men who organised rather than invented; or those, like Arkwright, who ruthlessly appropriated and forged together other men's ideas. The creators, indeed, got themselves a bad name among the right-thinking. The head of the house of Rothschild said there were three ways to lose your money: through women, gambling and engineers. 'The first two are more pleasant, the third more certain.' English snobbery played a devastating and destructive role. The machine-men were not welcome in society. To be so, they must first pass through the transfiguring and cleansing process of acquiring a landed estate. Because, in this sense, English society was so open, the industrialists never became an assertive, self-protective and influential caste: they could escape into respectability by buying broad acres, and the seats, peerages and political power which went with them. Money was the sole materialistic incentive of industrial pioneering; all others – social position, political influence, public esteem, intellectual approval – were non-industrial, even anti-industrial.

If England became an industrial country against the current of social approval, it is hardly surprising that its growth was unsystematic, haphazard and violently irresponsible. It emerged from the unplanned activities of small men, and it remained decentralised, composed of tiny or medium-sized firms, often highly specialised; there was no public or private pressure to produce an integrated national, or even regional, structure. We still suffer from these evils today. But the horrors produced at the time were far more obvious, though no one sought to relate them to causes. Though the English had learned the virtues of long-term planning in estate-management – it had become part of their moral code – industry was created without any forethought, or any consciousness of the need for it. The first generation of factories were built to employ steam only for spinning. Their enormous output brought into existence a quarter-million of hand-loom weavers, who were completely outside the factory system. When the power-looms arrived in quantity, rising from 2,400 in 1813 to 224,000 in 1850, the independent weavers were mercilessly starved out of existence. This human tragedy was an industrial tragedy, too, for these wretched families, by accepting depressed wages, allowed inefficient factories to operate on marginal profits long after they should have been replaced. Already by the 1840s, England had lost her technological lead in cotton: and even a diminished rate of expansion could only be maintained by exploiting protected colonial markets in the backward parts of the world.

The growth of population, indeed, did not cause the industrial revolution in England; it very nearly aborted it. Factories paid higher wages than domestic industries; all the same, they were very low, chiefly because most of the factory hands were women and children. Low wages kept home consumer demand down; worse still, they removed the chief incentive to replace primitive machinery by the systematic adoption of new technology. State limitations of human exploitation came too late, and were too ineffective, to make the quest for productivity a virtue: the English did not discover it until the twentieth century, by which time the trade union movement had constructed powerful defences against it. No attention was paid to management efficiency, cost accounting, development research or the planned relationship between skill and machinery. Profits could be made without such frills; and when trade declined and profits fell, the answer was simply to cut production and lay off labour. When world credit crises occurred in the 1820s, 1830s and 1840s, as they periodically did, without warning or apparent cause or cure, no one thought to compensate for the drop in exports by stimulating home demand. On the contrary, wages were allowed to fall and unemployment to rise: the men who controlled society, even enlightened and well-informed men like Peel, simply waited in a spirit of pure Micawberism.

Cotton as a prime motive force of national economic growth was a fading instrument even in the 1830s. The huge distress, and the consequent agitation, of these years made many people believe that the industrial revolution was on the point of collapse. And so it was. Cotton was too narrow and vulnerable a base to produce self-generating advance. The Chartists, rightly according to the evidence available to them, rejected economic solutions and concentrated exclusively on a political programme. Long before Marx, they thought the final crisis of capitalism was at hand; they wanted democracy to mitigate and civilise its consequences. Corn Law Repeal was the only widely accepted economic panacea – as it turned out, an irrelevant one. The economic schoolmen, and their political pupils, still clung to the subsistence theory of wages: if men were paid more than they needed to stay alive, they would simply work less. Not until about 1870 did anyone in England grasp the economic virtues of high wages. To end the Corn Laws, it was thought by the enlightened, would enable manufacturers to keep wages low and so profits high: thereby lay progress.

In the end the English economy was saved by accident, and with it the social system. It was as though, as with the most primitive system of agriculture, the possibilities of growth in one set of fields had been exhausted, and the tribe moved on to virgin lands. Demand for cotton fluctuated alarmingly; but demand for coal, a much older industry, grew steadily: by 1842 Britain was producing 30 million tons annually, two-thirds of the world's output, though most of it went into English fireplaces. And the needs of mining gave a violent propulsion to the capital goods industries. The stuff had to be moved in ever-increasing quantities: so steam was at last applied to transport. In 1825 the Stockton–Darlington line was built to get coal from the north-east fields to the ports of shipment. It was a primitive goods-line, run like a road, with coal-owners supplying their own trucks and even engines. But it made enormous profits, and it was seen as the future. There was a little railway boom in 1835–7, and an enormous one in the late 1840s. By 1850 over 6,000 miles of track had been laid in Britain. Railways were a much more spectacular development even than cotton: they were not concentrated in limited areas, but ubiquitous – they literally cut through all the delicate and traditional strands of a still rural society. They also absorbed colossal quantities of capital – they could not be self-financed like cotton factories – and were closely related to all the advanced technologies. They gave an enormous and sustained impulse to the British economy, and dramatically ended the first crisis in capitalism. They stimulated new ranges of metal industries, which in turn produced others. The railways led to the ocean-going steamships, and so to the great shipyards. These new industries paid high wages: they had to; and the percentage in the total work-force employed in them rose steadily. By 1914, indeed, engineering embraced the largest single group of male workers, and the highest paid. There were great leaps forward in steel, and an ever-growing demand for coal, which gave work to 1·2 million by the 1910s.

High wages and engineering created the social basis of mid-Victorian stability. Social problems which had seemed insoluble and menacing in the mid-1840s had vanished from sight a few years later. Chartism collapsed; the working class was de-politicised for a generation; the events of 1848, which terrified the Continent, were a fiasco in England, as lower-middle-class Special Constables kept a diminished and bewildered multitude at bay. The constitution, which had barely survived in 1830–32, was firmly re-established as the sole custodian of human felicity, and the parliamentary statute as the only engine of progress and justice. There was a genuine degree of reconciliation between classes. Peel laid down the maxim: 'Whatever be your financial difficulties or necessities, you must so adapt your measures . . . as not to bear on the comforts of the labouring classes of society.' This was high Tory doctrine, but increasingly endorsed by the bourgeois elements now fully integrated and identified with the property-state. 'The middle classes know,' said Shaftesbury, 'that the safety of their lives and property depends upon their having round them a peaceful, happy and moral population.' The workers turned to education and self-improvement. 'Denounce the middle-classes as you may,' an anti-Corn Law speaker told a Chartist gathering, 'there is not a man among you worth a halfpenny a week that is not anxious to elevate himself among them.'

But if the second phase of the industrial revolution brought the English social and political stability, and even a measure of widely spread prosperity, it ultimately dethroned them from their unique position in the world economy. In the first phase we exported exclusively consumer goods; in the second, machines and technology. The difference is important. Until the 1830s, England was the only industrialised country; the English had performed the miracle, and the key to it was still in their exclusive possession. At that stage, the English could still, by an act of conscious and deliberate policy, have set a double standard for themselves and the rest of the world. They might have circumscribed the major industrial developments within the limits of the British Isles, and denied the other developing countries, notably the United States, France and Germany, the financial capital, the capital goods, the expertise in knowledge and skilled personnel, to construct their own industrial bases. Such a policy was not beyond the bounds of possibility. It would doubtless have aroused intense foreign antagonism. But Britain was then in a position to apply the new industrial processes to military technology in a manner denied to the world beyond, and to achieve an overwhelming supremacy in the use of firepower over any other nation or group of nations. This was, indeed, English naval policy, maintained with increasing difficulty until 1914. But the two-power standard, as operated by Britain, was essentially defensive. It never seems to have occurred to the English even to consider the possibility of exploiting the new industrial power they had created to achieve and maintain a world hegemony of advanced weapons. Still less did it seem to them right or desirable to confine the industrial age to its insular base. Other nations might have taken a very different attitude. They might have applied logic and foresight and chauvinism to the problem of restricting heavy industry to a single, national power-base, and to erecting a high and perpetual barrier of technology between themselves and the rest of the world. But England was an open society. It lacked the psychology of a fortress-civilisation. The English saw themselves as superior: but it was a superiority of degree, not of kind. Even in their most manic phase, even when they saw themselves as the chosen people, they sought to be missionaries, not conquerors. They had no desire to be the master-race. The world was there to be evangelised, not conquered. Everything English was for export. They wanted the sincere flattery of imitation, not the servility of helots. The world could have English political and constitutional habits, free. It could have English industrial habits and experience too, at a fair commercial price. To the English, no other decision seemed possible. The matter was not even debated. It was just allowed to happen.

Thus, in the middle decades of the century, England became the workshop of the world; and, in the process, helped to create rival workshops throughout it. In the 1850s Britain produced two-thirds of the world's coal, half of its iron, more than half its steel, half its cotton cloth, 49 per cent of its hardware, virtually all its machine-tools. The industrialisation of half a dozen major economies took place by courtesy of British tools, patents, industrial know-how and skilled personnel; and it was largely financed by British capital. By 1840 Britain had £160 million invested abroad; by 1873 nearly £1,000 million. During this period international trade multiplied five times over, and passed the £2,000 million mark. The railway–steamship age created the modern world-market economy; the English device of the gold standard was generally adopted, and centred on London as the financial pivot of the liberal international trading system. For the English, it was the high-water mark of their fortunes relative to the rest of the world. English ideas, institutions, attitudes, tastes, pastimes, morals, clothes, laws, customs, their language and literature, units of measurement, systems of accountancy, company law, banking, insurance, credit and exchange, even – God help us! – their patterns of education and religion became identified with progress across the planet. For the first time, the infinite diversities of a hundred different races, of tens of thousands of regional societies, began to merge into standard forms: and the matrix was English.*

. . .

* Helped by the fact that Britain was also the leading exporter of people. Between 1750 and 1900 the population of Europe rose from 140 to 401 million, raising its share of

While England was evangelising the world – forging and exporting the matrix – the English were engaged in a protracted and in many ways unsuccessful struggle to civilise themselves. We regard the nineteenth century as the great age of reform, and indeed it was, in the sense that virtually all the institutions and attitudes of English society were re-examined and altered so that the structure which surrounds us today is in most ways the product of that time. But little that was entirely new was added; hardly anything which was essentially old was removed; and the net effect was rather to rationalise and strengthen the basic components of the eighteenth-century nation than to rebuild it from its foundations. All that England really got in the age of reform was an elaborate face-lift, a piece of cosmetic surgery which left the old bones of property and class virtually intact, in some ways stronger than before. We can learn a great deal about the English – and in particular about their inability to escape from their present-day predicament – by examining this process of conservation by change.

On Thursday 8 October 1885, a remarkable funeral took place at Westminster Abbey. The Earl of Shaftesbury, better known to the older generation as Lord Ashley, had been the most pertinacious and ubiquitous of all the great Victorian free-lance reformers. He was a Tory by birth and conviction; he opposed, though with increasing diffidence, all schemes of political reform. In some ways he was a reactionary. As a fundamentalist evangelical, he was a powerful enemy of secular education, and one of the architects of the Sunday Observance Laws. But he was not a party man. He declined office. He loathed organisational politics; he thought government was impersonal, cynical, materialistic and often grossly hypocritical – and he had a long and bitter experience of negotiating with British governments. Ultimately, he predicted government would take over the task of making a better world; but in the meantime he, and other voluntary do-gooders, had the sacred duty of standing between defenceless groups of humanity and the blind or malevolent forces which degraded them. He was, as he put it, 'the great *pis-aller*'. It is easy to sneer at Shaftesbury, from behind the ramparts of the Welfare State. He exuded paternalism, and

---

world population from a fifth to a quarter; within Europe, Britain's share rose from 5·7 to 9 per cent. The British Isles provided the lion's share of the European emigrants, especially in the first half of the nineteenth century; even in the period 1850–1900, 10 million out of 23 million European emigrants came from the British Isles. The chief target was the United States. Its first census (1790) revealed that 80 per cent of the white population was English, 92 per cent British. Between 1815 and 1914, out of 35 million fresh emigrants to the US, 20 million were British. In 1950 it was estimated that total world population of British stock numbered 140 million, half in the US, one-third in the British Isles. See D.F. Macdonald 'The Great Migration', in *Britain Pre-Eminent: Studies in British World Influence in the 19th Century*, edited by C.J. Bartlett (London, 1969).

the unctuous religiosity of the Sabbatarian. Yet in a real sense he represented all that is best in the English character. He not only loved justice and hated cruelty: he devoted his entire energies throughout a very long life to doing something practical about it. 'Practical', indeed, was the word he always applied to his religion. He denounced what he called the 'speculative Christianity' of men like Keeble and Pusey, its obsession with dogma and ritual. He honestly believed that there was a real struggle between good and evil going on in the world, and that the Christian must throw himself wholeheartedly into it. He thought the Devil was actively at work through slave-owners and factory bosses, through the men who forced opium on China and the politicians who abetted them, through fathers who beat children and a materialistic society which refused to educate them, through the massed, complacent ranks of the Victorian middle and upper classes, enjoying unprecedented opulence based on a morass of cruelty and deprivation. No worthy cause in his day went without his eager and energetic support. His mind was blunt and uninquisitive, but his heart was immense, and his purse (such as it was) always open. He had had a miserable childhood. His parents were cold, hard and merciless. His only comforter was a servant. He was brutalised at his private school, so that (as he said) he did not know which to fear most, term-time or holidays, and even the horrors of Harrow he regarded as blissful release. Such experiences were not uncommon, for the English have always believed in making life disagreeable for their children. Shaftesbury was remarkable in that he carried into adult life a blazing determination to lift some of the burden of cruelty, not only from children (though they had his first affection) but from all the weak and oppressed categories of mankind.

His funeral was, as it were, a physical reflection of his efforts. It was a gathering of Victorian philanthropy at its most impressive, bizarre and (to our eyes) comic, a roll-call of the better forces in a harsh society earnestly striving to make itself less intolerable. The pall-bearers were men and boys from the Shoeblacks' Brigade, the Industrial Schools, the YMCA, the Costermongers' Mission, the Ragged Schools, the National Training Ships and the London City Mission. Packing the Abbey, and lining the streets of Westminster, were curiously dressed and pathetic contingents from all over England. They represented the Society for the Relief of Persecuted Jews and the society to convert them to Christianity, the Society for Suppressing the Opium Trade, and the Association for International Peace, societies for providing drinking-fountains for the poor, and surgical appliances for the crippled; there was the Tonic Sol-fa College and the women from the Female Inebriates, the RSPCA and the Anti-Vivisection Society, the Aid for Cripples' Homes and the Association of Bradford Factory Workers; there were societies to prevent cruelty to children and to cure chest diseases; and, above all, a great multitude of clubs, associations, leagues, unions and missions to alleviate the vast spectrum of nineteenth-century misery – London Flower Girls, Unemployed Cab Drivers, Unemancipated Slaves, Poor Curates, Sons of Poor Clergymen, Turkish Refugees, Distressed Italian Immigrants, Blind Women, Fallen Women, Destitute Children, Abandoned Orphans, Distressed Seamen, Lifeboatmen's Families, Starving Chinese, Indian Females, Poor Parisians, Widows of Medical Men, Discharged Prisoners, Unemployed Railwaymen, Poor Irish, French Refugees, and Consumptive Girls. In all, delegations from 232 different groups were there, and as the coffin was carried out for its last journey to Dorset, the band of the Costermongers' Temperance Society played the March from *Saul*.

The gathering was an impressive record of endeavour and even achievement; and certainly Shaftesbury succeeded in placing on the statute book an astonishing variety of progressive laws, from his great Factory Act to the Act for the Protection of Merchant Seamen. But he died almost in despair: 'I cannot bear to leave the world,' he wrote at 84, 'with all the misery in it.' His diaries and letters constantly express his dismay at the little that years and decades of effort had been able to achieve, and the shock of discovering at every turn fresh pockets of horror clamouring for legislative action. They are a terrible indictment of the forces of indifference and reaction in Victorian society: manufacturers and economists, landowners and peers, insurance companies and religious pressure-groups; above all, the House of Lords, the Bench of Bishops, and the judges. Shaftesbury found governments of all complexions icily neutral, and more inclined to raise difficulties than to remove them. Occasionally, he won important converts: both the Radical J. A. Roebuck and the Peelite Sir James Graham eventually admitted publicly that they had been wrong to oppose factory legislation. More often, ministers and leading politicians dismissed him as a crank, a neurotic and a time-consuming nuisance, the source of what Lord Beaumont, spokesman of the Catholic peers, called the 'pitiful cant of pseudo-philanthropy'. Yet all Shaftesbury's measures were simply attempts to mitigate (scarcely end) abuses which would now – or even a decade or so after his death – be regarded as unspeakable barbarisms. His life was, and is, a discouraging illustration of the difficulties which confront anyone who tries to make the English change for the better. It took him 17 years of incessant agitation to secure the 1850 Factories Act, and many more to get it working effectively. Another 17 were spent on the Lunacy Act, a measure of simple humanity for the insane, and again another five years to operate it. Many of his bills were mutilated in committee, and so became dead-letters. As an MP he often fought his bills through the Commons, to see them destroyed in the Lords. As a peer, he sometimes successfully bullied the peers only to find that his support in the Commons had collapsed. Worst of all, having got a statute through, he discovered that the judges could make it wholly futile.

The story of the infant chimney-sweeps, or the 'climbing boys' as they were known, is a terrifying example of the massive resistance which English society presents to the reform of even the most spectacular and indefensible abuses. These boys formed a small group, perhaps never more than 10,000, but they were typical of many forgotten and brutalised classes, too weak to organise themselves, and therefore wholly dependent on philanthropic champions. They were recruited from workhouses, from the age of four up, and strictly bound as apprentices by the Poor Law Guardians; they could be imprisoned, and flogged, if they broke their articles by escaping. They not only swept the chimneys, but were used to put out fires; often they were forced up by the use of long pricks, and by applying wisps of flaming straw to their feet. They suffered from a variety of occupational diseases and many died from suffocation. What made the injustices from which they suffered more repellent was that the 'political nation' knew all about them: they were not tucked away in some obscure corner of the coal-fields or the London slums, but were regularly and visibly employed in the homes of the upper and middle classes. Everyone knew that tiny children (including a few girls) swept their chimneys. Indeed, the resistance to reform sprang from the unwillingness of the possessing classes to rebuild their chimneys, or to pay the higher fire-risk premiums which the insurance companies (who organised the opposition to reform in Parliament) claimed must follow if children were banned. Nor could anyone claim ignorance of the worst aspects of the system. As far back as 1760 two Sunday school teachers set up an agitation on the boys' behalf; and in 1785 one of them, Jonas Hanway, published a detailed account of the horrors in his *Sentimental History of Chimney-Sweepers in London and Westminster.* In 1788 an act was passed forbidding the employment of children under eight; it was totally ineffective. In 1804, 1807, 1808 and 1809, new bills were thrown out. In 1817 a Select Committee investigated, and published, a catalogue of sickening horrors – reinforced by a brilliant article by Sydney Smith in the *Edinburgh Review* – but a bill based on its findings was thrown out by the Lords. In 1834 a new bill was actually passed, limiting the age to ten, forbidding master-sweeps to send children up chimneys which were on fire, and stipulating that all future flues should be a minimum size of 14 inches by 9; but it was universally evaded, and became nugatory.

In 1840, aided by a passionate reformer from the Hand-in-Hand Insurance Office, Thomas Steven, Shaftesbury succeeded in carrying through yet another bill, against resistance that can only be called fanatical. Despite his efforts to bring test cases, it proved impossible to secure convictions in the courts. In 1851 he got another bill through the Lords, but it lapsed in the Commons for want of support. In 1853 he produced a third bill; but it was referred to a Select Committee, which reported that it was 'inexpedient to proceed further'. He got a fourth bill through the Lords in 1854, but it was voted down in the Commons. In 1855 he could not even get it read a second time in the Lords. In 1861 he got the sweeps referred to the Children's Employment Commission, and in consequence he persuaded Parliament to pass an act raising the age of employment to 16. This was in 1864; but two years later the Commission reported that the act was a failure. The fault lay not in the drafting, but in the general conspiracy of local authorities, magistrates, police, judges, juries and the public to frustrate the law. Boys continued to die as a result of glaring breaches of the act; Shaftesbury noted two cases in 1872, and in 1873 he referred to the Lords the coroner's inquest on a boy aged seven who suffocated in a flue. In 1875, following the death of a boy aged 14, Shaftesbury at last secured a conviction for manslaughter against a master sweep. The sentence was only six months, but the case caught the eye of *The Times,* and in the ensuing agitation Shaftesbury finally carried a draconian bill through what he called a 'very inattentive' Parliament. It had taken precisely 102 years to secure this elementary act of justice to defenceless children.

The agonising gradualness of reform reflects not merely the gross indifference of a supposedly enlightened English public, but the sheer frictional power of English institutions, most of which, throughout the nineteenth century, grew stronger as they relinquished untenable outworks. Between 1790 and 1840 the monarchy lost its direct political power, partly because of the decline of patronage, and partly because of George III's madness and his subsequent inability to discharge detailed business, the idleness and indifference of his sons, and the inexperience of his granddaughter. But Victoria made a tenacious recovery once she learned how to manipulate the system. She wholly declined to behave like a constitutional monarch; she did not accept the principle that the sovereign reigned but did not rule; the notion that her power was confined to 'the right to be consulted, the right to encourage, and the right to warn' was a fantasy of Walter Bagehot's, based on pure ignorance of how the British political system actually worked. His *British Constitution* became a best-seller and deluded the nation. If Victoria ever read it, which is doubtful, how her receding chins must have quivered with delight, how she must have clapped her plump hands, how her short, stocky legs must have drummed the floor with satisfaction that this sophisticated editor, polymath and know-all should have been so completely taken in! The subsequent publication of her letters, and other documents, tell a very different story. From the 1840s, she and her husband constantly intervened in foreign affairs, and invariably on the side of Continental reaction; both were fully paid-up members of the monarchs' trade union, as the Stuarts had been. After Albert's death, Victoria took an increasing and baleful interest in domestic politics, invariably on the side of reaction. After 1868 she was a straightforward party instrument, acting on behalf, and often with the advice and consent – even the active encouragement – of unscrupulous Tory leaders like Disraeli and Salisbury. Liberal ministers, and notably Gladstone, felt obliged to treat her with forbearance because they believed she was in danger of going insane, if crossed, and might confront Parliament with the intolerable difficulties of a regency crisis. They might hold fast against her on absolutely central issues; but in many appointments to the Church, the armed services, the Empire and even to the Government, they surrendered to her views. She rarely showed her hand openly (as over the Gordon telegram) but she was astonishingly active and relentless behind the scenes. She always threw her considerable weight against reform, and did her ingenious best to defeat Liberal measures or to secure Tory electoral victories. She could not halt reform; she could, and did, delay or emasculate it. She wore down her more progressive ministers by absorbing a phenomenal amount of their time on her selfish and trivial family concerns.*

* One example among many: in August 1872, during the fleet manoeuvres, the captain of the royal yacht, the *Victoria and Albert*, insisted on firing the fleet's sunset gun, on the grounds that the Prince of Wales was aboard and took precedence over the senior admiral commanding. This flouted all naval traditions, for the fleet was theoretically at war, and firing the gun was an operational command. There was a tremendous outburst of Service wrath; Mr Goschen, the First Lord, naturally sided with the navy, and the Cabinet with Goschen; but the Queen vociferously supported the royal party, and for nine angry months stuck, literally, to her gun. The affair ended in compromise, as the navy and the Government were anxious to resume work. See Frederick Ponsonsby: *Sidelights on Queen Victoria* (London, 1930), Chapter 1: 'The Fatal Gun'. Victoria's selfishness took many forms. Like George III, she liked under-heated houses, and her servants and guests suffered accordingly. She also insisted at meals that the plates should be removed the second she finished a course; as she was served first, and wolfed her food, many went hungry. The only man who successfully protested against such proceedings was the 8th Duke of Devonshire: 'I say! Give that plate back!'

Constitutional monarchy had to await the accession of her despised son; and even he, though genuinely anxious to act fairly between the parties, inherited her tradition of acting as a brake on radical change, a royal posture maintained to this day. For the reformer, the monarchy was always one more river to cross, and there were limits even to the energies of the great nineteenth-century pioneers.

Around the throne spread the formidable buttresses of the House of Lords. There had been proposals to abolish it as far back as the sixteenth century, but throughout the age of reform no motion to this effect even got so far as the table of the House of Commons, despite the fact that the Peers overrode the wishes of MPs on over 100 occasions either by rejecting outright Commons bills or amending them beyond recognition. During the nineteenth century the upper house contained an overwhelming majority of Tory peers, and their hostility to non-Tory governments was only marginally qualified by the deliberate policy of the Tory leadership – not always enforced – of restraining the backwoodsmen to avoid a dangerous conflict between the Houses. Long before the Lords were christened 'Mr Balfour's Poodle', they had scurried back and forth to the commands of the Tory high command. In 1832, Macaulay asked the Earl of Clarendon how Wellington would justify the Reform Bill to the Lords. 'Oh, that will be simple enough. He'll say: "My Lords! Attention! Right about face! Quick march!" and the thing will be done.' If the Lords lost their pocket boroughs in 1832, they retained direct personal influence over more than half the Commons seats. In 1866 37 eldest sons, 64 younger sons, and 15 grandsons of peers sat as MPs, plus a further 100 closely related by marriage or descent.

Membership of the Lords broadly embraced the landowning classes, as the monumental 'Return of Landowners, 1873', known as the New Domesday, revealed. John Bateman's analysis, *The Great Landowners of Britain and Ireland,* based on the 'Return', calculated that 2,500 people, the overwhelming majority peers or close relatives of peers, owned 45 per cent of all the agricultural land. The only parliamentary defeat the landed interest suffered throughout the period was the suspension of corn duties in 1846: this, indeed, was why it caused such a sensation, because it was so unusual. In fact, for a whole generation it operated in favour of landowners, for Peel was anxious to encourage 'high farming' and accompanied the measure with subsidies for improvement, particularly drainage (cheap, factory-made field drains came on to the mass-market after 1842); rentals rose steadily in the middle decades of the century, and peers enormously increased their incomes by the development of urban house property, mining, the railways and ports. They derived more financial benefit from the industrial revolution than any other class, including the manufacturers themselves. Justifying their existence in 1867, Bagehot characteristically underestimated their power: 'There is no country where a "poor devil of a millionaire" is so ill off as in England. The experiment is tried every day, and every day it is proved that money alone . . . will not buy "London Society". Money is kept down, and so to say, cowed by the predominant authority of a different power.' In fact peers married into the new-rich without the smallest hesitation or scruple provided the terms of settlement were right; Lord Rosebery, for instance, snapped up the leading Rothschild heiress, with £100,000 a year to add to his £40,000, although he had little time for Jews. More to the point, the leading peers were in fact much richer than even the most successful industrialists: the Derbys, the Bedfords, the Devonshires, the Westminsters and the Sutherlands, plus perhaps 40 other families, enjoyed regular incomes of between £100,000 and £400,000 a year.*

The Tory peers blocked the bills of Liberal governments in a skilful and systematic manner. But more insidious and damaging, in some respects, was the restraining pressure of the nominally Whig and Liberal peers. Although from the 1830s, Whig peers drifted steadily into the Tory camp, leaving gaps which new creations could not fill, enough remained to exert powerful blackmail on the Liberal leaders. After Gladstone's smashing popular victory in the 1880 election, Earl Fitzwilliam wrote meaningfully to Earl Granville, Gladstone's deputy:

> I believe it is mainly through my instrumentality that six Liberal members have found seats in parliament. My own political opinions are well known, and I have every reason to believe that it was confidence in the moderation of my views, which brought about this success – you will therefore understand that I must take a deep interest in the formation of a cabinet which I and mine will have largely contributed to place in power.

The hint was taken, to judge by the overwhelmingly aristocratic composition of Gladstone's 1880 Cabinet, which outraged rank-and-file Liberals. A few years before, indeed, Lord Hartington, a stupid and

* The Earl of Derby's rental was nearly £300,000 a year; in 1893 he employed 727 servants, gardeners and estate staff. The Dukes of Bedford, Portland, Devonshire, Sutherland and Northumberland, and the Marquess of Bute, were richer. See Randolph S. Churchill: *Lord Derby, King of Lancashire* (London, 1959). The richest of all was the Duke of Westminster, who had over £250,000 a year from his London properties alone. At his Cheshire house, Eaton Hall, he employed the following: 85 in the house and stables, 43 in the gardens, 32 in the stud and 168 on the estate. See the chart on p. 138 of Gervase Huxley: *Victorian Duke* (Oxford, 1967).

lazy man, chiefly interested in racing and women, had succeeded Gladstone during his temporary retirement, on the grounds that it was invidious to choose between the more plebeian contenders and that 'no one need feel affronted by being passed over for the eldest son of the Duke of Devonshire'. Rosebery got the leadership for much the same reason in 1894. In point of fact, subservience to the prejudices and interests of Whig peers availed the Liberals not at all: the Whig rump ratted almost to a man over Home Rule in 1886.

Much of the power of the Peers sprang from a misconception of their will to exert it, and from the self-deceptive hocus-pocus with which the English always surround matters of class. Too many Englishmen had swallowed Burke's pernicious nonsense about the 'chivalry' of the aristocracy, 'that generous loyalty to rank and sex, that proud submission, that dignified obedience, that subordination of the heart, which kept alive even in servitude itself, the spirit of exalted freedom'. Gullible visitors accepted this conventional view. 'Of what use are the lords?' asked Ralph Waldo Emerson; and replied: 'They have been a social church proper to inspire sentiments mutually inspiring the lover and the loved . . . Tis a romance adorning English life with a wider horizon; a midway heaven fulfilling to their sense their fairy tales and poetry.' Yet if one examines the legislative record of the Lords, one finds precious little chivalry and no inspiration. The attitude of the Lords, wrote Shaftesbury, was invariably 'hard and unsentimental'. The peers benefited from the extraordinary conspiracy of silence which surrounded the sexual misdemeanours of the great, and the professional expertise of solicitors like Sir George Lewis, who specialised in keeping the peerage out of the courts (his papers were, alas, burned according to his instructions after his death). Most of them were selfish, irresponsible, short-sighted and mean.

They benefited, too, from the raging middle-class snobbery which was such a feature of Victorian England, and from the grotesque cult of the 'gentleman', assiduously propagated by renegade-progressives like Thackeray. Macaulay had raged against this in 1833, writing to his sister:

> The curse of England is the obstinate determination of the middle classes to make their sons what they call gentlemen. So we are overrun by clergymen without livings; lawyers without briefs; physicians without patients; authors without readers, clerks soliciting employment, who might have thriven, and been above the world, as bakers, watchmakers, or innkeepers.

Yet Macaulay was himself both the victim, and the willing accomplice, of the system: though he was the greatest parliamentary orator of his generation, and a faithful party man, the Whigs excluded him from senior office purely on social grounds. Taine, writing his *Notes sur l'Angleterre* in 1872, rightly pointed out that English snobbery was a grave source of economic weakness: the money-makers could easily obtain admission to the ruling class, but only at the price of abandoning their commercial attitudes – quite different industrial *élites* were by then growing up in Germany and America. It was true, as Charles de Montalembert argued in *The Political Future of England* (1856), that the anxiety of the new men to join the gentry was a source of political strength and stability. But strength for whom? Stability for what?

On the whole, the direct and indirect power of the Lords survived all the legislative assaults of successive governments: the extensions of the suffrage in 1867, 1869 and 1884, the Ballot Act (1872), the Corrupt Practices Act (1883) and the Redistribution Act (1885), which taken together effectively created a mass-democracy of males in Britain. As legislators, they were more successfully aggressive in the 1890s than in the 1840s. Perhaps their neatest trick of all was to frustrate the intentions of the Local Government Act of 1888, which replaced the old Quarter Sessions of JPs by democratically elected County Councils. In 29 out of 48 counties, the lord-lieutenants, who had presided over the old sessions, were elected chairmen of the new councils. Peers dominated nearly all the rest, a notable exception being Durham, which fell into the grip of rurally based mining communities. As the Earl of Harrowby wrote to Salisbury, the Prime Minister: 'We shall all have to live in the country for the next three years, to keep things straight.' Peers are still keeping things straight in most of the counties in the 1970s. At the urging of the Marquess of Abergavenny, Salisbury adopted the policy of giving the lieutenancies to young, 'reliable' peers, who would outlive any risky Liberal interregnum. Thus, the Earl of Powis, whom he made Lord-Lieutenant of Shropshire in 1895, held the job until 1951, triumphantly surviving one Liberal and three Labour administrations, two wartime coalitions, a national government and 13 prime ministers.

After 1875, as we shall see, the landed strength of the English aristocracy was progressively eroded, not so much by legislative action as by external events; but in some important ways the class system was systematically reinforced, and new fields in which it could operate were opened up. What the English deplored in foreigners as servility they applauded in themselves as 'deference', or 'beneficent snobbery' as G. M. Trevelyan unsmilingly called it. J. S. Mill wrote despairing to Mazzini in 1858:

> The English, of all ranks and classes, are at bottom, in all their feelings, aristocrats. They have the conception of liberty, and set some value on it, but the very idea of equality is strange and offensive to them. They do not dislike to have many people above them as long as they have some below them . . .

Such attitudes were found even among the new Labour MPs (in time to be in the market for peerages themselves), and their humourless Fabian mentors. In 1906, on the morrow of the great radical landslide, Beatrice Webb's diary contains revealing notes on two dinner-parties, one at the Asquiths', the other at the Lord George Hamiltons'. There was no doubt which she preferred:

> The Tory aristocrat and his wife were, in relation to their class, living the simple life; and the Yorkshire manufacturer's son was obviously 'swelling it', to use the vulgar expression for a vulgar thing.

The Webbs ended up as peers themselves; so, for that matter, did the Asquiths. If the Parliament Act eventually removed the Lords' veto, it did not seriously damage their political power, still less their social influence, based as it was, and is, on the hereditary principle. And the last attempt to demolish the hereditary basis, in 1968, was characteristically frustrated by a rabid, cross-party coalition of Tory troglodytes and Labour neanderthals, the latter group led by a member of one of England's leading establishment families, which already includes two life peers and a Knight Bachelor. How William Pitt would have laughed – or, rather, sneered, for he despised peerages himself and rightly considered his father made a fatal error in taking one.

One important way in which the Victorian English underpinned the class structure, even exported it, was by inventing sport, and then organising it on a class basis. All kinds of ancient pastimes, and some new ones, were drawn into the net. Racing was codified in the 1840s, with gentlemen as owners and arbiters (through the Jockey Club), the middle classes as trainers, and the proles as riders (except in steeplechasing, where a mixing of the classes was thought desirable to underpin support for fox-hunting). When cricket was organised in 1846, this once-classless game was subjected to a rigid division into gentlemen and players, with apartheid dressing-rooms and (until very recently) a distinction even in the way names were listed on the score-cards. Big-game hunting was introduced for the upper classes, mountaineering for the upper middle class, rugby (1846) for the middle class and Welshmen, the Boat Race (1856) for the upper classes and clergymen, Association Football (1858) for the workers, open golf (1861) for the middle class and Scotsmen; finally in the 1860s and 1870s, prizefighting became boxing, for the gentry to watch and control, and the workers to perform, and tennis, badminton and croquet were graciously bestowed on respectable ladies. There was something for everyone in Victorian sport, and everyone firmly in his, or her, place.

Yet if there was a powerful current in society tending to institutionalise the class system, there was an equally powerful current of opinion demanding that society should become more efficient. How could the two be reconciled? The Victorians devised a neat, pragmatic solution, which deserves examination because it illustrates, in classic form, the genius of the English for fossilising change, and rendering it socially harmless, even while it is taking place. There were three elements. The administrative reorganisation of central government carried through by Pitt and others working in the Shelburne tradition had transformed a corrupt, oligarchic and court-orientated muddle into the nucleus of a modern machine. It was now beginning to be capable of discharging tasks which hitherto had been beyond the resources of any State. Secondly, the Benthamite tradition of utilitarianism indicated strongly what these tasks should be: Parliament should use the weapon of the statute to remodel the offensively inefficient jungle of life into a neat and ordered structure, and civil servants should see that the law was carried out. But this meant not merely an enlargement of government activities but also a progressive expansion of its personnel. The experience of the early attempts to regulate factories, before 1833, showed that they were futile without an inspectorate.

This was only one of many examples. From the early decades of the century onwards there was an irresistible impulse to investigate the ills of society and propose remedies, using the machinery of the parliamentary committee and the Royal Commission. Charles Darwin was later to write: 'My mind seems to have become a kind of machine for grinding general laws out of large collections of facts.' His method brilliantly summarises the reforming process. The committee examined witnesses, accumulated volumes of facts, summarised its conclusions and reported; Parliament acted; government enforced. Dickens might deplore, in *Hard Times,* the undue importance the Victorians attached to facts, but their massive and effective deployment was the only way to overcome the inertia of society. The reformers usually won in the end because they won the argument; and they won the argument because they marshalled the facts in such a way as to appeal simultaneously to men's reason and their hearts, to their commercial sense and to their moral sense. Shaftesbury and those in the evangelical tradition could put the emotional case; but it required also men like Sir Edwin Chadwick, from the very different Benthamite tradition, to put the factual case. He was able not merely to state, but to prove, that reform saved money. The real argument of his famous Report on the Sanitary Condition of the Labouring Population, the most impressive of all the Victorian reforming documents, was that it cost less to spend money on the preventive measures of public health works, than to cure epidemics. The Report, published in 1842, was a best-seller; naturally it met resistance, but, as if by an act of God, an outbreak of cholera intervened, rammed home Chadwick's point, and propelled the Public Health Act of 1848 through Parliament. Thereafter, the fact-men, working in bizarre but effective harness with the evangelical moralists, found the going easier. In his *Apologia Pro Vita Sua,* Newman sneered at the prevailing philosophy:

> Virtue is the child of knowledge, and vice of ignorance. Therefore, e.g. education, periodical literature, railroad travelling, ventilation, drainage, and the arts of life, when fully carried out, serve to make a population moral and happy.

This was good for a laugh, and it is true there was no particular logic in the Victorian approach. All the same, the combined factual–moral analysis appealed strongly to enlightened English society, and it produced impressive legislative results. But how were the new laws to be enforced? Or, rather, who was to enforce them?

Here we come to the third element: the staffing of an expanding civil service. The clerks of the minimum state were the junior offspring of the ruling class; they were appointed by ministers from among their acquaintance and families, and by parliamentary nomination. Such people were incapable, and often unwilling, to undertake the new tasks. In any case there were too few of them. The census of 1851 registered less than 75,000 public employees (there were 932,000 in France in 1846), the vast majority working in the customs, excise and post office. Only 1,628 manned the central departments of civil government. Where to get the new men? The answer was obvious. The 1832 Reform Act had already begun to integrate the middle classes with the political nation. They must now be brought into the administrative nation. Obviously, in this case, the aristocratic principle of selection could not work. So selection by merit, by competitive examination, must be applied. India had already shown the way. Running a subcontinent of over 100 million people demanded a professional administrative class, and competition had been employed to recruit it. The system threw up men of genius like Macaulay, Grote and James Mill, and had thoroughly justified itself by results; in 1853, it was extended to all Indian posts. The lessons of India were applied to England: the Northcote–Trevelyan report in 1853 was followed by the Civil Service Commission two years later, and Sir Gregory Hardlines was firmly in the saddle. The mess of the Crimea, like another act of God, intervened to point the factual moral about the armed services, too; and it was only a matter of time before commission by purchase went and the army likewise embraced the examination panacea.

The triumphant vindication of the survival of the fittest in government service, neatly coinciding with Darwin's *Origin of Species* of 1859, was a famous and crushing victory for the enlightened elements in Victorian society. Opponents of the reforms were made to seem obscurantists, and no doubt they were. Yet with the benefit of hindsight, some of their arguments seem surprisingly impressive. Trollope doubted whether the ability to excel in exams was in fact the best criterion of a good civil servant; and in an age when the whole principle of written examination is being challenged, we tend to sympathise. But some more fundamental arguments were put, about the very nature of English government and society. Giving evidence in 1857, Earl Grey made a fool of himself by suggesting that exams for army officers would induce a high rate of brain disease as, he said, had already happened in France. He also, however, voiced a valid fear. If, he said, promotion was by merit, the army would 'always be desirous to force the country into war'. He was thinking of Napoleon and his marshals. The English had always disliked standing forces, controlled by career-officers. They accepted the need for a large navy, officered by men who knew their business: it posed no threat to their liberties. But the army, they felt, should reflect the political nation, should be in a literal sense identical with it. Its officers should come from broad acres, should constitute not a military caste but the defensive arm of property. To fight was a last resort. Under purchase, soldiering was a social rather than a professional career. Officers disliked fighting, above all abroad; they were liable to resign their commissions if subjected to wartime inconvenience. Beau Brummell's public career as an arbiter of taste began in 1799 when he left the 10th Hussars in disgust because it was moved to Manchester; England was then engaged in a European war, but no one thought the worse of him. Regimental officers despised the 'Indians' as vulgar careerists, always on the lookout for wars and promotion: they expected to become a colonel for nothing, merely through efficiency and valour. They might just as well be Frenchmen! Lord Cardigan may have been an arrogant fool, but he undoubtedly had the English constitution on his side. And (others might add) if the old system did not work in the Crimea, that was an added reason for avoiding Crimeas.

But such arguments could not repel the gathering impetus of

Victorian enlightenment. The army became professional, after a fashion. By the 1880s and 1890s young officers, having sweated through Sandhurst and Woolwich, daunted by the long columns of the Army List barring their way to the top, were desperately anxious to get into action, as Churchill's *My Early Life* eloquently testifies. Professional ambition became a powerful driving force behind the spread of empire. A characteristic novel of the period, A. E. W. Mason's *The Four Feathers*, centres on a hero who is branded a coward, because he resigned his commission to avoid service in the Sudan; England had come a long way since the days of Beau Brummell. Moreover, as soldiering became a science, the envious eyes of ambitious English officers began to turn ever more frequently to the great and growing Continental armies, and to speculate on the role Britain might play in a vast European conflict. From speculation grew planning, and from planning grew secret international staff-talks, integration of plans, the concept of an expeditionary force. The transformation of the English army officer is one of the many sinister chains of events which led to the catastrophe of 1914, burying the Northcote–Trevelyan efficiency in a sea of Flanders mud. Oddly enough, the reforms did not dislodge the gentry from command, especially in the cavalry; and the cavalry got the top jobs. In the Crimea, cavalry commanders merely killed their own men and horses; in the Great Professional War they killed a generation.*

In the civil departments the reforms brought about a more subtle, but equally sinister and far more permanent change. The old constitution had a narrow franchise, but within its limits it was supremely representative because ministers accepted absolutely their day-to-day responsibility to Parliament. They feared 'bureaucracy' as they feared a standing army; and if their fears seemed atavistic by the nineteenth century, they were based on sound instinct, as the twentieth century was to prove. When the young Queen Victoria asked Palmerston what was 'bureaucratic influence', he replied:

> In England the Ministers who are at the head of the several departments of the State, are liable any day and every day to defend themselves in Parliament, and in order to do this, they must be minutely acquainted with all the details of the business of their offices, and the only way of being constantly armed with such information is to conduct and direct those details themselves.

* What is more, cavalry officers, like Lord Trenchard, created the structure, and inspired the military philosophy, of the RAF, above all the concept of strategic bombing, which the RAF first elaborated and then passed on to the Americans (and the Russians). The thermo-nuclear devastation of cities by intercontinental rockets is thus ultimately derived from the notions of nineteenth-century English cavalry commanders. When General Curtis Lemay, head of Strategic Air Command, spoke of bombing people 'back into the Stone Age', he voiced the authentic tones of a modern Lord Cardigan.

He contrasted this with the loose organisation of Continental governments, where executive power fell into irresponsible hands. Now Palmerston was speaking in the Pitt tradition: Pitt ran the Treasury like a private office, initiating and controlling all the details of government finance. This was the way Palmerston ran foreign policy. Sir George Shee, one of his Under-Secretaries, wrote to another, John Backhouse, in 1832:

> Lord Palmerston, you know, never consults an Under-Secretary. He merely sends out questions to be answered or papers to be copied when he is here in the evenings, and our only business is to obtain from the clerks the information that is wanted.

The clerks were literally clerks, and even Under-Secretaries, as we know from Palmerston's papers, had occasionally to turn their hand to copying when the clerks were overworked.

It was against this background that some of the older civil servants resisted reform. One of them, Sir R. M. Bromley, told the Select Committee on Civil Service Appointments in 1860 that the virtue of the nomination system was that it integrated Parliament with the administration. Civil servants were usually closely connected with members of both Houses of Parliament; they were part of the ruling class, in no sense a separate caste structure. Open competition, he continued, would eventually produce a very powerful bureaucracy, not easily subjected to political control. To this we might add the shrewd predictions of Sir Henry Taylor in *The Statesman* (1836). Reflecting on the consequences of a wider suffrage, he pointed out that powerful civil servants, who willingly took orders from a Ministerial duke or earl, would be less inclined to obey 'a man raised from a lower rank in society to a high official station'. As we shall see, these predictions were tragically fulfilled, especially in the Foreign Office and the Treasury. The assertion of bureaucratic control which marked the beginning of the twentieth century coincided with the period when the new type of civil servants got to the top. These men thought they were a cut above the politician because they had risen through the survival-of-the-fittest process of competitive examination and promotion by merit – as opposed to Ministers, who had merely exploited the workings of democratic choice by an ignorant electorate. Thus real power tended to slip away from the grasp of the masses at the very moment when they appeared to be acquiring it at last. It was a new version of the old English vanishing trick.

In any case, steps were taken to ensure that the new mandarins of the State came from the right catchment area, and were subjected to an appropriate social conditioning. How, we asked, could class be reconciled with 'efficiency'? The answer was through the education system. At the beginning of the nineteenth century this presented an extraordinarily confused picture: over 800 charitable foundations, many of great antiquity, whose original purposes had become distorted and which had often been taken over for the exclusive use of the middle or the upper class; a number of highly effective Dissenting Academies, which had produced the leaders and the skilled personnel of the first industrial revolution; and thousands of charity schools, run by Anglican voluntary societies, which gave a rudimentary education to a section of the workers. Topping it all were the Oxbridge colleges, exclusively staffed by Anglican clergymen, and for the exclusive use of Anglican communicants.

There had always been, in England, a tradition of comprehensive, and classless, education, which indeed went back to Alfred. The charities had originally been founded to promote it, and had been immensely strengthened in the sixteenth century in the wake of the Reformation. These institutions made possible the achievements of the English in the seventeenth century, and lay at the root of their world predominance. But the property-state had allowed them to be absorbed in the class structure, and interclass schools had virtually disappeared by 1800. Men like Henry Brougham, who led the movement for the reform of charitable abuses, were anxious to amalgamate the educational charities with the new movement for educating the poor. He believed, quite explicitly, in comprehensive education. If his views had prevailed, Britain's education today would be entirely different; vastly more efficient, non-sectarian, and classless. But he was in a tiny minority. Even educational 'progressives', like Dr Andrew Bell, thought the purpose of education was to reflect, and reinforce, the social divisions. As he wrote in one of his books:

> It is not proposed that the children of the poor be educated in an extensive manner, or even taught to write and cypher. Utopian schemes . . . for the diffusion of general knowledge, would soon . . . confuse that distinction of ranks and classes of society, on which the general welfare hinges. . . . There is a risk of elevating by an indiscriminate education, the minds of those doomed to the drudgery of daily labour above their condition, and thereby rendering them discontented and unhappy in their lot.

The 'risk', indeed, was not very serious; a quarter of a century later, the Select Committee on the Education of the Poorer Classes (1837–8) calculated that only 1 in 12 had any kind of education, only 1 in 24 an education of any use; in the great industrial cities the proportion fell to 1 in 38 (Birmingham), 1 in 35 (Manchester) and 1 in 41 (Leeds). To Bell's argument, Brougham put, in effect, the modern comprehensive case:

> Is it contended that persons of a certain yearly income engross among them all the natural genius of the human race? . . . If, then, among two millions of persons in the lower ranks, who now receive no education at all, there are a certain proportion of fine understandings, utterly buried and forever lost to the world for want of cultivation, would it not be worth while . . . to give that matter a certain degree of attention?

The answer of society was a decisive 'No'. Not only Oxbridge, but the seven leading public schools were specifically exempted from the operations of the Charities Commission, which Brougham eventually got established. He was no more successful with the town grammar schools, which had been taken over by the middle class. In answer to his campaign, Dr Butler, headmaster of Shrewsbury, put the class objections to the comprehensive principle rather more nakedly than its opponents would dare today. Middle-class parents, he argued, did not object in principle to the idea that poor boys and girls should get the same education as their own children: they objected that they should get it *together*. 'Their children can learn no improvements in manners and morals by associating with all the lowest boys of the parish, and they will feel it necessary and inevitable to forgo one of these two advantages, either the preservation of their children's minds from the contagion of vulgar example, or the benefits of an institution which they cannot enjoy without exposing them to so great a risk. Now, Sir, do you think that any sensible and affectionate parent will hesitate a moment which of these two he should choose?'

Dr Butler undoubtedly reflected prevailing opinion, and the results were foreseeable. The class basis of the system was systematically strengthened and institutionalised. The leading public schools were reformed, after a fashion, and became exclusively the property of the upper class: already, by the mid-century, it cost £200 a year to send a boy to Eton. New public schools were founded, modelled in all essentials on the old ones, but carefully graded in price and social pretensions to accommodate the sons of the upper middle class.* Examination

* One such, veering uneasily between upper- and upper-middle-class station, was Radley. In the 1850s, its headmaster, William Sewell, explained to the boys in a sermon that he could not invite all of them to dine with him. 'This college, and all its system, was planned, as I have often told you, for the sons of the upper classes. . . . Shall I tell you, my boys, how a Christian gentleman, such as all of you without exception are here, feels towards rank and blood, especially to noble blood? . . . A gentleman, then, and a Christian, whether boy or man, both knows, and is thankful that God, instead of making all men equal, has made them all most unequal.' He then announced that he intended to invite to dinner all boys of noble blood, in which he included the sons of bishops!

requirements of the civil service and the forces were adjusted to fit in with the curricula of these schools, and the same relationship was maintained between them and Oxbridge. Thus the road to the top, even in a 'reformed' and 'efficient' society, lay exclusively through, as it were, a toll-gate geared to the class system. The triumph of the exam, indeed, reinforced the structure. Wealthy dissenters, who were ambitious for their children, found themselves compelled to send them to Anglican public schools. In this respect, the lifting of the ban from Dissenters at Oxbridge made no difference. (Catholics remodelled their own public schools on the Anglican prototypes.) Worse, the grammar schools did their best to ape the public schools; and the new secondary schools, finally created in the last decades of the century, tried to ape the grammar schools. At the bottom of the heap, elementary schools, established in 1870 and made compulsory after 1880, did their best to reflect the educational and organisation patterns of their betters. The whole structure was underpinned by the all-powerful schools inspectors, recruited of course by examination. The first generation were a mixed lot, and included some originals and eccentrics, for the newer public schools had not yet begun to process a regular middle-class intake. But the second and subsequent generations uniformly enforced the educational attitudes of the *élite* establishments.

The effect was not merely to reinforce the class structure but to deny knowledge. The muddle of English education in the eighteenth century had some surprisingly healthy features. Some of the schools for poor children were very good indeed, with a wide curriculum and enterprising teachers. Many grammar schools taught science a hundred years before Eton. The Dissenting Academies were usually admirable. The very lack of system had its virtues, because it allowed experiment to flourish. When reform came, and the class matrix was imposed, it brought with it the absolute paramountcy of the classics (plus Euclid mathematics), whose survival had been ensured by the 1660 Restoration, and which Oxbridge and the leading schools had perpetuated. Thus in some crucial respects, English education actually took a step backward in the later nineteenth century, at a time when the United States, France and, above all, Germany were organising mass-education on modern lines.

The great public schools, as standardised in the 1840s, were an extraordinary combination of the barrack-room, the utilitarian prison, the medieval monastery, and the Athenian academy. The flogging was only one degree less severe than in the Wellingtonian army. Without exception, all the 'great' headmasters were floggers, though Moss of

Shrewsbury held the record with 88 lashes delivered personally on a single boy (others exceeded this, but only when they flogged whole classes). A great variety of weapons were used, some of them weighted with lead. Christ's Hospital (forsooth!) was the worst: one boy related that he spent part of the night by the bedside of a victim, easing the shirt from his back and extracting 'at least a dozen pieces of birch-rod, which had penetrated deep into the flesh'. But all schools were savage – and this at a time when in hapless Poland, partitioned between three empires, it was already illegal to use corporal punishment on children. The institution of fagging, a form of bondage modelled on what was supposed to be the 'feudal system', spread to all these schools. The gang-mentality of these brutalised boys was canalised into the House system. Team-games were enforced to tame the arrogance of the intelligent. In some ways the schools were run like a medieval kingdom: the birth of a son to the headmaster was usually followed by 'an amnesty of penals and punishments'. Extraordinary efforts were made to eliminate sexual activities, some of them clearly inspired by the experience of the Prison Commission.* At Wellington, Dr Benson placed wire entanglements over the sleeping cubicles. His instruction about bedtime procedure for younger boys read:

> While they are undressing, steward and matron to walk up and down in the middle of the dormitory and report any boy who goes out of his own dormitory to another. . . . Door of cubicles as at Eton to be incapable of fastening on inside, but may be locked on outside, every door to be commanded by a master key.

Most of these attitudes, including fagging, were endorsed by the report of the Public Schools Commission in 1864.

Still more extraordinary was the rigorous exclusion from the curriculum of virtually all knowledge which had not already been available to a well-educated citizen of imperial Rome. Despite increasing pressure, as the century advanced, the headmasters vigorously resisted the encroachment not only of the natural and physical sciences, including geography, but modern mathematics and history. A sixth-form exam paper, entitled 'Modern History', which Palmerston preserved in his papers, presented at Harrow in 1829, contained nothing beyond the reign of Alfred. How angry the old king would have been! The 1851 Exhibition may have impressed headmasters, but it did not induce them to admit the teaching of science: Hawtrey

* Later, the process was reversed. In 1929, a Labour Home Secretary, J.M. Clynes said (or rather was told by his officials to say) that 'His Majesty's Prisons embody some of the best features of our Public Schools'.

of Eton set it as the subject for a Latin prize essay (rather as, in the 1960s, the Vatican strove to find Latin terms with which to denounce the contraceptive pill). Moberley of Winchester summed up their attitude to modern knowledge before the Public School Commission:

> . . . a boy who has learned grammar, has learned to talk and to write in all his life; he has possessed himself for ever of an instrument of power. A man who has learned the laws of electricity has got the facts of science, and when they are gone, they are gone for good and all.

Under pressure, headmasters argued that what they taught must reflect the demands and standards of Oxbridge. But the argument was circular, for Oxbridge replied that they must build on what was taught in the schools. The truth is that Oxford and Cambridge had contributed virtually nothing to English education since the early seventeenth century; since then, indeed, they had actually impeded it. The industrial revolution had been made possible by the fact that Dissenters like Priestley had been excluded from the endowed Anglican system. During the nineteenth century, Oxbridge and the public schools produced very few men of distinction in the sciences, engineering and the organisation of industry. Their products dominated politics, the Bar and the Church, and towards the end of the century the home and imperial civil service – as well, of course, as the armed services. But they did not turn out wealth-makers or creators. Of the scientists, Lord Rayleigh stayed one half at Eton, and Sir John Herschel a few months. Darwin was at Shrewsbury, but wrote in his *Autobiography*: 'Nothing could have been worse for the development of my mind than Doctor Butler's school.' Without exception, the rest of the great figures in British industry and science went to grammar schools or private academies or, like Faraday, had virtually no formal education. Not until Dr Sanderson went to Oundle towards the end of the century did any major public school take science seriously (though it is true T. H. Huxley was made a fellow of Eton). More ominous, in the long run, was the way in which the anti-science bias spread downwards from the *élite* schools, to embrace virtually the whole of the educational system. Not only did self-made engineers, scientists and industrialists send their children to public schools (Brunel's boys went to Harrow), but some of the more adventurous establishments accepted the classical bias to conform with the Oxbridge requirements. Even in conservative circles, there was some awareness that a great industrial nation like Britain was storing up trouble for herself. The *Quarterly Review* warned in 1867:

England, at least as far as the natural and experimental sciences are concerned, seems in danger of sinking to the condition of what in political language would be called a third- or fourth-rate power. Our greatest men are perhaps still greater than those of any other nation; but the amount of quiet, solid, scientific work done in England is painfully less than that done in Germany, less even than that done in France.

Oxbridge resisted the advance of science just as doggedly. Until the last decade of the century the amount of scientific work done in either university was negligible and the number of graduate engineers produced by Britain was already very small by comparison with the United States and Germany.* The relative decline of Britain as a great industrial nation was already apparent by the 1870s, and pronounced by the 1880s; and it has, of course, continued ever since. It was produced by a number of factors, but by far the most important was the backwardness of the English educational system. And for this the Anglican Church, with its incorrigible belief in the classics, was almost wholly responsible. One might say, indeed, that the triumph of Anglicanism in the sixteenth century set England on the road to becoming a world power, and that the Indian summer which the Church enjoyed in the nineteenth century set in motion the slow process of decay, which continues, relentless and remorseless, as I write these words. The English have paid a terrible price for Eton and Harrow, for Oxford and Cambridge.

Curiously enough, this Indian summer, which allowed the Anglicans to establish a vice-like grip on all the *élite* institutions of education, came at a time when on every other front the established Church was on the retreat and its very foundations were being undermined by the spread of unbelief. If the High Church revival allowed it to capture the fellowships and the headmasterships, it lost the theological battles decisively: the retreat of Manning, Newman and so many others to Rome was a symptom of disaster. In 1800 the Church still retained a massive and satisfying array of temporalities: the annual incomes of the Archbishop of Canterbury and the Bishop of Durham were £19,000 apiece; one rectory was worth £7,000 a year; pluralism was universal and unchallenged – from 1780–1829, the Reverend the Earl of

* Equally, Victorian Oxbridge contributed little to medical science. Medical academics were, next to the clergy, the most violent opponents of university reform: in 1852, the Regius Professor of Clinical Medicine at Oxford strongly supported the Archbishop of Canterbury in maintaining the privileges of noble undergraduates. In 1877, the *British Medical Journal* engaged in a protracted controversy about the proposition that hams cured by menstruating women go bad. One correspondent thought that the matter 'might be decided by experiments made in lunatic asylums or prisons, under the direction of the medical officer'.

Bridgewater held a Durham prebend and two benefices in Shropshire, while living in Paris surrounded by cats and dogs dressed as humans.* Fifty years later, the loot had been rationalised and equalised; there were complaints of clerical poverty from all sides; and in the last three decades of the century there was an abrupt decline in the number of sons of aristocrats and gentry seeking ordination, a sure sign that the game was up. The religious revival, the last England was to see, antedated the Victorian era by nearly a quarter of a century; it was already levelling off by the 1840s. Coinciding, as it did, with the enormous increase in wealth which marked the first industrial revolution, and at a time when Parliament would only grant public works relief for ecclesiastical building, it left a spectacular legacy of 20,000 churches – next to the railways, the greatest physical memorial of the time. But by mid-century, it was becoming increasingly difficult to fill them. The Census Report on Public Worship, taken in March 1851, showed that little over 7 million people out of nearly 18 million in England and Wales went to any kind of religious establishment on Sunday, and for the Anglicans the trend was much more sinister. They had already lost the struggle in the towns and cities, and were beginning to lose it in the villages. None of the other Churches benefited much from this erosion: the only real victor was indifference. The lifting of disabilities in the 1820s allowed the inter-Church warfare to be carried out on roughly equal terms, and education was inevitably the chosen battlefield. The only effect was a further brake on the spread of education, for the difficulty of getting the various sects to agree to any proposal imposed delays up to and beyond the turn of the twentieth century. The attempt to evangelise children meant, therefore, that they got an inferior education, but it did not succeed in making them Christians.

The bottom began to fall out of the Anglican world in the 1850s. Charles Lyell's *Principles of Geology* (1830–33), vulgarised in Robert Chambers' *Vestiges of Creation* (1844), undermined the Biblical account of the origin of the planet. The ground was thus already prepared for Darwin's explosion, in the next decade. The outstanding ecclesiastical figure of the high Victorian period was Bishop Wilberforce of Oxford and, later, Winchester, whose life was punctuated by a series of lost

* The Anglican clergy also exercised power directly. Though barred from sitting in the Commons, they controlled a homogeneous block of 25 seats in the Lords, and, until 1832, formed a significant proportion of the electorate, especially in certain seats, such as Oxford University, which carried particular kudos: it was the clergy who punished Peel in 1830 by ejecting him from Oxford. More important, they contributed more than half the JPs and could, when they chose, dominate Quarter Sessions and other organs of local government. For further details of clerical temporalities, see W.O. Chadwick: *The Victorian Church*, Vol 1 (London, 1966).

rearguard actions, from which he emerged with increasing ridicule.* In 1860, when the Association for the Advancement of Science met in Oxford, Wilberforce, already known by the devastating name of Soapy Sam, was ill-advised enough to challenge Darwinism in open debate with T. H. Huxley, an encounter which his biographer wisely glosses over. The same year, he attacked a harmless volume of advanced theology, *Essays and Reviews,* in the pages of the *Quarterly* (for which he was paid the handsome sum of 100 guineas). As a result, two of the contributors were persecuted for heresy, but the finding against them was overthrown by the Privy Council. A third contributor, Benjamin Jowett, was prosecuted in the Chancellor's Court of Oxford University, a piece of Laudism which merely evoked sneers; and the fact that Jowett's High Church enemies took their revenge by blocking his salary as Regius Professor of Greek merely covered the establishment as a whole in contempt. The truth was, the Protestant Church, unlike Catholicism, did not claim to embody a tradition and interpret it on the basis of authority; it was a documentary faith, and its documents, the Bible, were now seen to be losing their validity. The actions of a Wilberforce or a Pusey were essentially those of men in a panic.

Pious Victorians did their best to accommodate the new knowledge within the framework of their assumptions. On reading *The Descent of Man,* Augustus Pitt-Rivers commented: 'The thought of our humble origins may be an incentive to industry and respectability.' The establishment stretched its increasingly elastic doctrines to embrace Darwin. When he died in 1882 he was buried in Westminster Abbey. Three years later his bust was unveiled in the Science Museum; the Archbishop of Canterbury was present and, according to an observer, 'gazed steadily for half an hour at the marble image of his victorious foe'. But the younger generation of intellectuals was opting out. John Morley, Leslie Stephen and Frederick Harrison, the historian, all lost their faith in the early 1850s, and these were only three examples from a multitude. All three were cut off by their families, in consequence, and the experience left them shaken and distraught. Morley was one of many Victorians who feared that loss of faith, of a framework for life, would lead to melancholia and eventual madness. William Ward,

* He made a sensational exit from life. In 1873, while riding with the Foreign Secretary, Lord Granville, he was flung from his horse and died instantly. The Victorian public was stunned by this arbitrary act of God against their leading ecclesiastic. Lord Shaftesbury noted: 'This event struck me like an earthquake. I was all but horror-struck . . . absolutely thunderstruck with amazement and terror.' Granville, however, assured the bishop's son that even the manner of his father's death was essentially prelatical: 'He must have turned a complete somersault; his feet were in the direction in which we were going, his arms straight by his side – the position was absolutely monumental.' – Lord Edmund Fitzmaurice: *The Second Earl Granville* (1905), ii, 270.

of the Oxford Movement, was told by an eminent doctor: 'The chief causes of insanity in England are the pressures of the commercial system and the uncertainty of religious opinions.' The need for a framework explains the popularity of Auguste Comte and other systematic philosophers: both Morley and Harrison, for instance, became Positivists. It also accounts for the cult of George Eliot (which persists to this day) among intellectuals not normally accustomed to taking novels seriously. Though an agnostic, and unconventional enough to live with another woman's husband, she contrived to preach a moral law which made sense to the new, rationalist conscience. Frederick Myers described a conversation with her in which, 'taking as her text the words God, Immortality, Duty, she pronounced, with a terrible earnestness, how inconceivable was the first, how incredible the second, and yet how peremptory and absolute the third'. Not everyone could find such lines of escape from the dilemma. The problem deeply disturbed the early manhood of many born in the second half of the century. It is quite normal now for people to go through life without an ultimate object, but to the Victorians it was new and daunting. No wonder so many of them were such odd fish – Kitchener, Rosebery, Salisbury, Dilke, Curzon, Carson, Randolph Churchill, Fisher, Rhodes, Milner. In many cases certitude was replaced by a streak of violence, and the loss of God contributed to the afflatus of English imperialism.

Of course, where the Church could repel boarders, it did so, and its chief victim was, needless to say, the weaker sex. But here it had all the instincts of society behind it. Women had fewer rights in Victorian than in Anglo-Saxon England. This is not surprising: in the inventory of the property-state, the wife was a valuable item. Until 1870 she had no property rights at all: her husband could steal her earnings for drink with legal impunity. Even the Married Women's Property Act, which caused an uproar, made little difference; it was heavily watered down in the Lords and, until 1882, did not cover most forms of property. No woman had the parliamentary vote, of course (in this, as in other, respects, Britain had fallen behind the most advanced countries by the end of the century), and the few entitled to vote in local government lost their right if they married. If marriage involved surrender, getting unmarried was virtually impossible, at least for a woman. The Church was responsible both for the vast expense of divorce and the delay in extending it. The 1857 Act brought only marginal relief, and was deliberately weighted against women. A man could divorce his wife for adultery alone; the wife had to prove that her husband's adultery involved incest, bigamy, rape, sodomy or bestiality, or adultery plus legal cruelty. There were still only 800-odd divorces in 1901, as against more than 25,000 in the US the same year. The Church proved wholly reactionary over marriage reform because its moral theology was defective: it still could not make up its mind what a valid marriage was. That ancient conundrum from Leviticus, marriage to a deceased wife's sister (on which the English Reformation had hinged), was still unresolved at the beginning of the twentieth century. It was, indeed, the hardy perennial of Victorian parliaments. In the years after 1850, bills to legalise it were defeated on 29 occasions, chiefly in the Lords, and one finally scraped through only in the wake of the Liberal landslide in 1906.*

Fear of the encroachment of women upon male preserves undoubtedly lay behind efforts to suppress the public manifestations of sex. It was one aspect of the resistance to reform, and antedated the Victorian era by several decades. The sexual patterns of the nineteenth-century English did not differ from those of any other age, as newspaper reports of divorce cases, and much other evidence, make clear. Victorian public men did not want to suppress vice. They did not choose, for instance, to clear the London streets of prostitutes, or even to shut the brothels: the chief motive of the smear campaign against Gladstone was to deter him from his efforts to close an 'introducing house' in St George's

* In 1901, one such bill was killed in the Commons by Lord Hugh Cecil, the leader of the Tory roughs, by crawling through the division lobby on his hands and knees, thus ensuring that the vote could not be completed in time. Supporters of the bill protested that such behaviour was not cricket. He replied: 'I am not playing cricket but preserving the transcendental sacredness of the marriage tie.' But was he? One of the greatest practical failures of Christianity was its inability to work out a satisfactory canon law of marriage. This was a central weakness: it affected everyone, since the Church insisted on basing its social theory on the monogamous family as the basic unit of society. Few couples could have the absolute assurance that they were validly married; and since the Church did not permit divorce, like other religions, an assault on the validity of the marriage was the only road to separation, and the confusion of canon law often made it a possible one. The higher the social class, the more likely it was that this course would be pursued; and where affairs of State, or inter-State relations, were involved, the uncertainty of the law posed a real threat to the unity of Christendom. It was thus no accident that Roman Christianity came to grief over Henry VIII's marriage problems. The legal confusion remains to this day a weakness in Roman Catholicism, which finds itself fighting damaging and ultimately futile rearguard actions against civil divorce. What is extraordinary is that the Anglicans, too, failed to solve the problem. In 1898, the Bishop of London, Mandell Creighton, admitted in his report to the Committee of Convocation on Divorce: '... there is no point on which the Western Church displayed such incompetence, for I can call it by no other name, than in its dealings with the question of marriage ... the State had to interpose, because the Church had reduced matters to such extraordinary confusion ... it is a matter of fact that the Church found exceeding difficulty, and showed exceeding reluctance, in defining what marriage was ... and how a valid marriage could be contracted.' Even today the Anglican clergy are still bitterly divided on whether divorced persons can, or should, be remarried in church. If the Anglican Church had produced a clear doctrine of marriage, the world would never have known civil divorce. Thus the Church has shown itself not the defender of the marriage tie, but its enemy.

Road catering exclusively for MPs and peers. The Victorian house-party was geared to adultery, which was taken for granted provided it did not end in public scandal. But 'sex' like democracy, socialism, atheism, pacifism and equality, was a subversive word. It raised questions about society too fraught with peril to permit public debate. Thus the Victorians used a double vocabulary, as the English stage had learned to do since the early seventeenth century.* And sexual reform, identified with vice, was subjected to the familiar English battering of guilt by foreign association. When Francis Place advocated birth-control in the 1820s, a magazine was published to oppose the campaign: it was called the *Bulldog*. France, as usual, was cast in the role of villain. Rubber contraceptives were called French letters, though they first, in fact, came from America. Sir Charles Dilke, celebrated for his encyclopaedic knowledge of foreign affairs, and his links with French republicans, got the xenophobic works during the Crawford divorce case. Mrs Crawford claimed 'he taught me every French vice', and her counsel, Henry Matthews, thundered: 'He was charged with having done with an English lady what any man of proper feeling would shrink from doing with a prostitute in a French brothel.' (In point of fact, the English ruling class, led by the Prince of Wales, were the best customers of the famous establishments in the Rue de Provence. One of them, for the benefit of English travellers, included a mock-up of a *wagon-lit*, mounted on machinery to simulate motion, and with a canvas panorama of the chateaux of the Loire rolling past the windows.) One might argue that the public silence the Victorians maintained on the subject of permitted sex, and the public excoriation of vice, sprang not least from the belief of the English that they were a race apart, maintaining themselves in splendid isolation from the contaminations of the Continent, and purifying their energies for the dedicated task of running a world empire. Sex had destroyed Rome; it should not destroy Britain. Beneath the public façade, however, all the evidence shows that the English tended to treat sex much less seriously, as indeed a prime subject for laughter. The lengthy newspaper accounts of Victorian divorce actions (which shocked Queen Victoria so much that she complained to the Lord Chancellor) were punctuated with the phrase 'laughter in

* An act of 1606 (*3 James 1, Chapter 21*) prohibited profane language in public plays: substitutions for words like *zounds* can be detected in the Shakespeare folio of 1623. The dual system persisted until 1968, when theatre censorship finally collapsed. Puritanism is the parent of linguistic invention, and the Victorians used an illicit sexual vocabulary which has never been equalled in size and vividness. Among synonyms for whore were: academician, biddy, bobtail, bunter, bung-up, cat, cock-chafer, frow, flymy, pave-thumper, trooper and blowen; brothels were variously referred to as academics, corinths, peggers or pegging-cribs, swells'-kens and convents. See Ronald Pearsall: *The Worm in the Bud – the World of Victorian Sexuality* (London, 1969), Chapter 8.

court'. And, within the limits of certain conventions, the music halls, the Victorian equivalent of television, showed a consistently ribald approach to sex. The leading practitioner of comic innuendo was the Great MacDermott, who brought the houses down with his ditty 'Jeremiah, Blow the Fire'. He was also, it should be noted, the man who made famous 'We Don't Want to Fight, but by Jingo if we Do'.

The theme of this section has been the agonising slowness with which the English were induced to reform and improve their society, and the manner in which changes, when they came at last, often served chiefly to reinforce the existing structure. Certainly, the pace of improvement – in an age dedicated to improvement – must have seemed almost unendurable to the enlightened. Take the comparatively small but significant matter of the compulsory payment of church rates, an indefensible anomaly which rightly enraged not only progressives but most apolitical men and women. A test-case was brought in Braintree. It was fought over 16 years, before 28 judges and in eight courts, four deciding in favour, and four against, until, in 1853, the House of Lords gave a complex judgment which, in effect, made it extremely difficult to enforce payment; but not until 15 years later were compulsory church rates abolished by statute. In some cases the ancient citadels of horror withstood all attacks. In 1845, *Punch*, appalled by the public execution of a woman, and by the unctuous reverence with which prison chaplains gave a Christian blessing to the act of judicial murder, gave a confident assurance: 'Still, have we this comfort: whether the men of God assist the goodly work or no, the gallows is doomed, is crumbling, and must down – overthrown by no greater instruments than a few goosequills.' Alas, even the public spectacles continued for another generation, and hanging was not finally abolished until 1965, 120 years later. English reformers, too, had the mortification of seeing other countries, once derided for their social backwardness, catch up and overtake Britain. In 1884 the Germans got accident insurance, and State insurance against sickness, followed five years later by old-age pensions. In 1905, before even the Liberals had accepted the principle of State welfare, nearly 19 million German workers were insured against accident, 12 million against sickness, and 14 million against old age and incapacity.

All the same, we must not underestimate the magnitude of the English achievement during this remarkable century. If some countries were beginning to progress more rapidly in certain directions, it was the English, for the first time in history, who contrived to harness the idea of progress to the immense engine of the popular consensus. Man had been martyred through the ages; now he was free, if he chose to use his freedom. The historian Henry Thomas Buckle, in many ways the Victorian archetype,* summed up his *History of Civilisation* in a triumphant paean of Baconian optimism:

> The powers of man, so far as experience or analogy can guide us, are unlimited; nor are we possessed of any evidence which authorises us to assign even an imaginary boundary at which the human intellect will, of necessity, be brought to stand.

Later Victorians, more sophisticated and critical, more disillusioned too, found these sentiments crude. It was said that Buckle saw history as 'a sort of vast anti-Corn Laws agitation, with the substitution of knowledge for cheap bread'; that he reduced progress to an infantile ditty:

> I believe that all the gasses
> Have the power to raise the masses.

Writing of the 1850s, Leslie Stephen laughed at an age 'when people held that the Devil had finally committed suicide upon seeing the Great Exhibition, having had things pretty much his own way until Luther threw the inkstand in his face'. Of course Buckle's optimism was crude: but all truth emerges first in crude forms. History teaches that the terrible predicament of mankind can never be improved by resignation, and that the self-confidence of the species is the pre-condition of all progress. In the nineteenth century, the English made a great act of faith in the future, and communicated it to many peoples: that faith has been shaken but not destroyed, and it has permitted the human race to survive calamities which to the Victorians would have been unimaginable.

Moreover, during this period, the English achieved their maturity as a people. They learned to smile at the darkness, and to respect the resources of the intellect. For the first time, they conquered the violence in their natures decisively, and for good. The chaotic and frightening society of the 1830s, which Disraeli described in *Sybil,* had vanished 20 years later; by then it must have seemed as forgotten as the Lancashire of *The Road to Wigan Pier* seems to us today. Late-Victorian England was profoundly shocked when a crowd of unemployed smashed London windows in 1886, though no one was killed or even seriously hurt. By then the tradition of non-violence had already been firmly established, regarded indeed as immemorial, taken for granted. The

* Including the fact that he secretly kept a mistress, Mrs Faunch. See Giles St Aubyn: *A Victorian Eminence* (London, 1958).

fact that for the first three decades of the century England had often been on the brink of bloody revolution, and that the mob was then the prime instrument of political change, was already seen as part of the debris of history, as remote as the Wars of the Roses.* By the 1880s, a marriage had taken place between the political nation, and the nation as a whole – an imperfect marriage, to be sure, marked by bickering and occasional threats of divorce – but strong enough to permit the dialogue of change to be conducted through legal and constitutional forms. Meanwhile, elsewhere in the world, the experience had been very different. The United States had fought a merciless and prolonged civil war, which raised as many problems as it solved, indeed institutionalised its own traditions of violence; France had undergone three revolutions in the vain quest for stability; Germany was moving towards military dictatorship, and Russia was preparing to perpetuate a bestial autocracy in the name of a mythical people. The English, as we have seen, paid a high price for domestic peace. But in terms of human happiness who can say that the purchase was not a shrewd one?

The honest broker in this bargain, the celebrant at the marriage between the two nations, was Mr William Ewart Gladstone. The story of this extraordinary man's political pilgrimage, of his transformation from the youth Macaulay called 'the rising hope of the stern, unbending Tories' to the old democratic eagle whom the world acknowledged as 'the people's William', is essentially the story of how political maturity was reached. The majestic debate in his own mind both instructed and echoed the debate in the nation; like all great politicians, he both led and followed. Now other men had done this, notably Peel and even, to some extent, Palmerston. What made Gladstone unique was the triple combination of a conservative temperament, a radical intellect and an insatiable conscience. The conscience quested, the intellect resolved, the temperament harnessed it to tradition. Thus what was in fact the perpetual motion of the times seemed as stable as the earth spinning on its axis. The development of Gladstone's political

* In September 1838, at Wymondham in Norfolk, an incident occurred which might have figured in *The Paston Letters*. Following a lawsuit over the possession of Stanfield Hall, one of the claimants, a Mr Larner, took possession of the hall by force with an army of 80 followers, wearing laurel leaves in their hats by way of livery. They were eventually dislodged by the 4th Dragoon Guards. See Owen Chadwick: *A Victorian Miniature* (London, 1960). The inability of the authorities to enforce the law in remote parts, even in the 1840s, is the theme of R.D. Blackmore's novel, *Christowell*, set on Dartmoor. But the coming of the railways and the electric telegraph made civil disorder increasingly easy to suppress, or anticipate. Modern technology is the enemy of the mob: today, in advanced countries, successful revolution is impossible without the intervention or acquiescence of the armed forces; and even in backward states it is becoming vastly more difficult.

personality could not have been more revolutionary and complete; yet it is hard to point to any particular episode which involved a qualitative change in his views; the process had the inevitability of gradualness. As a young man his instinct was to be a churchman (as Cardinal Manning's was to be a politician) and for many years he saw public life more in ecclesiastical than in political terms. In October 1832 he breakfasted with his patron, the Duke of Newcastle, who had offered him a safe seat, and the conversation he recorded shows what a long way the young man was to travel:

D. of N: I confess I have a great notion of the horrors of enthusiasm.

W.E.G: Your Grace, I think we must expect to see enthusiasm in the present day, for where, after a long period of prosperity and ease, men's minds are disturbed ... it naturally happens that opinion starts forth in every variety of form which it can possibly assume.

D. of N: Yes, it is so. There can be no doubt that, if we desert God as a nation, he will desert us.

W.E.G: Yes, my Lord. And we seem to be approaching a period in which one expects events so awful that the tongue fears to utter them. ... All seems to be in preparation for the grand struggle between the principles of good and evil. The way to this seems to be in preparation for the approaching downfall of the Papacy.

D. of N: Yes, Popery is attempting to rally its forces, but I think only preparatory to its utter defeat and destruction.

W.E.G: The Roman Catholic religion is so bad, and yet the prospect after its overthrow is so very dreary, that one scarcely knows whether to wish for its continuance, or destruction.

D. of N.: The question as to what is to succeed is full of interest beyond calculation.

W.E.G: I fear that infidelity must succeed – for a time at least.

D. of N: I think there can be little doubt that we ought to wish for its destruction.

W.E.G: It appears to me that those are right who think there are great evils in the state of society – but wrong when they think them so superficial that they can be cured by legislation.

D. of N: Yes, all depends upon individuals; the matter cannot be reached by Act of Parliament.

Many light-years of intellectual experience later, Gladstone was to believe, more strongly perhaps than any other statesman, that political action was itself capable of becoming a moral force, and that the parliamentary statute was the supreme instrument of public elevation. For a man to set his sights so high was, of course, to invite failure; and Gladstone's life was a failure in terms of his stupendous objectives. All the same, the statute became in his hands a formidable weapon;

there is no parallel to his record of achievement in English history. More important, he discovered that democracy is not a monster to be contained, still less excluded, but a moral force to be unleashed. He learned to trust the people; he sought to teach the lesson to others, not always successfully. In one sense this became his considered view of Christianity. Give every man his freedom, and God's light will shine in that man's mind, however humble he may be, as clearly as in the mind of a man born to rule; of course he may reject the light, but the exercise of his free will cannot be denied without denying the potency of God; and experience increasingly shows that in the majority the light will be admitted. This Christian theory of democracy was, in essence, Pelagianism taken to its ultimate conclusion; and thus, as the English finally achieved political maturity, we can see in their evolution an admirable continuity and symmetry. Gladstone had the rare capacity to admit error without losing faith in his judgment. His abandonment of the élitist view of politics left his confidence not merely undiminished but increased: he saw the people as a source of strength. What particularly struck him was the behaviour of the Lancashire cotton operatives during the American Civil War: not merely did they reject the apparent self-interest which dictated support of the South, but they believed (unlike Gladstone) that the North would win, and events confirmed their prescience as well as their rectitude. It was, he said, 'a great lesson to us all, to teach us that in those little tutored but reflective minds ... opinions and sentiments gradually form themselves . . . which are found to be deep-seated, mature and ineradicable'. From this episode sprang the germ of the Midlothian campaign, the conviction that the people could be brought right into the centre of the public stage and express themselves as a moral chorus to which the world would listen; and from this point it was but a short further step to grasp the principle of self-determination, and to seek to apply it to the Irish people. So, in an age of rising empires, not least their own, the English gave birth to the idea of a liberated world.

Yet there is a melancholy coda to this story. To his theory of Christian democracy, Gladstone added a final qualification. Right at the end of his life, he wrote a testament to the young heir who would inherit the Hawarden estate, on which he had lavished so much care and anxiety, and so many copious draughts from the bottomless well of his conscience. The document is not always clear; Gladstone had many of the mental habits of the schoolman, and he was often least intelligible when he was most in earnest. One passage, on the social power of the landed estate, reads:

The influence attaching to [properties] grows in a larger proportion than mere extent, and establishes a natural leadership, based upon free assent, which is of especial value at a period when the majority are, in theory, invested with a supremacy of political power which, nevertheless, through the necessities of our human nature, is always in danger of slipping through their fingers.

Now what Gladstone meant, I believe, is this. Englishmen born to wealth or privileges have a special duty to society, to supply the defects of a mass-democracy which arise from the political consequences of inequalities which no human wisdom can finally eliminate. The English people will, from time to time, be deceived, and be their own worst enemies, for they will fail to exploit the potentialities of the power given to them. An *élite* is needed, not to govern, but to enable the people to govern themselves. These last words of Gladstone, written in June 1897, provide an important clue to the history of the English in our own times. They suggest one way in which the splendours of progress can be made to outweigh the miseries. But first we must examine how the miseries overcame the English.

PART SIX

Hubris and Nemesis

[1870–1972]

IN 1870 England was universally regarded as the strongest and richest nation on earth, indeed in human history. The English aroused little affection. In general, they were cordially disliked; and Lord Palmerston, who had died five years before, had taken this for granted: as he frequently told his ambassadors, it was only to be expected that a wealthy, fortunate and successful country like England should arouse envy and criticism; so long as such feelings were confined to words, and tempered by respect, or if necessary fear, there was no cause for concern. England operated from motives of self-interest, which happened to coincide (by the disposition of a benign providence) with the long-term interests of the civilised world, in fact of the entire human community. England was moving in the direction of progress, and pulling the world along in her wake. England did not need nation-states as allies, because her true allies were the forces of enlightenment, moral, economic and constitutional; by their very nature they were ubiquitous, permanent and immutable. The English had grasped the salient truth that international morality and the pursuit of wealth were not merely compatible but, in a sense, identical: as Locke had put it, 'virtue is now visibly the most enriching purchase, and by much the best bargain'. This the English had discovered for themselves, and were now teaching the world. God was the Great Book-keeper, the Ultimate Accountant, the Chairman of the world liberal economy, and His instruments were free trade and the Royal Navy.

Exactly a hundred years later, there is absolutely nothing left of this vision. England is now the weakest, and in many respects the poorest, of the industrial nations. The signposts to the future no longer point the way she is travelling; on the contrary. The English are still criticised, not least by themselves; but the tone is no longer envious or indignant, but rather impatient and admonitory. Fear has yielded to indifference, respect to pity, and admiration to contempt. The arrogance of the English has gone, and with it their self-confidence. The world suddenly seems a vast and alien place, and the English to occupy a very tiny portion of it. The God that Palmerston worshipped is revealed as an old wooden idol, blind and impotent. English prayers fall on deaf ears, and the cheering crowds turn elsewhere. Such historical transformations have occurred before, but never with such speed and decision – and never, certainly, to the English. The common fate of nations appears to them a unique and devastating blow of providential injustice, unforeseen, undeserved, irrational and, in the deepest sense, immoral. What went wrong? How did it happen? Who is to blame? When did progress cease to move at an English rhythm? The answer is really very simple. It is the old story of hubris and nemesis.

On 8 February 1870 Oxford's first, and newly appointed, Professor of Fine Art gave his inaugural lecture. John Ruskin was then 51, his mind barely clouded by approaching madness, at the height of his enormous powers, a national celebrity. His books sold in tens of thousands, his prose was universally admired and frequently emulated. As a polymath he was without a rival, even in an age of polymaths; and many people thought he was the greatest man alive. So many, indeed, that the theatre of the Oxford Museum, where he was due to lecture, was filled to capacity over an hour before the appointed time, with hundreds outside clamouring to get in; and the chairman, Sir Henry Acland, decided to adjourn to the Sheldonian, where there was more room. Through the icy streets of Oxford, the bizarre and spiky figure of Ruskin, like the Pied Piper, led the eager and academic mob. What he eventually had to say, as it happened, was well worth hearing. It struck a new note for the times, though the theme was an ancient and (to English ears) a tempting one. Well might Ruskin complain, afterwards: 'My University friends came to me, with grave faces, to remonstrate against irrelevant and Utopian topics of that nature being introduced in lectures on art.' For Ruskin seemed chiefly concerned not with art in an academic guise, or in any guise at all, but with a call to racial heroism:

> There is a destiny now possible to us, the highest ever set before a nation to be accepted or refused. We are still undegenerate in race; a race mingled of the best northern blood. We . . . still have the firmness to govern and the grace to obey. . . . Will you youths of England make your country again a royal throne of kings, a sceptred isle, for all the world a source of light, a centre of peace; mistress of learning and of the Arts? . . . This is what England must either do or perish; she must found colonies as fast and far as she is able, formed of her most energetic and worthiest men; seizing every fruitful piece of waste ground she can set her foot on, and there teaching these her colonists that their chief virtue is to be fidelity to their country, and their first aim is to be to advance the power of England by land and sea . . .

It is always a serious matter when pundits, scholars and academics feel inspired to stray outside their chosen disciplines and lend their authority to vast, portentous and mystic pronouncements about the human race – and still more about any particular race. Oxford dons had hitherto bent their energies to resisting the spread of education, and their influence, though almost wholly bad, had at least been merely negative. Ruskin began a new fashion, and opened the era of the mad professor. It is true that he had a John the Baptist: the gentle and reputedly saintly John Henry Newman. In the 1840s, Newman had lectured to the students of Dublin on 'The Idea of a University', and had told them, among other things, that the chief virtue of the English public schools was their ability to breed

> ... heroes and statesmen, literary men and philosophers, men conspicuous for great natural virtues, for habits of business, for knowledge of life, for practical judgment, for cultivated tastes, for accomplishments, who have made England what it is – able to subdue the earth.

Within a few years, the Great Famine had struck Ireland, and the 'heroes and statesmen', at least in Irish eyes, had not been conspicuous for 'practical judgment', had indeed confessed their inability or unwillingness to do anything to mitigate the greatest natural catastrophe in Ireland's history. But since then Darwin's ideas had been bandied around for a whole generation, had been absorbed, vulgarised and perverted, and seemed to many Englishmen to give a new lease to the concept of a chosen race. The race, their own, owed its appointment with destiny not to spurious historical documents, or the supposed activities of first-century apostles, but to the processes of natural selection which allowed only the fittest to survive and rule: the English had been chosen not by God but by nature herself, as could be demonstrated by irrefutable scientific theory. Of course Darwin had said nothing of the kind; he had always been careful to insist that natural progression was morally neutral. But the 'survival of the fittest' seemed to describe so accurately the facts of English history, and the dominant position of the English race, that English pundits naturally assumed that the laws of science endorsed England's global policy. It was, at any rate, an appealing idea to clever young men, and Ruskin had many enthusiastic followers. Convinced that the moulding of the chosen race demanded physical as well as intellectual discipline, Ruskin set his students to work digging roads, under the guidance of his gardener, Downs. Among those who toiled away was Alfred Milner; and another disciple, Cecil Rhodes, failed to take part only because of his weak health. Thus the age of imperialism was born in the home of lost causes.

It is important to realise, if we are to understand the history of the English in the last hundred years, that this new imperialist concept –

born in Oxford, bred in Westminster and then shipped to the colonies – was in all essentials alien to the spirit in which England's overseas territories had been acquired and administered. The English of the sixteenth and seventeenth centuries had been missionaries as well as traders, concerned to liberate the natives from the darkness of heathenism, or still more of Spanish Catholicism, as well as to win raw materials and markets. If this current had been submerged in the vast expansion of the eighteenth century, the strident materialism of an international trading empire, with its vile and profitable instrument of slavery, it still flowed beneath the surface. And, in the closing decades of the eighteenth century, idealism became once more a governing motive in the activities of the English across the oceans. It is significant that, at the time of the Act of Union with Ireland, George III, no less, firmly declined to assume the title of Emperor, on the grounds, according to Canning's secretary, A. G. Stapleton, that 'he felt that his true dignity consisted in his being known to Europe and the world by the appropriate and undisputed style belonging to the British crown'. The lessons of the American revolt had been learned. The English began to see their rule as essentially a transitional phase, in which they were to act as trustees rather than freeholders. Burke saw the Empire as one of ideas rather than military occupation: 'As long as you have the wisdom to keep . . . this country as the sanctuary of liberty, wherever men worship freedom they will turn their faces towards you.' The English taught in many ways, notably through religion and commerce, but no one doubted that the pupils would eventually emerge from tutelage, and that the bonds of mutual interest would thereby be stronger, because voluntary.

The process was confused because many territories were acquired in a haphazard manner, as a result of disputes which were purely European in origin. But certain strong and ubiquitous principles emerged. What lay at the heart of the long debates on Warren Hastings was the growing conviction that the Indian sub-continent could not be ruled through the corrupt methods which English governors had acquired from Indian princes. Administration by bribery was no longer acceptable in England; it could not be practised elsewhere. Hastings was not a scoundrel: he was an anachronism, using devices no longer endorsed by English parliamentary and public opinion. Inevitably, as the Walpolean system retreated, 'efficiency' rushed in to fill the vacuum. The interests of the overwhelming majority, the natives, must be paramount. By Waterloo, England was responsible directly for 87 million Indians: indirectly for 43 million more. When Lord William Bentinck went out as Governor-General in 1828, he wrote to Bentham: 'I shall govern in name, but it will be you who will govern in fact.' Thus India was rapidly exposed to western ideas, and to honest, systematic and relentless methods of government. On the basis of Macaulay's majestic minute on Indian education, Bentinck declared in 1835 that 'the content of higher education should be western learning, including science, and that the language of instruction should be English'. The decision was momentous. It pushed Indian history, and indeed the history of most of Asia, and later of Africa, in a radically new direction: hundreds of millions of people were to follow, economically, politically, socially and technologically in the path of the West. With the ideas came machines: less than half a generation separated the railway age in India from its climax in England, and the time-gap narrowed with each fresh wave of invention. The strains of this forced development of an ancient civilisation were enormous. The Mutiny of 1857 was not a nationalist revolt against alien rule, which was largely beneficent, and at all events minimal. It was a protest by the conservative forces in Indian society against unrestricted penetration by the western way of life. It led to many readjustments in English administration, notably a curbing of the pace of westernisation by the deliberate reinforcement of traditional Indian institutions. But there was no immediate revision of the ultimate object. Macaulay predicted that the moment when the Indians, 'having become instructed in European knowledge ... demand European institutions' would be 'the proudest day in English history'. This remained the common supposition until the imperial age.

The end of the Napoleonic wars found England the residuary legatee of the crumbling European empires. The English were almost alone on a world stage, of which the navy was the custodian. They did not seek to exploit this situation in any imperialist sense, but to apply to the entire planet the principles of moral improvement and self-betterment which were already being preached in their homeland. If anything, the English ruling class was notably more liberal overseas than in the British Isles. The Quebec Act of 1774 gave official status to Roman Catholicism and French civil law, ending disabilities which British and Irish Catholics had to endure for another half-century. The loyalist refugees from the United States who settled in Upper Canada did not have to wait long for a measure of self-government: both Canadas received model constitutions in 1791. What was appropriate for European settlers could not ultimately be denied to anyone else. The English emphatically rejected Aristotle's notion that 'many men are born ignorant and slavish and therefore ought to be slaves'. Two years before the French Revolution, the English evangelicals founded the Society for the Abolition of the Slave Trade, and the year the Canadas got their constitutions it sponsored a settlement for freed slaves in Sierra Leone. The influence of Wilberforce and the Clapham Sect penetrated the Colonial Office, now emerging as a great department of State. It was the creation of two earnest and pious gentlemen, Lord Bathurst and his under-secretary, Henry Goulburn; but its real driving force was Wilberforce's brother-in-law, James Stephen, who remained the permanent under-secretary until 1847. He did not seek to administer, but to improve. Enormously hard-working, he had the faults of a doctrinaire and a prig; but the virtues, also, of a crusader and a reformer, inspired by a profound sense of moral obligation to the subject races. He did not ask: 'What can the natives do for us?' but 'What can the English do for the natives?'

Well, what could they do? There were two rival theories, usually advanced by men who shared a common Biblical inspiration. At the end of the nineteenth century, in *The Man of Destiny*, Shaw was to sneer at the hypocrisy with which the Englishman acquired his Empire: 'When he wants a new market for his adulterated Manchester goods, he sends a missionary to teach the natives the Gospel of Peace.' But he was writing in the experience of the new, brutal and cynical imperialism. A hundred years earlier, English missioners and colonists were inspired by wholly different motives, whose very altruism produced tremendous clashes. The four great overseas missionary societies, founded around 1800, repudiated any connection with government: the secular arm, they feared, would bring oppression and pollution, whereas they aimed solely to elevate the natives:

> Must we not endeavour to raise these wretched beings out of their present miserable condition, and above all, to communicate to them those blessed truths, which would not only improve their understandings and elevate their minds, but would, in ten thousand ways, promote their temporal well-being, and point out to them a sure path to everlasting happiness?

In New Zealand, in particular, the Church Missionary Society tried hard to prevent any form of colonisation, as fatal to its objects:

> Only let New Zealand be spared from colonisation and the Mission have its free and unrestricted course for half a century or more, and the great political moral problem will be solved – of a people passing from a barbarous to a civilised state, through the agency of Europeans, with the complete preservation of the Aboriginal race, and of their natural independence and sovereignty.

But the colonisers were equally earnest in their anxiety to promote moral welfare by practical means. Moreover, they brought into the equation the additional factor of the British and Irish poor, living on islands universally held to be overcrowded and menaced by Malthusian catastrophe. What of the felons? Was it not better, instead of hanging them, to offer them a chance of hard-working repentance in territories which were virtually empty? At any rate, Lord Sydney, a well-meaning Home Secretary, began the process in Australia, in 1786, thus anticipating the missionaries by a decade; and before the system was ended in 1868, nearly 200,000 had been transported, mostly with success. As for the natives, would not their progress to civilisation be hastened by the example and assistance of industrious white folk, released from unemployment in the cities, and from subsistence living in the exhausted fields of Ireland and the Scottish Highlands, to employ their frustrated energies in creating new worlds? Could not the hand of God be seen in bringing together these two compatible objects? With hindsight, we can see the terrible fallacy. But, apart from the missionaries, every dedicated improver in early nineteenth-century England believed in the merits of colonisation. The flow of immigrants began after 1815, and from 1830 it was subsidised. Behind the movement, directed by the National Colonisation Society, were upright, God-fearing men like Charles Buller and Edward Gibbon Wakefield, who believed wholeheartedly that colonisation was the road to freedom, to the enlargement of the human spirit. White men would raise their stature overseas, and carry the natives with them. It was no accident that Buller and Wakefield went with Lord Durham to Canada, in the aftermath of the Canadian revolt, and helped to write the Durham Report of 1838. With all its contemporary limitations, the Report was the blueprint for a future community of self-governing dominions, and by the end of the 1860s not only Canada, but Australia, New Zealand, Newfoundland and the South African territories were for all practical purposes administering themselves: the change was marked by the withdrawal of English troops, and noted at Westminster with almost universal satisfaction.

Indeed, it was possible, at that time, to foresee the culmination of the Empire in universal self-government not in a remote future, but in a matter of decades. The Empire was cultural, not military. Captain Cook's first expedition in 1768 had been essentially a scientific venture, organised and supported by the Royal Society, to bring the Pacific and the South Seas into the orbit of knowledge: the process continued for a century, with generations of naval officers and scientists exploring and mapping the entire Indian and Pacific oceans. In the 1840s the drive was extended inland, to Africa: but the motive was consistent with the aims of what was fundamentally a pacific and humanitarian Empire. English army and naval officers tended to know more about cartography than firepower. The naval and trading stations were acquired and garrisoned: Trincomalee (1786), Mauritius (1810); Singapore (1819), Aden (1839), Hong Kong (1842); but they provided merely a minimal framework of security within which, it was assumed, commerce, education and religion could be safely left to transform backward and barbarous societies into progressive and free ones. The thinking may have been naïve; it was certainly not hypocritical.

Indeed, one can detect definite signs of impatience in London that the maturing process of empire was not proceeding more rapidly. Adam Smith, not Chatham, was the guiding spirit. Colonies had originally been acquired, in part at least, to embody mercantilist principles, to provide exclusive access to raw materials, and exclusive markets for English exports. The industrial revolution made nonsense of mercantilism, at least from the point of view of the English, and once the French wars were over, the old system was scrapped, in favour of free trade, with remarkably little argument. Huskisson dealt the fatal blow in 1825, and in the Forties and the Fifties free trade was extended to the whole Empire. This being the case, and with ultimate independence for all within sight, there seemed no reason to suppose that the Empire should have any formal future at all; informal arrangements based on mutual commercial interest, and common culture, were both more durable and less expensive. 'Those wretched colonies,' said Disraeli in 1852, 'will all be independent in a few years and are a millstone round our necks.' This was an acrimonious, perhaps extreme, example of what was becoming the prevailing mood. Lord Derby rejected Australian demands to annex Pacific islands; he asked an Australian delegation 'whether they did not want another planet all to themselves'. When, indeed, the navy became active in the Pacific in 1872, its object was to stop the Australians running the Kanaka labour trade, rightly regarded as a form of slavery. In South Africa, the English willingly allowed the Boers to separate themselves from British rule, and confined their military activities chiefly to defensive actions against the Kaffir confederations, pushing down from the north. In 1872, Mr Gladstone thought it expedient to accelerate the winding-up process of empire. The year before, reflecting the received opinion of the age, Bismarck had pronounced: 'Colonies are of no more use to us than a fur to a Polish count with no shirt.' What was the point of empire? J. S. Mill, in writing *Considerations of Representative Government*, approached the question with some diffidence. He finally concluded that there *was* a point, thereby adumbrating the modern theory of the Commonwealth, in phrases which have become the clichés of Commonwealth Prime Ministers' conferences:

There are strong reasons for maintaining the present slight bond of connection.... It is a step, so far as it goes, towards universal peace and generally friendly cooperation between nations. It renders war impossible among a large number of otherwise independent communities ... it has the advantage ... of adding to the moral influence and weight in the councils of the world, of the Power which, of all in existence, best understands liberty.

It is astonishing to reflect that these words were published exactly 99 years before Harold Macmillan found it necessary to deliver his 'winds of change' speech. In the 1860s, the English seemed to have reached the last chapter of the colonial epoch, and were about to close the book with satisfaction and the consciousness of duty done. How was it that the book was rudely reopened, and new, bloody and catastrophic chapters written – chapters catastrophic not merely for the English, but for the entire world?

We cannot simply blame the ideas-men, like Ruskin. Ideas, after all, are impotent unless they both reflect and reinforce physical events. But we cannot entirely exonerate Ruskin, either. A scholar must give some cautionary thought to the probable use, or misuse, of his ideas by desperate and unscrupulous men of action. Ruskin's lectures were promptly published and widely circulated. One copy certainly fell into the hands of Benjamin Disraeli. In 1872 he was an ageing and frustrated politician, coming to the end of a very long road without having once glimpsed the promised land of power in all its plenitude. The great parliamentary majority still eluded him. Brief and tantalising spells of office had invariably deposited him back on the opposition benches. The huge extension of the suffrage which he had himself carried in 1867 had brought advantage only to the Liberals. He was in the market for any idea, any issue, which might propel the new voters in a Tory direction. Now Disraeli was not a Darwinian; on the contrary, he was 'on the side of the angels'. But Ruskin's transmutation of Darwinian concepts was quite another matter. The voters did not like to be told they were descended from apes. But they might welcome the news that they were in process of becoming gods. At any rate, Disraeli decided to give it a try, and at the Crystal Palace on 24 June he electrified a great congregation by scrapping all his previous views on the colonies and unveiling a new vision of empire.

Why, he asked, had he promoted the 1867 Reform Act? Because that act had been based on his belief that the working classes were proud of belonging to a great, imperial country, and wished to maintain its Empire. For 40 years the Liberals had sought 'to effect the disintegration of the empire of England'. Of course, he, like everyone else, supported self-government. But the donation of such 'ought to have been accompanied by an imperial tariff', by a 'military code for the defence of the colonies', by provision for aid from the colonies for the mother country, and 'by the institution of some representative council in the metropolis'. Self-government, yes; but only 'as part of a great policy of imperial consolidation'. But the tragedy was that the Liberals had viewed 'everything in a financial aspect, and totally passing by those moral and political considerations which make nations great, and by the influence of which alone men are distinguished from animals'. He then enunciated a new doctrine of secular racialism:

> The issue is not a mean one. It is whether you will be content to be a comfortable England, modelled and moulded on Continental principles and meeting in due course an inevitable fate, or whether you will be a great country, an Imperial country, a country where your sons, when they rise, rise to paramount positions, and obtain not merely the esteem of their countrymen, but command the respect of the world.

The consequences of this speech were curious. There is no evidence that Disraeli's new line had any particular effect on the 1874 election, which the Liberals lost for a variety of other reasons. Nor did Disraeli, once in office, show any inclination to apply his doctrines in a systematic manner. Indeed, he was incapable of system. He was old and tired by the time he found himself in power with a coherent majority. The legislative business of his government was inspired and carried through by other men, notably Richard Cross. Disraeli, savaged by gout, and weakened still more by the loss of his wife, could still occasionally concentrate his brilliant intellect on particular issues which captured his imagination; but to carry through a conscious and long-sighted policy of imperialism was beyond him. His purchase of the Suez Canal shares was an instinctive response to one of those financial 'opportunities' which had tempted him to disaster in his youth – though now he had the credit of the British Treasury and the pound sterling to play with. In the long run it drew Britain into an 80-year occupation of Egypt, periodic wars, huge military expenditure and political embarrassments which culminated in Sir Anthony Eden's humiliating venture in 1956; it must therefore be counted an unfortunate speculation, the true Disraeli touch. He made the Queen, it is true, Empress of India, something which appealed strongly to the vulgar streak in her nature (and which her grandfather had soberly resisted); but this was a piece of Disraelian theatricals rather than a carefully considered attempt to reorganise the basis of British rule. Disraeli's passionate interest in the Near East led him to play a star role at the Congress of Berlin. But this, too, was more show than substance; most of the time, said his Foreign Secretary, Lord Salisbury, he did not really grasp what was going on, partly because he was ill, partly because he did not understand French.* In any case, he failed entirely to comprehend Salisbury's scheme to follow up the settlement by vigorous British action. Salisbury wanted to use British military consuls to create an Indian-style empire on the ruins of Turkish Asia or, as he put it, 'to promote that pacific invasion of Englishmen which is our principal reliance for the purpose of getting power over the country'. This, one presumes, was exactly what Disraeli had in mind when he spoke of 'paramount positions' for 'your sons' in his Crystal Palace speech. But he took little interest in Salisbury's plans, and failed to use his influence with the Treasury to provide the money, for lack of which the enterprise foundered – fortunately, no doubt. Disraeli was a verbal imperialist, no more. His words, indeed, sank into the consciousness of his party; his vision became in time Tory orthodoxy, and remained such for 70 or 80 years. But the imperialism of fact arose from quite different causes.

The decade 1870–80 was a key one in the history of the world, and from its tragic events flowed momentous consequences, not least for the English. For the first time, the new, interlocking world economy, which had been expanded at breathtaking speed for 30 years, suffered a major breakdown. The first crisis of modern capitalism had reached its climax in the late 1830s, when the British economy still entirely dominated world trade. Britain saved herself, and so the world, by expanding out of crisis through the explosion of railway technology, by creating, and then exporting, the matrix of heavy industry based on coal and steel. The United States and Germany became great industrial powers. Other countries – France, Belgium, Austria, even Russia and Japan – began to follow. The modern economy took shape in the middle decades of the century, and for a time it seemed possible that this shape would be essentially English, with London as the financial pivot, and unrestricted free trade, by treaty or unilateral action, as the dynamic of unlimited, self-sustaining growth. The world was going England's way: hence the almost crazy optimism of the 1850s and 1860s.

But England, as we have seen, was in many respects a grossly inefficient and mismanaged country. The cotton revolution had been

* 'What with deafness, ignorance of French, and Bismarck's extraordinary mode of speech, Beaconsfield has the dimmest idea of what is going on – understands everything crossways – and imagines a perpetual conspiracy.' (Salisbury to Lady Salisbury, 23 June 1878.)

botched, at huge human sacrifice. The railway revolution was botched, too. No one seems to have sat down and thought out a philosophy of railway economics. No one planned a national network. The lines were simply built, at great speed, often in senseless competition, financed by an almost limitless flow of capital from men and women who believed all railways were bound to make money. In fact, some made no profits at all. In the 1860s the average yield settled down to no more than 4 per cent, and there were a number of spectacular failures. Abroad, a few of the lessons of the English railway-expansion were learned, but by no means all.* In Europe and the United States, giant engineering operations were launched in a spirit of boundless optimism, often on the sketchiest financial framework, and without any systematic attempt to relate costs to profits. The engineers were in control, eagerly followed by a greedy and gullible public.

The first blow to confidence came in 1866, when the great banking house of Overend and Gurney went broke. The City weathered this storm, with some difficulty. Then, in 1873, there was a financial panic in Vienna, as the result of an orgy of company flotations, riding the crest of the German railway mania. It spread rapidly through central Europe. Worse, it coincided with a railway boom-and-bust in the United States. By 1876 the slump had become general. Two years later, the impregnable City of Glasgow Bank collapsed. Recoveries were possible, and were indeed staged. But the glad confident morning which the coal-steel age heralded was gone for good. Panic firecrackers were now liable to burst, almost without warning, at any of the great financial centres, with unpredictable effects throughout the world. The collapse of a great Paris bank in 1882 brought a further downturn in the world economy: all the indices now showed jagged variations, instead of the smooth upward curve which had seemed to men the unassailable certitude of the modern age. Moreover, these recurrent panics not merely robbed the well-to-do and the middle-class investor: they brought vast factories to a standstill, turned thousands on to the streets, provoked riots and demonstrations and exerted entirely new pressures on governments which, whatever their complexion, now had to respond to mass public opinion. Liberalism no longer seemed to have all the answers.

The collapse of the great mid-century boom coincided with a crisis in European agriculture, which the new technologies themselves provoked. By throwing open the ports in 1846, Peel had stimulated

* The high price paid for land, and legal expenses, made the cost per mile of line in England and Wales three times as high as in Prussia, and five times as high as in the United States; E.J. Hobsbawm: *Industry and Empire* (London, 1968).

agricultural growth throughout the Continent. English farmers and landlords had responded as eagerly as anyone: there had been heavy investment, encouraged by the high and rising price of land, and a rapid increase in agricultural productivity. Throughout Europe, free trade was seen to be working: cheap food meant an improved diet and higher consumption for populations which were themselves expanding fast, thus in turn raising the demand for agricultural products. All the farmers of Europe were producing, and selling, more, at prices which the industrial explosion kept buoyant. But the cycle of growth could endure only so long as geography protected Europe from the full effects of freedom. By the 1840s, the Americans were opening up the great wheatfields of the mid-west. Shortage of labour led American farmers to demand machines; American industry and technology supplied them; the rise in American food-production was phenomenal, and the cost began to fall with unprecedented speed. The railways followed the farms as fast as the track could be laid, which was very fast indeed. To stimulate development and settlement, the railway companies transported the grain to Chicago at cost price. The new steamships brought it to Europe at rates which fell steadily. In the 1850s, even in the 1860s, it was not yet profitable to bring American food in bulk across the Atlantic. But by 1873 to ship a ton of grain from Chicago to Liverpool cost only £3·35. Ten years later it was £1·2 and still falling. Technology was catching up with free trade, and revealing the tragic distortions of a world economy run on pure liberal principles.

Throughout Europe, the American grain invasion terrified every farmer serving markets within reach of the railways. In England, the effect was catastrophic, for the arrival of cheap American food coincided with five consecutive wet summers, culminating in 1879, the worst in the memory of the oldest rustic. In the past, bad weather had brought its compensation in high prices: but now they fell rapidly. From the decade beginning in 1877, English wheat dropped from 56 shillings a quarter to 31 shillings, and farming incomes were cut by up to 75 per cent. On the Continent, there was similar distress, followed by vigorous government intervention. Unlike England, Continental countries had millions of landowning and clamorous peasants; they also had conscript armies, which the peasants' sons manned. There could be no question of permitting clearances, and driving the peasants into emigration, as had already happened in Scotland and Ireland. As the German ruling class put it: 'Agriculture must provide our soldiers, and industry must pay for them.' It was a view Continental ministers and parliaments took to be axiomatic (indeed, it is still the principle on which the European Common Market is based). There was an ominous rattling throughout the Continent as the tariff shutters went up. By 1880, free trade as a world system was dead. The industrialists, alarmed by the end of cheap food for their workers, and seeing governments bend to the pressure of the farming interests, sent up their own yelps of fear; and they, in turn, got tariffs on imported manufactures. This, of course, angered the Americans: they had never really abandoned tariffs, and their system of government was peculiarly susceptible to protectionist demands from powerful lobbies. In 1890 they erected the McKinley tariff structure, and this provoked further Continental retaliation.

The rapid retreat from free trade left Britain isolated on a lonely sandbank. The immense conservatism of the English, their unwillingness to contemplate radical change without decades of investigation, the huge built-in barriers to reform which existed at every level of the political system, united to inhibit any sharp response. It had taken more than half a century for Adam Smith's doctrines to win acceptance and implementation. By 1875, however, they were the supreme orthodoxy. Free trade was traditional, had existed (in spirit if not in fact) since time immemorial, was almost a supernumerary article in Magna Carta. It was what England was all about. Abandon free trade, merely because some frightened foreign governments had lost faith in it? One might as well propose to abolish the monarchy, or the established Church, or the public schools, or even the navy. No leading politician of either party was prepared even to contemplate such a proposal. The depression of the 1870s exposed the English public mind at its worst: drugged by a dogma which had once enshrined empirical truth.

It also exposed Disraeli as an ageing conjurer who had run out of tricks. It should have been the culmination of a career of remarkable prescience. He might have told Parliament: 'This is what I predicted in 1846. The catastrophe I foresaw was delayed, but it is now upon us. We must act fast.' In fact he did nothing of the kind. Indeed, he did nothing at all. Perhaps he felt himself too old to fight the consensus again. Perhaps he was not fully aware of what was happening. He noted dolefully in 1878 that many great aristocratic London houses were not being opened for the season. His letters reflect the grumbles of his old friend the Duke of Rutland, and the ravages of his own estate in Buckinghamshire. But he does not seem to have realised that, by failing to act, he was murdering the agricultural interest he had once championed. He had no ideas, and no policy. He told the Lords in December 1878:

> Her Majesty's Government are not prepared – I do not suppose any Government would be prepared – with any measures which would attempt to alleviate the extensive distress which now prevails.

As for the future, like Mr Micawber, he vaguely saw

> ... symptoms of amelioration and general amendment which must in time – and perhaps sooner than the country is prepared for – bring about those advantageous results which, after periods of suffering, we have before experienced.

This philosophy of impotence was, indeed, the general view. As one Liverpool Tory MP put it:

> No one can see clearly when 'the good times will come again'. But that they will come, 'ere long, is just as certain as that the light of day follows the darkness of night. Prosperity and adversity move in cycles; and the one is simply the reflex of the other, and has nothing to do with politics.

To put it bluntly: the English political nation abdicated during this key decade. When a tiny group of protectionists, early in 1880, called for a Select Committee to investigate 'the one-sided so-called Free Trade', very few MPs bothered even to attend the debate. The Government spokesman remarked fatuously that 'one-sided Free Trade was better than no Free Trade at all', and the motion was lost by 75 to 6.

Why did the English landed class, which had defended itself so cunningly through the centuries, which was still so powerful, and indeed had its hands on all the levers of government, accept such a devastating blow almost without protest? The episode is a complete mystery.* But the facts are clear enough. In the mid-nineteenth century the aristocracy had practised high farming, on a massive scale, for the first time since the Black Death. All that now came to an end. England's rulers had watched with indifference the plight of the hand-loom weavers and the Irish peasantry; now they stood idly by while their own homelands, their own dependants and kith and kin were devastated by blind economic forces. Over a quarter of the area under wheat – more than a million acres – went out of production. Estates were sold off or consolidated. Upper-class capital drifted into the City, industry, mining, or migrated. Nearly 100,000 labourers were driven off the land;

* Even the spokesman of the agricultural interest, Henry ('The Squire') Chaplin, admitted that the free trade issue, 'whether for good or evil, was settled during the last generation with the deliberate sanction and approval of the nation'. The young A. J. Balfour, while stating that the case for a duty on bounty-supported foreign sugar was obvious, added: 'Of course I know well enough that there are unanswerable reasons, administrative and political, which make the imposition of such a duty perfectly out of the question.' See Paul Smith: *Disraelian Conservatism and Social Reform* (London, 1967). Of course there were no 'unanswerable reasons', merely prejudice against change. The agricultural depression undermined the self-confidence of the landed gentry. Mark Girouard: *The Victorian Country House* (Oxford 1971) prints a graph showing the number of large country houses being built during the period 1835–90: it rises steeply until the mid-1870s, then falls away, never to recover.

in the 1870s alone, emigration topped one million. Throughout recorded history, England had always been an advanced agricultural country, with high rates of productivity and a genius for 'improvements'. The mid-century had seen spectacular achievements in this field, and had been marked by a sharp improvement in agricultural wages and rural incomes generally. Now, all progress on the land was halted, or even reversed. Productivity, investment, standards of farming fell. Britain produced 75 per cent of her wheat in the 1860s, less than 35 per cent by the 1890s. As the volume of food imports rose, so the burden on the balance of payments increased. The great agricultural counties acquired an air of seediness, even of despair. The true poet of the age was not the strident Kipling but Thomas Hardy, who caught the sad, autumnal note of the betrayed countryside:

> The land's sharp features seem'd to be
>   The Century's corpse outleant,
> His crypt the cloudy canopy,
>   The wind his death-lament.
> The ancient pulse of germ and birth
>   Was shrunken hard and dry,
> And every spirit upon earth
>   Seem'd fervourless as I.

In this poem, 'The Darkling Thrush', Hardy fancied the bird had 'some blessed Hope, whereof he knew And I was unaware'. But hope was not restored for many long decades. By 1913, a third of the land under cereal, some 3 million acres, had gone out of cultivation; by 1932 it was less than half the 1872 figure. Until 1914 most of the big landowners tried to hold on to their property, for social reasons; but after the wartime boom temporarily raised profits, they sold out to their tenants. Over a quarter of the land of England changed hands, the biggest transfer of ownership since the Norman Conquest. The landless peer became the norm, the squire a rarity. Between the wars, English agriculture reached its historic nadir: the land of idle acres. The war against Nazi Germany rescued it from virtual oblivion; but it was the post-war Labour governments, by extending and improving the wartime system of government intervention through cooperative boards and subsidies, which placed British agriculture once more on a stable foundation. The rebirth was as rapid and spectacular as the fall. By 1971 British agriculture was again the most productive and efficient in the world, the most highly mechanised and scientific, with an unparalleled record in experimentation and research. It was also highly profitable, and brought rich returns to those landowners who had held on to, or bought their way back into, the land. Thus, by a characteristic English paradox, the representatives of the industrial workers put the rural *élite* back on its pedestal.* But, of course, the pedestal was now purely functional: agriculture had simply become an efficient industry. It had ceased to be the underpinning of English society.

The events of the 1870s dealt a devastating blow to the English agricultural community, undermining a sector of the economy which had hitherto been highly efficient. But they also added to the problems of British industry, which was already exhibiting many sinister features of backwardness. Ancient and inefficient plant, a low rate of investment, outmoded ideas, dogmatic and complacent management, lack of interest in new technologies, or refinements of old ones, the defensive conservatism of the work-force expressed in restrictive practices – all these characteristics were already apparent to shrewd observers, not least Britain's chief competitors, the United States and Germany. American industrialists, responding to the challenges of an immense continent, thought in terms of bigger and bigger capitalist units, whose very size made possible the investment of vast sums in research and development, and the recruitment of industrial scientists and engineers which America's forward-looking universities produced in growing quantities. In Germany, too, heavy industry rapidly consolidated itself in vast combines, linked to banks which supplied a constant supply of finance for investment in new equipment. The later a power industrialised, the more likely it was to achieve economies of scale. In Germany, and still more in Russia and Japan, the State intervened to force the pace of industrial growth, to underwrite credit, and impose rationalisation. American industry was highly organised to exploit the resources of the State through Congressional lobbies. In other industrialising countries, the State was a forceful and aggressive partner. Only in Britain did government leave industry entirely to its own devices and uphold sternly the liberal consensus that trade and industry 'had nothing to do with politics'. The structure of British industry reflected its pace-setting origins: a multitude of small or medium firms, highly specialised and provincial in outlook, usually controlled by a single family, administered almost like landed estates and handed down from generation to

* The success of leading landed families in retaining their position, and indeed increasing their wealth, is well illustrated by a table comparing the holdings of 79 families in 1967 with their holdings in 1873, published by Roy Perrott: *The Aristocrats* (London, 1968), pp. 151–6. Some of these families are vastly more wealthy than in Queen Victoria's day. The Duke of Bedford's 30 London acres were worth (1967) 'at least £20 millions'; the Duke of Westminster (Grosvenor Estates) owned 300 London acres, presumably worth in the region of £200 million.

generation. Although London was the biggest capital market in the world, British investors played only a tiny role in the financing of British industry, chiefly because the openings did not exist. Very few firms went public. Even in 1914, over 80 per cent of British industrial firms were still privately owned. British money went elsewhere, notably into the ambitious investment plans of Britain's industrial competitors.

By the 1880s, Britain had ceased to be the leading industrial power, trailing behind America and Germany, and with new challengers coming up fast. The collapse of British agriculture meant a steady increase in the volume of food Britain had to import, and a corresponding need to export more manufactured goods. But this was now becoming difficult, as British exporters came up against foreign salesmen in more and more markets which had once been the exclusive property of 'the workshop of the world'. There was a growing volume of complaints about the price and quality of British goods, poor delivery dates, and the complacency and indifference of British salesmen in the face of highly organised and determined competition. The supposed excellence of British workmanship, enshrined in tradition by the Great Exhibition of 1851, was already regarded with cynicism in some quarters. In May 1887, a new class of 24 British torpedo boats went on a trial cruise. By the end of the first day, eight had been disabled. Engines broke down, crank-brasses were fused, wire cables parted, the top of a feed-pump blew off, main valves leaked, propeller blades snapped; in one ship, to quote the official report, 'the boiler furnace crown came down, the engine-room and stoke-hole staff were scalded, and three men subsequently died'. There were ten delays or accidents from defective steering-gear, one collision, one grounding on the rocks. An American reporter, George W. Smalley, sneered in the *New York Herald:*

> And how do you suppose the English, who have a turn for philosophy, console themselves? Why, that bad as they are, their rivals are probably worse. Defective iron, brittle steel, bad workmanship, imperfect designs – all these exist in the English Navy. Let us hope they exist among our neighbours, too, responds the indomitable Briton. Does he think they exist with the Germans, for example? The arithmetical statement of this torpedo expedition is simple indeed. There were in all, and during rather less than a fortnight, 27 accidents to 19 boats.

There was an even more sobering story from New Zealand, one of many that could be quoted. In 1883, New Zealand ordered 20 locomotives from Britain. At the end of 18 months, only two had been delivered. In despair, the New Zealanders switched the order to a firm in Philadelphia which completed deliveries in less than four months; each engine cost £400 less than its British counterpart.

The cost-advantage of Britain's competitors slowly became decisive in many fields. Between 1883 and 1910, German and US steel prices fell by 20 and 14 per cent respectively, while British prices rose by a third. The explanation lay in size and technology. As Andrew Carnegie told British steelmasters in the 1890s: 'Most British equipment is in use 20 years after it should have been scrapped. It is because you keep this used-up machinery that the US is making you a back number.' But there were other factors in Britain's shrinking share of export markets, all widely commented on at the time: lack of salesmanship, and especially of trained salesmen; goods not marked in kilos and metres; brochures sent out only in English; the failure of British sales staff to speak any language but English; lack of credit facilities, especially in comparison with the Germans. All these weaknesses, so drearily familiar to British newspaper readers of the 1960s and 1970s, were already in evidence nearly a century before. The economic decline of Britain has deep roots in the past.

Once the edge of Britain's industrial thrust had been blunted, other conservative factors in British life came into play. The economic distress of the 1880s ended the long truce between the working classes and the men of property. From this decade we note the first use of such terms as 'unemployed' (1882), 'unemployable' (1887) and 'unemployment' (1888). There was a rebirth of the working-class movement, to some extent on a socialist basis, marked by the founding of H. M. Hyndman's Social Democratic Federation (1881), and the Fabian Society (1884). But the response to the slump, in practical terms, came from the trade unions, which fought bitterly and blindly to protect the jobs and living-standards of their members. Many of these unions were already ancient. Some could trace misty origins back to the seventeenth century. They reflected the chaotic structure of British industry, in their multiplicity and anomalies, in their fear of change, and in their complex and irrational methods of business, using procedures for calculating wage-rates and defining occupations which went back to the late eighteenth century, sometimes beyond. The trade union movement was already the House of Lords of the British working-class, waving historic banners, defensive in outlook, resisting innovation on principle. By the turn of the century, Britain's relative economic decline began to inspire a literature of self-reproach, and the blind conservatism of organised labour came in for heavy criticism. S. J. Chapman, in *Work and Wages* (1904), noted the willingness of American workers to accept technical change, whereas 'an English workman finds it almost impossible to imagine that the adoption of labour-saving methods could result in higher wages or more employment'. Their memories were long and bitter: they still tended to regard new machines as a potential threat; their unions were organised to conserve, not to elevate. When British manufacturers introduced new machinery, they often found it impossible to adjust the piece-rates to the extent needed to pay for it.

Nevertheless, the critics were unanimous in identifying the conservatism of British management as the overriding cause of decay.* As F. A. McKenzie put it in *The American Invaders* (1902):

> If our workmen are slow, the masters are often enough right behind the times. In spite of all recent warnings, there is a stolid conservatism about their methods which seems irremovable. Even great houses which have the name of being most progressive, often enough decline to look into new improvements.

British industry, wrote the German sociologist Veblen in *Imperial Germany and the Industrial Revolution* (1915), was burdened with 'the restraining dead hand of past achievement'. The result was dolefully reflected in the statistics. In coal, American productivity rose by half in the period 1880–1914, almost entirely due to the introduction of cutting machinery. In Britain, where there seemed to be no shortage of labour, it remained virtually static. Even in 1924, less than 20 per cent of British coal was cut by machine, against 70 per cent in the US. There is not much point in quoting other examples – they are too numerous, and all emphasise the same trend. By 1914 the only basic industry in which Britain was leading in technology was pottery. The most spectacular failure was in chemicals, where Britain had once been absolutely dominant. Virtually the entire export business was handed over to the Germans. By 1913 Britain contributed only 11 per cent of world production, against 34 per cent for the US and 24 per cent for Germany. Germany's export trade was now more than twice Britain's, and in some areas she had a near-monopoly: 90 per cent of Britain's synthetic dyestuffs were now imported from Germany.

Chemicals was a key case, for there the connection between innovation, export performance and scientific research could be most clearly traced. The failure of British industry was the failure of British management, but this in turn was essentially the failure of British education. The sons and grandsons of English industrial pioneers were nurtured in the Anglican public schools and universities and taught to despise science. British manufacturers not only made little effort to exploit scientific

* See D.H. Aldcroft: 'The Entrepreneur and the British Economy, 1870–1914', *Economic History Review*, August 1964.

education, they often distrusted men with technical degrees. Some of the scientists and engineers produced by British universities were forced to emigrate: there was a brain-drain even in the 1890s. Britain had plenty of skilled workmen; she continued to produce theoretical scientists of immense distinction and originality. But between these two categories there was a fatal inability to find work for applied scientists. In 1872 a British deputation visiting Germany found that there were more students engaged on chemical research in Munich than in all the universities and colleges of England. After 1900, in response to growing criticism, a group of provincial universities and polytechnics were opened in Britain. Even so, in 1914 Germany had 58,000 full-time and 16,000 polytechnic students, as against 9,000 and 4,000 in Britain. It was not just a question of numbers but of industrial attitudes. E. D. Howard, in *The Cause and Extent of the Recent Industrial Progress of Germany* (1907), put it flatly: '. . . one of the most fundamental and important causes of the present prosperity of the German nation is the close relations which exist in that country between science and practical affairs.' S. J. Chapman quoted the director of a German steel works as saying: 'We can compete and make profits because of the scientific basis of our manufacture and the technical education of our workpeople . . . every one of our foremen and managers has had two years special education at the cost of the firm – a technical and scientific education.'

The Americans placed less emphasis on science than the Germans, but considerably more on industrial organisation, mass-production, standardisation, cost-control and marketing. Around 1900 the modern factory system was taking shape, based on thousands of uniform machine tools and the production line: this made nonsense of traditional British methods, but it embodied the type of industrial thinking in which the Americans had been pioneers. As early as 1880, America had been producing certain standard machine-tools at half Britain's prices. By 1913 her exports in this category were four times as big. With the advent of the motor-car, and the third phase of the industrial revolution, America was way ahead right from the start. Characteristically, the British car industry operated in tiny units, to a multiplicity of designs, some created for individual customers. By 1913 Britain had put nearly 200 different makes of car on the market, more than half of them unsuccessfully. In 1914 no British manufacturer had succeeded in producing more than one car per man per year; even ten years earlier, Henry Ford was turning out 1,700, with a work-force of 300 men. The quality of Britain's industrial performance could be measured not merely in export sales but in other, and more sinister, ways, notably in the naval competition with Germany. Admiral Jellicoe commanded the Grand Fleet in 1914 in the sombre knowledge that his battleships were in many respects technically inferior to their German counterparts: it explains the caution for which he was roundly abused. At Jutland, Admiral Beatty watched two of his battle-cruisers blow up and turned to his flag-captain with a bewildered but apt comment on Britain during the third phase of the industrial revolution: 'Chatfield, there seems to be something wrong with our damn ships today.'

Of course, the distribution of national effort is not a matter of blind chance. To some extent it reflects the conscious choice of a people. As I have emphasised before, the English have never taken industry very seriously; indeed they have taken their achievements in this field almost for granted,and have never systematically sought to reinforce them. Industrialists have not, on the whole, been rewarded by English society, either with place or privilege. For an industrial nation, they have occupied a remarkably inconspicuous place in public affairs. Success in business, especially industry, has never been regarded as a qualification for high office. Peel based his political career not on his father's factories, but on his own inherited wealth and acres. Men like Cobden and Bright, who spoke for the manufacturing interest, were notable for their lack of business acumen (Cobden left his family penniless). Joseph Chamberlain, the apostle of Birmingham industry, in fact sold out his business interests in 1874, before embarking on politics: the £120,000 he thus realised gave his public career an essentially rentier financial base. An examination of the financial status of British twentieth-century prime ministers reveals a curious pattern. Some, like Churchill, Eden and Wilson, have made money as writers, using their privileged access to State documents. Lloyd George made money through honours-skullduggery, and invested it, unsuccessfully, in farming. Only one, Macmillan, had a successful business career – significantly in publishing, where Britain has held on to a commanding lead. A. J. Balfour inherited some £4 million at the age of 21. He died a comparatively poor man, and his house had to be shut up. He lost, it seems, £250,000 in backing a process to turn peat into powdered fuel, and there were, presumably, other ill-judged speculations.

Baldwin's case is particularly instructive. He spent his youth in the family iron-and-steel firm, a typical Midlands industrial concern, medium-sized, localised, specialised, conservative, well meaning, lacking both ruthlessness and dynamism. Baldwins was involved in odds and ends of business: a South Wales colliery, worsted spinning mills, a carpet manufactory, a tin-plate works. It sounds a ramshackle and inefficient set-up, but cosy in an English way. Baldwin himself described the main plant in glowing terms, in a speech he made as Prime Minister in 1925:

> . . . a place where I was able to talk with the men not only about the troubles in the works, but troubles at home and their wives. It was a place where strikes and lockouts were unknown. It was a place where the fathers and grandfathers of the men then working there had worked, and where their sons went automatically to the business. It was a place where nobody ever 'got the sack', and where we had a natural sympathy for those who were less concerned in efficiency than is this generation, and where a large number of old gentlemen used to spend their days sitting on the handles of wheelbarrows, smoking their pipes.

It says a lot for the state of competition within British industry that this firm, far from going to the wall, actually expanded and became a major producer. Baldwin found himself worth over half a million pounds in 1919, largely due to wartime excess profits, and nobly gave one-fifth to the nation. But he, like Balfour, ended his life in straitened circumstances. Bonar Law did rather worse. He made his money in the Glasgow iron market, as a partner in William Jacks. At his death, the preference shares he held in this firm were almost worthless, and the £71,000 he left reflected legacies from relatives which totalled £60,000; not a notable business career. Neville Chamberlain began active life by losing his father £50,000 in a Bahamas sisal business, which was a complete write-off. He later won a modest competence in various small Birmingham concerns, the chief of which made ships' berths and ammunition racks for warships – a perfect example of the limited and highly specialised outlook of British industry.

The truth is, the English had a particular order of priorities in the way in which they invested their brain-power, and industry certainly came low down the scale. The *élite* education system was geared to produce, above all, politicians, lawyers and churchmen.* It inculcated habits of thought peculiarly well adapted to these professions. It deliberately and systematically encouraged the ablest young men to aspire to be prime ministers, lord chancellors, archbishops. And, within the limitations of its terms of reference, it was conspicuously successful. It is no accident that England was able to move from oligarchy to democracy, and then to social democracy, without revolutionary

* For statistical analyses, see T.W. Bamford: *The Rise of the Public Schools* (London, 1967), especially Chapter 9. Of the 5,034 MPs in the period 1734–1832, 1,714 came from the seven *élite* public schools. Even in the period 1918–35, the public schools produced 43 per cent of MPs, and 30 per cent in 1945. During the period 1918–55, the public schools produced 30–40 per cent of the ruling class: W.L. Guttsman: *The British Political Elite* (London, 1963).

violence – the only country in the world to do so. It is no accident that England had, and has, the most stable political system of any major country. Equally, England acquired, and in time dismantled, the largest empire the world has ever seen, with the minimum of bloodshed. The English created, and still maintain, a uniquely resilient and efficient judicial system, distinguished both for its fairness and its dispatch. They have contrived to avoid religious warfare, and to confine doctrinal battles to the realm of scholarship. None of this came about by chance. It reflects the extent to which the English were prepared to invest their abilities in these particular fields of endeavour, and in the institutions which dominate them. To become a Member of Parliament, Anthony Trollope correctly observed, was the height of ambition of every decent Englishman. This helps to explain the vigour and flexibility of English political life. The same remark could not conceivably have been made in the United States or Germany. In neither was politics a uniquely attractive and honoured career, absorbing a regal share of the best brains in the country. And the consequences were, and indeed are, evident. The German nation, then and now the best educated in the world, industrious, dutiful, splendidly organised and equipped, twice surrendered itself to political imbeciles who led it to disaster.* America, also, has under-invested her talents in politics: this explains her long failure, from 1900 to 1941, to accept the world democratic leadership which her physical power made desirable, and her very indifferent performance since she has reluctantly shouldered the task. Much of what happened in, and to, Germany, much of what is now happening inside America, and to her efforts overseas, is the consequence of a particular set of priorities in the allocation of *élite* human resources. The history of modern Japan reflects a similar choice. The English put stability and non-violence before industrial performance: and they got what they paid for, no more, no less. The price they paid in wealth has been a heavy one. When Lord Birkenhead was negotiating with both sides of the coal industry in 1921, he remarked that he would have thought the miners' leaders were the stupidest men in the country, had he not had occasion to meet the owners. What did he expect? He himself, to use Lord Beaverbrook's phrase, was 'the cleverest man in the kingdom'.

* In 1914, the German illiteracy rate was only 4 per 1,000, as opposed to 30 in France, 300 in Italy. But German political education was stunted. Bismarck, as Max Weber said, 'left behind him a nation lacking in all and every political education, far below the level they had reached in this respect 20 years earlier'. The point assumed direct practical importance in the Great War. The British system produced Lloyd George; the German system removed Bethmann Hollweg, to replace him first by an unknown civil servant, George Michaelis, then by the elderly and senile Count Hertling. Recently published documents on the inner conflicts in Germany, 1914–18, edited by Wilhelm Deist, confirm this analysis.

As such, he would never have contemplated going into the coal-mining industry. Naturally, and inevitably, he sought the glittering prizes in the law and politics; a century before he might have aimed, equally, at Lambeth Palace. The English education system was not designed to produce a happy and prosperous coal industry: it scarcely taught the miners how to read and write, and to the owners it gave, at best, a modest fluency in dead languages.

In the light of this, it is not surprising that the English, confronted by growing evidence that they were no longer the world's leading industrial power, sought redress and relief not in an economic solution but in a political one. They did not use the State to become more efficient. They used the State to enlarge the area in which their inefficiency would matter less. In short, they invented modern imperialism. This was, or at any rate seemed to be, the easy way out. But it was a choice directed not by strength, but by weakness. Unable to compete successfully in the developed markets, the English turned to a further exploitation and enlargement of their sphere of influence in the backward parts of the world. This was essentially a reversion to the pattern of trade in the eighteenth century – colonial raw materials in return for British manufactures. Not surprisingly, it was in India that this second wave of exploitation was operated most successfully. In the last quarter of the nineteenth century, Britain ran a steady, often a huge, deficit on visible trade. The gap was closed partly by a rise in earnings on 'invisibles' – banking, shipping, insurance and other services, and yield from overseas investments – and partly by the huge profit Britain made on the Indian Empire, which in 1914 covered up to two-fifths of her trading deficit. Other colonial ventures were less profitable. Indeed, in the 20 years up to 1900, exports of British manufactures to the colonial territories actually declined relative to other areas. But this point was not grasped, at any rate at the time. There was a general belief that painting places red on the map must be good for trade.

The economic motive behind late nineteenth-century imperialism was reinforced by others. Among the Tories in particular there was a conscious attempt to find and exploit issues which would appeal to the new working-class electorate. The number of voters doubled in the 1880s, reaching 60 per cent of the adult male population. Redistribution equalised constituency areas, the ballot box and the Corrupt Practices Act evened the party odds. The number of contested elections rose from two-thirds in 1868, to three-quarters in 1874, to four-fifths in 1880: by 1885, 19 out of 20 seats were contested. Fighting general elections on this scale force-fed the growth of national party machines, involving hundreds of thousands of party workers on both sides. The Tories had to compensate for the undoubted edge the Liberals possessed on most social issues, and they chose increasingly to play the cards of patriotism and empire. Here, indeed, Disraeli sounded the note of the future: but it was left to Lord Randolph Churchill to make it practical politics in electoral terms. In 1884 he did a deal with Lord Salisbury to ensure that the new mass-movement of voluntary workers became the docile instrument of the party brass; the same year he brought in the Tory ladies through the Primrose League. By 1900 the earliest psephologists noted the skill with which class deference and Empire loyalty was employed to turn lower-middle-class and working-class Tory voters into activists. Inevitably, the right wing of the Liberal Party were tempted to toy with the notion of imperialism, much to Gladstone's disgust. In 1882, chiefly as a result of Disraeli's commitments, he found his government bombarding Alexandria, occupying Egypt, and getting deeply involved in the politics of the entire Nile valley. Struggle as he might, he could not extricate the Liberals from this commitment, which brought the angry and anguished resignation of John Bright, a man whose Liberalism ante-dated even his own. Gladstone announced that Britain's military presence would be 'temporary', a pledge repeated and broken 66 times by British governments over the next 60 years. What saddened Gladstone most was that his youthful protégé, Rosebery, who had planned the great humanitarian campaigns in the Midlothian, became the most outspoken, articulate and relentless of the Liberal imperialists. In 1892, it was Rosebery who prevented Gladstone from evacuating Egypt. The old man came bitterly to regret that he had made Rosebery Foreign Secretary, but he was powerless to prevent Rosebery's promotion to Prime Minister when he himself retired; and Rosebery promptly annexed Uganda, the first of a new wave of African depredations.

The new imperialism lacked a unifying theory, but it rapidly acquired a distinct racial tinge. Where Ruskin had led, a tribe of eccentric professors eagerly followed. In 1881–2, J. R. Seeley, Regius Professor of Modern History at Cambridge, gave an immensely successful series of lectures, published as *The Expansion of England*: a greater Britain was arising to rival the emergent world empires of America and Russia. In 1885 James Anthony Froude wrote *Oceana, or England and Her Colonies,* in which he appropriated some of the racial theory of history popularised by Treitschke in Germany: Henry VIII was, to Froude, England's man of 'blood and iron'. A great deal of pseudo-Darwinian nonsense was written, and believed, by amateur sociologists like Benjamin Kidd, who thought imperial war and struggle purified the race by natural selection; and by J. A. Cramb, who argued, in *The Origins and Destiny of Imperial Britain* (1900), that universal peace was unthinkable in a living, developing planet. Academic imperialism was thus already inclined to see war as inevitable, even useful. Of course, few Englishmen were willing to take racial theories of world politics to their logical conclusion. Houston Stewart Chamberlain, who married Wagner's daughter, transferred himself to Germany to develop his full-blooded Aryan doctrines, and during the Great War became a violent critic of Britain's 'race betrayal'. Moreover, English racial imperialists had to meet volleys of ridicule and abuse from Fabians, socialists, nonconformist Liberals and other well-entrenched and articulate intellectual groups. Nevertheless, in the 1890s at least, race-imperialism captured the English consensus and lingered on for a good while afterwards. Rosebery gave it a high-minded gloss, in a speech to the Colonial Institute in 1893, neatly combining an instructive contemporary metaphor, a blatant distortion of history, and some characteristic English humbug:

> . . . we are engaged at the present moment, in the language of mining, in 'pegging out claims' for the future . . . it is part of our heritage to take care that the world . . . shall receive an English-speaking complexion, and not that of other nations . . . we should in my opinion grossly fail in the task that has been laid upon us did we shrink from responsibilities and decline to take our share in a partition of the world, which we have not forced on, but which has been forced on us.

This was a far cry indeed from the liberal internationalism of J. S. Mill 30 years before. The notion of England as a trustee of others had yielded to a much more narrow, selfish and racist vision of what Rosebery termed 'the future of the race of which we are at present the trustees'. The idea that the Empire must have a natural terminal date lapsed. Instead, emigration was taken to grant the English a natural right to annex wherever they settled. Other apologists of empire were less sophisticated than Rosebery. Cecil Rhodes, in 1877, had put the point with schoolboy crudity: 'I contend that we are the first race in the world and that the more of the world we inhabit, the better it is for the human race.' Some of the most violent imperialists had a Germanic background and training. Alfred Milner, for instance, was born in Germany and one of his grandparents was German (the radical press always insisted on printing his father's name as 'Karl'). He entered politics as the secretary of George Joachim Goschen, the son of a German merchant, and left the Liberals when Goschen resigned over Home Rule. From first to last, he insisted that politics should be shaped by race-factors. He repeatedly and emphatically denied that he was a 'cosmopolitan' (a term of abuse already linked to anti-Semitism). In *The Nation and the Empire* (1913), he asked: 'What do I mean by the British race? I mean all the people of the United Kingdom and their descendants in other countries under the British flag.' This was for public consumption. But after his death, among his papers was found a document entitled 'CREDO – key to my position'. It is worth quoting:

> I am a British (indeed primarily an English) Nationalist. If I am also an Imperialist, it is because the destiny of the English race... has been to strike fresh roots in distant parts... It is not the soil of England, dear as it is for me, which is essential to arouse my patriotism, but the speech, the tradition, the spiritual heritage, the principles, the aspiration of the British race ... follow the race. The British State must follow the race... we cannot afford to part with so much of our best blood.*

Against this background, part economic, part racist, the English engaged, for the first time, in a deliberate, vast and world-wide land-grab. In South-East Asia, in the Pacific, in South, Central, East and West Africa, enormous territories were directly annexed to the Crown or placed under its 'protection'. British governments proclaimed piously that they had not started the scramble, and had merely reacted defensively in formalising what were already British spheres of interest. This is very largely nonsense. America can be blamed for detonating the movement by her tariff policy, and she was herself, of course, engaged in 'manifest destiny' imperialism, in the Mid- and Far-West, and against Mexico and Spain. Russia also had practised systematic Asian imperialism for 300 years, and had latterly been imitated by Japan. But in the tropics and sub-tropics, Britain was the pacesetter, and by far the greediest and most successful of the imperial powers. In less than 20 years she appropriated 3·5 million square miles, with a further 1·5 million of protectorates. It was the lion's share, in quantity and still more in quality. Some English imperialists, indeed, did not bother to deny it. Writing of Lord Salisbury, Rhodes thought it extraordinary that 'a man who never travels further than Dieppe or the Riviera should have found out all the places in South Africa where an Englishman can breed, reserved them for Great Britain, and rejected all the

* *The Times*, edited by one of Milner's disciples, Geoffrey Dawson, published the 'credo' on 27 July 1925, described it as the 'conscious political faith of the best and most thoughtful patriots of the Empire', and reprinted it as a pamphlet for distribution in the schools. It provoked a sharp retort from the Aga Khan: 'To call India a Dominion in an empire based on British race patriotism would be an absurdity.' See A.M. Gollin: *Proconsul in Politics* (London, 1964), Chapter VI.

others'. This was the view most foreigners, notably the Germans, took. They observed that small powers like Belgium and Portugal, traditional allies of Britain, had somehow been permitted to absorb large and valuable territories, that France had got nothing much worth having apart from the south Mediterranean littoral, and that late-comers like themselves and the Italians had been fobbed off with large slices of damn-all.

Imperialism inflicted lasting damage on the English people in a number of different ways, not least in producing a certain coarsening of the national spirit. Thus, by a just irony, a movement expressing the exaltation of a race which believed itself superior to all others produced in fact marked signs of moral degeneration. The point, of course, must be heavily qualified. The standards of the colonial service were high ones. Britain ran a quarter of the globe in an autocratic, but also a judicious, spirit. There was no parallel in British-controlled territories to the unspeakable cruelties practised by the Dutch, the Belgians, the Portuguese, the French, the Germans and the Italians in the areas they occupied. The British Empire reflected some of the salient characteristics of the domestic minimum-state. It was administered by a handful of people, backed by a tiny army and small, locally recruited police forces. It therefore involved a large degree of acquiescence, indeed consent, on the part of the governed. In some ways, the Empire resembled eighteenth-century England: power confined, in practice, to the political nation of the governing *élite*, but many important freedoms enjoyed by all. Freedom of movement and communication, freedom of speech and the press, free access to an impartial system of justice, freedom to enjoy life and property under the rule of law – these were very substantial benefits. In addition, the Empire was conducted on rigorous principles of Gladstonian public finance, and corruption at any level was mercilessly punished. Perhaps most important of all, within its authoritarian framework it enforced a high degree of racial and religious tolerance. Indeed it had to. The British Empire was the greatest Moslem power and the greatest Hindu and Buddhist power; it was the greatest Protestant power, but it embraced many majority Catholic communities. It protected free-thinkers; it also, for instance, appointed the Catholic Dean of Malta. Its officials, troops and police made it possible for many scores of antagonistic races to cohabit in peace. Unhappily, indeed, its insistence on freedom of movement, and its rigorous suppression of sectarian or racial violence, allowed alien communities to establish themselves in many territories, notably in the West Indies, East Africa and East Asia. Again, its imposition of western standards of public health led to a revolution in life-expectancy, an assault on tropical disease, and a sharp fall in the infantile death-rate. From 1898, Britain began to devote increasing sums to imperial schemes of economic development, beginning in the West Indies, but gradually spreading to all parts of the Empire. Thus Britain, in a high-minded manner, and from the most altruistic motives, is largely responsible for the three salient characteristics of the Third World today: inter-racial conflict, the population explosion, and the illusory panacea of 'overseas aid' – what might be termed the International Poor Law.

But if the Empire conferred mixed blessings on its subject peoples, it brought nothing but evil to the English themselves. The problem of India, said Annie Besant, was India's pride and Britain's arrogance. The new imperialism enormously increased this arrogant spirit, the national hubris which the industrial revolution had generated and which, by the last quarter of the nineteenth century, bore less and less relation to the facts of life. As a world economic, political and military power, Britain was already in relative decline by the time the imperial spasm took place. Imperialism concealed the truth of decadence, inhibited inquiry into its causes, and blocked the search for remedies. It led the English to forget that they are a small people, inhabiting a tiny island not rich in natural resources, and that in a hostile and competitive world they must always live on their wits and their realism. It persuaded them to overestimate their strength and ignore their weaknesses. In this sense it was a true opium of the people, a drug which bred ruinous fantasies and produced a dismal awakening when the dreams were over. It was a classless drug, or rather one which all classes were encouraged to imbibe according to their station. The ruling class took heady and copious draughts, as it shared out amongst its members the glittering satrapies. There was a notable orgy on 28 October 1898, when Lord Rosebery spoke at a farewell dinner of Old Etonians, in honour of three *alumni* who were departing to take up office as Governor-General of Canada, Viceroy of India and Bishop of Calcutta. Even in the history of that fateful academy, there can have been few such scenes of rotund complacency and burping self-congratulation. 'We belong', intoned Rosebery,

> ... to the school that with an everlasting current of eternal flow turns out the Viceroys and the Bishops and the Ministers of the Empire that the Empire requires.... What, for example, would Canada have done without Eton, when out of the last six Viceroys all but one are Etonians? ... You are sending out three eminent men on three vitally important missions to different parts of the Empire ... when the battle is won, they will have a tale of stewardship which is nobly undertaken and triumphantly achieved, one

which has helped to weld the Empire which we all have it at heart to maintain, one which will redound to their own credit, and which will do if even but a little – for there is so much to be added to – to add to the glory and the credit of our mother Eton.

At almost every level of society, the same false note was struck. From Balmoral came a vulgar and debased echo of the age of Elizabeth, in Queen Victoria's notorious exchange with Cecil Rhodes. 'What were you doing, since I last saw you, Mr Rhodes?' 'I have added two provinces to Your Majesty's dominions.' At the polling-stations, Rosebery claimed that the new working-class voter 'feels that his personal honour and name are implicated in the honour and name of the Empire'. In 1896 Alfred Harmsworth launched popular journalism with the slogan: 'The *Daily Mail* stands for the power, supremacy and greatness of the British Empire . . . it is the embodiment and mouthpiece of the imperial idea.' The following year, the Diamond Jubilee brought to London, from all over the world, a Roman carnival of fancy-dress puppets, who paraded through the metropolis in an anachronistic display of antique militarism and tribal deference. Thus the high idealism of the great Victorian society collapsed into raucous bombast and tinsel theatricals. Gladstone was still alive, and watched in dismay the process whereby the bulk of the nation became intoxicated by a posthumous Beaconsfieldism, which he had once defined as 'an odious system of bluster and swagger, and of might against right'. He realised, with fearful prescience, that imperialism was linking trade to a military framework, and threatening to make industry the slave of violence. It was the negation of the liberal ethos, the denial of everything which the Great Exhibition had promised. When he attended the opening of the Kiel Canal, he burst out in anger at the scene: 'They told me it was a purely commercial enterprise. And the armed fleets of the world are gathered together.' There was never any doubt in his mind that imperialism must eventually lead to war.

Imperialism also played on the ugliest aspects of the English character. In India, Egypt and elsewhere, private soldiers were encouraged to carry swagger canes, to emphasise their ruling status in the eyes of the natives. Despised at home, they were lords abroad. Middle-class Englishmen became *bwanas* and *sahibs,* perched crazily on a bamboo caste-pyramid whose ultimate foundation was the unrelieved poverty of the dark-skinned masses. The illusion of racial superiority was bred and sustained by the grim reality of white fire-power and destructive technology. Sometimes Britain hunted with the combined white pack, as in China; more often, alone. The Maxim-gun, the mule-carried howitzer, the armoured river-boat were the chosen instruments of imperialist sportsmen, pushing forward the frontiers of civilisation under strict gun-room rules. 'Empire is commerce,' said Joseph Chamberlain, and believed it; Gladstone did not agree. As a rule, empire was 'Bobs' Roberts and Herbert Kitchener, professional exponents of the imperialist military equation: white-man's hardwear against the softwear of dark skins. Roberts had spent 41 years educating Indian and Afghan tribesmen in the arts of limited war. Kitchener, a bachelor who surrounded himself with elegant and well-connected young officers, had set himself up as a new Pharaoh of the Nile. They met briefly in South Africa, where they invented the concentration camp. Kitchener epitomised the dark side of the imperial impulse. He was a trophy-hunter, with the instincts of a looter; even imperialist Britain could not always approve of his treatment of prisoners. But he was efficient, an imperial engineer. His campaigns were exercises not in strategy, or valour, but in logistics, skilfully and systematically, and above all remorselessly, employing the railway and the steam-boat and the electric telegraph to effect the concentration of fire-power which ensured automatic success against barbarous armies. Moreover, Kitchener was cheap. His wars were based on principles of (to use a later term) cost-effectiveness. He 'costed' them in advance, and conducted operations within the strict framework of Treasury control.

Kitchener encouraged no illusions about the even-handedness of imperial justice. This led to a notable encounter in India, where he served as Commander-in-Chief under Curzon as Viceroy. Curzon was one of the three satraps whose departure Rosebery had toasted. He was in many ways the most thoroughgoing imperialist of them all, who deplored what he regarded as the weak-minded defensive policies of Salisbury. But he had a strong sense of imperial obligation. He disgraced a famous regiment when it failed to punish the murderers of a penniless native.* The Anglo-Indian community, and especially the white army, hated him. Curzon, who had a masochistic streak, rejoiced: that showed he was doing his duty. Kitchener, once installed,

* The incident occurred in 1902, and the regiment was the 9th Lancers. Curzon had earlier disgraced the Royal West Kents for the mass rape of an elderly Burmese in Rangoon. But often Curzon was powerless to act. In 1900 a private in the Royal Scots Fusiliers battered to death a punkah-coolie with a dumb-bell, but was acquitted. European juries usually declined to convict: only two Europeans had been hanged for the murder of natives since the Mutiny, though such cases were increasingly frequent. Curzon wrote angrily to the Secretary of State: 'That such gross outrages should occur in the first place in a country under British rule; and then that everybody, commanding officers, officials, juries, departments, should conspire to screen the guilty is, in my judgment, a black and permanent blot on the British name. I mean, so far as one man can do it, to efface this stain while I am here.' See David Dilkes: *Curzon in India* (London, 1969), Vol. i, Chapter 8.

had no intention of playing second fiddle to a civilian, especially one who disparaged the army. He quickly exploited his own prestige, and the enmity of the Anglo-Indians, to force Curzon's resignation. To Kitchener's mind, the inequality of the sword was incompatible with equality before the law. His whole career was based upon the deliberate exploitation of inequality. At Omdurman he had fought the set-piece battle of the entire imperialist age: the rapid, cheap and utter destruction of the Mahdi's host by the skilled use of white fire-power. It was more a *battu* than a battle. Echoing across two millennia, it bore an ironic resemblance to the Roman legionary destruction of the British after Boudicca's rebellion. There was nothing particularly glorious about it. But such bloody preludes to the advance of civilisation were necessary. It was all very well for Curzon to uphold theoretical ideas of empire: he did not have to do the dirty work. If whites and blacks were equal, what were the whites doing in India? How did they get there? Evidently because they were not equal. If you employed white troops as professional killers of subject races, you could not hang them for an occasional murder on the side, if the victim was black, that is. Curzon had the law behind him. But Kitchener reflected majority white opinion throughout the Empire. And it was Curzon who resigned.

The truth is, imperialism as a matter of practical necessity corrupted the master-race. It forced ordinary white men to behave like gods – false gods. A tiny *élite*, governing multitudes, had to acquire, or assume, a Jupiter-complex simply to get through the business of decision-making. The thunderbolts had to be rained on the heads of the wicked, the good rewarded with miracles. To many of the *élite* the illusion – the performance – became reality: they thought themselves gods. To others, the divine dispensation of rewards and punishments became a hollow and cynical routine. In either case, the imperialist made himself into a lesser man. The Empire did not so much shape 'character', as deform it. Moreover, it tended to imprison the rulers as well as the ruled. The English became slaves of their fantasies, both as individuals aping deities and, collectively, as a race aspiring vainly to a world role beyond their power. They found themselves constantly doing things, not by desire, but from a self-generating necessity which was, indeed, imperious. In his remarkable essay, 'Shooting an Elephant', George Orwell describes how, as a junior imperialist, he first became conscious of this predicament. As the man with the gun, vested with official authority, he ought to have been in command of the situation. His reason told him the elephant was harmless, and ought not to be shot. Nevertheless he shot it, because the natives expected him, as the ruler, to perform the decisive act appropriate to his role:

The people expected it of me and I had got to do it; I could feel their two thousand wills pressing me forward, irresistibly. And it was at this moment, as I stood there with the rifle in my hands, that I first grasped the hollowness, the futility of the white man's dominion in the East. Here was I, the white man with his gun, standing in front of the unarmed native crowd – seemingly the leading actor in the piece; but in reality I was only an absurd puppet pushed to and fro by the will of those yellow faces behind. I perceived in this moment that when the white man turns tyrant it is his own freedom that he destroys. He becomes a hollow, posing dummy, the conventionalised figure of a sahib. For it is a condition of his rule that he shall spend his life in trying to impress the 'natives' and so in every crisis he has got to do what the 'natives' expect of him. He wears a mask, and his face grows to fit it.

Thus at a number of levels, and in many subtle ways, the English became prisoners of their empire, and in the end its victims. The process was, indeed, at work even before the age of imperialism. For there was a spectre at the imperialist feast, a Banquo's ghost shaking gory locks matted with the congealed blood of centuries: England's oldest colony, Ireland. Ireland was the joker in the English pack. It always had been. It was a living, ocular refutation of the English claim to have the genius and temperament for an imperial mission. English statesmanship had been periodically exercised in Ireland for seven and a half centuries, and had invariably been found wanting. The English liked to think they were a tolerant people, easy-going, not inclined to press logic to unreasonable lengths, pacific unless provoked, adept at compromise, skilful in devising political solutions as an alternative to violence, constructive, unrivalled in fashioning institutional frameworks which canalised passion and reconciled the irreconcilable; above all, just. This was the image the English sought to present to themselves and to the world: they had built it up from innumerable examples culled from their activities in countries scattered all over the globe. But the moment the history of Ireland was mentioned, the whole shaky structure collapsed in ruins. Seen from an Irish perspective, the English were ruthless, but also irresolute; tyrannical, but cowardly; inconsistent even in their violence, oscillating wildly between repression, indifference and appeasement. Whatever the English genius was, in Ireland it had conspicuously failed. There, the English had imposed neither justice nor stability; they had created no durable institutions; they had generated no wealth. A long gallery of English potentates had impinged on the Irish scene: Henry Plantagenet and Richard of Bordeaux, Elizabeth and Cromwell, William of Orange and Walpole, Pitt the Younger and Robert Peel. Ireland had baffled them all, left them, indeed, angered at their own impotence. Confronted with Ireland, the English usually in the end took refuge in the bitter consolation that the Irish were incorrigible and worthless: thus attributing the consequences of their own criminal incompetence to the inborn nature of a hapless people who had never, in all those centuries, sought an English connection.

Ever since the Papacy had 'given' Ireland to the Crown, the English had, at various times, tried every solution (save one): outright conquest, partial conquest, the cordon sanitaire, genocide, assimilation through settlement, settlement without assimilation, apartheid, military rule, civil rule, plantations, puppet parliaments. Every experiment had ended in violence. At the beginning of the nineteenth century, the English tried Union, that is, direct rule from Westminster. The one thing they never tried, though to all but the English it seemed the obvious way to terminate this long tale of misery, was outright and total evacuation: it is still untried as I write. Direct rule proved the least satisfactory method of all, and inflicted terrible injuries on both countries. When the Act of Union was passed, Ireland's population was rising fast. It stood at over 8 million in 1841. The Famine ripped aside the arrogant pretensions of the English master-race, revealing them as bewildered pygmies. But what was a moral defeat for the English was a physical disaster for their subjects. By 1847 three-quarters of a million Irish were employed on relief works: 3 million were being fed the barest necessities at public expense. About a million are believed to have died. By 1851 the population had fallen by 20 per cent. It dropped to 5·8 million in 1861, and continued to fall until well into the twentieth century. Ireland must be the only country in the world whose population has dropped consistently and dramatically in the last century. That is the indictment the English, as the 'responsible power', have to face.

But Ireland took its revenge, in more ways than one. In the eighteenth century, Ireland had made a massive contribution to English prosperity: at least £750,000, on average, had been transferred, each year, from Irish tenants to their absentee landlords in England. And, of course, the cost of administration was borne by the Irish Exchequer. In the nineteenth century, the balance of economic advantages shifted decisively. A growing number of Irish estates yielded no revenue at all; or, if actively administered, tended to swallow capital. The shrewder landowners consolidated heavily or got out altogether, and after the general agricultural depression set in during the 1870s, it became politically possible for successive Conservative governments to pass a series of acts (1885, 1887, 1891, 1896, 1903) enabling Irish tenants to buy their lands: 13 million acres changed hands, and by 1917 there were over half a million individual holdings in Ireland. But this by no means diminished the Irish burden on the English Exchequer, which rose steadily throughout the century and beyond.

Ireland, indeed, became a serious political, financial, military and social liability. The Union was never accepted by the majority of the Irish people. At no point was it possible to administer the country, and keep the peace, within the ordinary framework of the law. The law either had to be, in effect, suspended, and political and economic crime go unpunished; or it had to be reinforced with coercive measures which were bitterly resented by the Irish people and abhorrent to a section of English opinion. Until 1829, of course, the Irish as a whole were excluded from the political processes of the Union. After it, they sent to Westminster a growing number of Members who formed a group hostile to the workings of the English parliamentary process. The decisive moment came with the secret ballot in 1872, which enabled tenants to vote against their landlords with impunity; and from then onwards, Charles Stewart Parnell was able to weld together a highly disciplined block of about 80 MPs who in abnormal times held the political balance of power, and even in normal times could disrupt parliamentary business. The effects on the House of Commons were serious and lasting. Throughout the century, the increase in government legislation, and the rise in the number of MPs who took an active part in parliamentary life, had forced governments to strengthen the rules of procedure, and so limit the freedom of back-benchers. In 1881–2 Gladstone met the threat of Parnellite disruption by bringing in the guillotine and other devices to permit essential business to get through the House. These changes, together with the rise of the party machines, killed the independent Member, and transformed back-benchers on both sides into regimented party units. Thus the initiative of power passed from the Commons as a body, and resided, in ever-increasing measure, with the Ministerial executive. This momentous development, which demoted Parliament to the status of a debating chamber, was brought about very largely by the effort to swallow Ireland. The question arose: was Ireland the prisoner of England, or was England the prisoner of the Irish problem she had created?

In 1885 Parnell brought matters to a climax by demonstrating that the Irish, by their votes at Westminster, and even by their votes in English and Scottish constituencies, could exercise a decisive influence on the central government of the whole Empire. It was a masterly display of purely destructive power, the first victims of which were the Liberals. Gladstone was now convinced that the moment had come to save the English political system by a decisive reversal of policy, as in

1829, 1832, 1846, or 1867. But the Irish demand for a large measure of self-government, which he now believed to be irresistible, ran directly counter to the swelling tide of English imperialism. Whichever party took the responsibility of absorbing the shock of impact between these two dynamic forces must be broken. Which was to take the risk? Lord Salisbury declined the poisoned chalice which Gladstone proffered him. Home Rule, he thought, might in some ways be desirable; it was not inevitable. For him to carry it would certainly break up the Conservative Party, and probably consign the rump to his arch-opponent, Lord Randolph Churchill. He did not believe that Irish self-determination was a moral necessity: 'There are no absolute truths or principles in politics.' Gladstone disagreed, with every fibre in his being. Home Rule would salvage the English Parliament. But it was also, he now saw, an end in itself. Christianity and democracy were self-sustaining. If an individual could redeem himself through the exercise of political freedom, did this not apply *a fortiori* to an entire nation? Nationalism, not imperialism, was the true, God-appointed road. Thus Gladstone willingly, almost eagerly, ran his party on the rocks, in the cause of something which he believed, rightly, to be infinitely more important than any mere political organisation. 'What?' asked Sir William Harcourt. 'Are you prepared to go forward without either Hartington or Chamberlain?' 'Yes. I am prepared to go forward without anybody.'

The year 1886 was the last great watershed in the evolution of the English political system. Hitherto, the great parties had been able to reach a consensus (sometimes a rather ragged one) whenever a crisis arose over issues fundamental to the constitutional structure. Henceforth they drifted apart, and the modern monoliths of Left and Right took shape. Hartington took over to the Tories the last of the Whigs, Chamberlain the radical industrialists, who soon ceased to be radical – became in fact imperialist and eventually protectionist. The parties began to organise on class lines, lines determined increasingly by income and occupation. A political chasm opened on the floor of the Commons.

The change expressed itself in many ways, all lamentable. Gladstone had long been worried by the lower moral tone of politics. 'Democracy,' he said, 'has not saved us from a distinct decline in the standard of public men. For this deterioration one man and one man alone is responsible – Disraeli. He is the Grand Corrupter.'* And the old charlatan, in Glad-

* Gladstone was right in the main, but his judgment of Disraeli was inflamed by personal hostility, dating probably from Disraeli's budget of 1852, which Gladstone thought quackish. When Gladstone succeeded Disraeli as Chancellor of the Exchequer, they had an undignified row over payment for the furniture in 11 Downing Street, and over the

stone's view, had found an apt pupil in Churchill. His Tory democracy, Gladstone said,

> . . . is no more like the Conservative Party in which I was bred than it is like Liberalism. In fact less. It is demagogism, only a demagogism not ennobled by love and appreciation of liberty, but applied in the worst way to put down the pacific, law-respecting, economic elements which ennobled the old Conservatism, living upon the fomentation of angry passions, and still in secret as obstinately attached as ever to the evil principles of class interests.

The crisis of 1886 poisoned English political life with fresh hatreds. Public men now held each other in contempt. London society, which had survived the convulsion of 1830–32 with remarkable aplomb, broke into warring camps. Social boycotting (a word, significantly, of Irish origin) was not entirely new. Gladstone was felt to have broken the conventions of the ruling class by his Midlothian campaigns, and Rosebery, who sponsored them, was blackballed from the Travellers. 'One of the minor results of the Midlothian election,' he wrote, 'was to close to me any London Club of which I was not a member. As I was already a member of eighteen this was of the less consequence.' But the Irish shadow added a new and venomous flavour. The Duke of Westminster, who had been elevated by Gladstone and sat in his government, ostentatiously took down the portrait of the GOM he had commissioned from Millais, and sent it to the sale-room. He electrified London society by cancelling a dinner in honour of Lord Spencer's brother (Spencer had stuck to Gladstone) on the grounds that he had dined with Parnell. The party whips circulated grandee hostesses with lists of guests to be vetoed. The Queen, needless to say, joined in the campaign with gusto: no Liberal, except Rosebery, was asked to Windsor or Balmoral. Excluding the Liberals and radicals meant excluding the intellectuals, and London society began to take on its inflated, vulgar atmosphere of wealth and snobbery – an atmosphere called Edwardian but which in fact dates from late Victorian times.* Plutocracy and the aristocrats were now in naked and open alliance.

Over this debased political scene, Lord Salisbury presided with a cynical appetite for power which he took pains to conceal. From his

---

Chancellor's robes, which Disraeli believed had belonged to Pitt, and insisted on keeping. Gladstone felt what he termed 'a strong sentiment of revulsion from Disraeli personally a sentiment quite distinct from that of dislike'. See *HMC: W.E. Gladstone, Autobiographica* (London 1972). Disraeli fully reciprocated the hostility. When he died, he was at work on a novel in which he hoped to destroy the victor of Midlothian by ridicule. See Robert Blake's Leslie Stephen Lecture, *Disraeli and Gladstone* (Oxford, 1969).

* Lord Derby had remarked to Disraeli in 1875 that the Conservatives 'are weakest among the intellectual classes: as is natural' (Disraeli papers).

Jacobean palace at Hatfield and his vast, yellow-coloured mansion in Arlington Street, he ran the country and the Empire in the interests of party and family. His taste for nepotism, and that of his nephew Balfour who succeeded him, was notorious: his government was the 'Hotel Cecil'. Gladstone had once liked, even respected, him. But when he heard that Salisbury had rescinded the ruling that Cabinet ministers should give up their directorships, he noted sadly that Salisbury showed 'an indifference on certain questions of honour that I cannot understand'. But Salisbury did not believe politics had anything to do with morals. In public, he and his family affected a devotion to High Church principles. In private, he regarded mankind as corrupt, and politics as the art of management by all appropriate means. Behind an edifying Victorian façade, he was Walpolean. Offices were handed out without scruple purely on party grounds. His Lord Chancellor, Halsbury, appointed judges and magistrates according to political colour, with Salisbury's full approval.* Where Gladstone had agonised for years over the choice of a poet laureate, Salisbury simply picked a Tory, Alfred Austin. He claimed to despise titles, but took a Garter. He was a white supremacist. In his speech against the Home Rule Bill he said openly that some races, 'like the Hindoos and Hottentots', were unfit to govern: the Irish were one of them. He encouraged Balfour to sneak on his political friends. He told the Queen Gladstone was insane, and cooperated with her in debauching the constitution. Salisbury has been gently handled by historians, partly because he said and wrote (at least in private) exactly what he thought: he had a clever, if shallow, wit, and his sayings were relished. But, on close examination, they betray an unpleasant nihilism, and a crude streak of snobbery. He did not want, he said, 'government by grocers', presumably a reference to the Liberal leader Campbell-Bannerman. He failed to recognise his chief lieutenant, W. H. Smith, at the dinner-table, an anecdote his daughter, in his official biography, relates with relish. But of course Smith was a tradesman; one doubts if Salisbury would have cut Lord Lansdowne. But snobbery did not prevent the sale of honours under the Salisbury regime.

The honours scandal merits a detour because it has been persistently

* As his daughter admitted: 'With regard to many non-political posts, he would be frankly partisan in his selections. Legal promotions did not come under his direct appointment, but he would never apologise for the practice of making them a reward for political "right thinking"' – Lady Gwendolen Cecil: *Life of Lord Salisbury* (London, 1931), Vol. iii. A reasoned defence of the Halsbury appointments is given in R.F.V. Heuston: *Lives of the Lord Chancellors, 1885–1940* (1964), Chapter v. For a more sympathetic view of Salisbury as a man and politician, see the excellent chapter, 'Lord Salisbury', in Kenneth Rose's study of the young Curzon, *Superior Person* (London, 1969).

foisted on to Lloyd George's admittedly broad and culpable shoulders. The question is shrouded in mystery, for documents dealing with it are still withheld from the public gaze. But it is clear that the systematic sale of honours for party purposes was already flourishing in the 1890s, and that both parties employed it freely. It took many forms. A Tory MP was bribed with a baronetcy to vacate his seat and let a Liberal in at the subsequent by-election. Rhodes paid £5,000 to the Liberal Party boss Schnadhorst in an attempt to influence policy on Egypt. But for the most part, honours went simply to party contributors. It was the plutocratic skeleton in the closet not of any group or individual, but of the whole political structure. It probably goes back to the Tory alliance with the brewers in 1874, provoked by Gladstone's efforts to reduce drunkenness. As Lloyd George later, and correctly, observed: 'the attachment of the brewers to the Conservative Party was the closest approach to political corruption in this country'. The intensification of party warfare produced by the Irish crisis deepened and widened the influence of money, and the afflatus of the Great War turned a closed scandal into an open one. Paradoxically, it was Salisbury's family who led the pack against Lloyd George. But what the Tories chiefly objected to – it was a major factor in the break-up of the Lloyd George coalition in 1922 – was not the association of honours with party contributions but the fact that most of the cash went into Lloyd George's funds, and not their own. A particularly sore point was the behaviour of Sir Horace Farquhar, to whom Lloyd George gave an earldom in his resignation honours. Farquhar was the Tory Treasurer, and had been given £200,000 by Lord Astor (who himself got a viscountcy in 1917). When Farquhar died, the Tory bosses angrily discovered the kitty was bare. In the words of Lord Beaverbrook, 'Horace had spent the lot', and 'L.G.' had got £80,000 of it. Under Baldwin, as the Davidson papers reveal, the system was changed. The private brokers were put out of business; Maundy Gregory was imprisoned, and on his release was whisked off to France by a Tory agent, and paid a quarterly pension from Tory funds as the price of silence. Money was still accepted by the Tory Party from honours-seekers, but on the strict understanding that the goods would be delivered only on merit. These transactions were still going on during the comparatively recent times. And today? Who knows?*

* The activities of Lord Farquhar are related in Lord Beaverbrook: *The Decline and Fall of Lloyd George* (London, 1963). For the handling of honours under Baldwin, see Robert Rhodes James: *Memoirs of a Conservative, J.C.C. Davidson's Memoirs and Papers, 1910–37* (London, 1969). For Maundy Gregory, see Gerald Macmillan: *Honours for Sale* (London, 1954), which has a useful appendix on the history of the honours system. H.J. Hanham: 'The Sale of Honours in Late Victorian England', *Victorian Studies*, March 1960, gives some interesting details.

Under the conflicting strains of imperialism and Ireland, the decline of English political life took other and more dangerous forms. The High Victorian consensus was built around the acceptance of a progressive extension of the suffrage, in the belief that the consequences would not be revolutionary, but could be accommodated within the existing parliamentary system. It was a typically English arrangement, an unspoken, but mutual, disarmament pact. The democrats undertook to advance with discretion, the *élite* to retreat with grace. Both forswore the use of ultimate weapons. Each paid a price. The result was a slow progression within a framework of stability: the English solution. But Ireland jammed the machinery of conciliation, because Home Rule seemed to many Tories a fundamental alteration in the geographical structure of the nation, and therefore outside the rules of the game. In 1893 the Tory Lords vetoed Home Rule, though passed by a Commons majority returned at an election when Ireland was a chief issue. They were in effect saying: there are limits to democracy and we decide when they have been reached. They got away with it, and for some years the Liberal leadership allowed Home Rule to become a dead issue in English politics. Thus encouraged, the Tory Party carried the theory further. In 1909 the Lords were used to veto the budget. After two elections, and the threat to create peers, the veto was removed by the Parliament Act. But the net result of the crisis was a return to 1885, with the Irish holding the parliamentary balance, and the passage of Home Rule essential to the survival of the Liberal Government; but now, of course, the Lords were no longer available as a longstop.

The Tories thus had to face some very serious questions. Did they support the existing system of government in England? Would they accept the democratic verdict, however distasteful? Was the parliamentary statute still the ultimate basis of law? Or would they go beyond the law, and resort to direct action? It was a question the English ruling class had never had to face before, except in the context of a struggle with the royal executive, because they had always controlled the constitutional machinery. But now it was in the hands of a mass-electorate. If the political struggle was allowed to escape from legal restraints, where would the process end? The women militants had already resorted to direct action: but then they were excluded from any other. The unions, at a time when prices were rising, and their living-standards actually falling, were seriously considering the political weapon of the general strike. The Commons had become a battlefield, with Lord Hugh Cecil and his Tory 'Hughligans' leading the forces of anarchy. The English had always put stability and continuity first. It had been bought at a price, usually paid by the poorer classes and the under-privileged.

Now the *élite* were asked to fork out a long-overdue subscription – in the form of the Protestant landed gentry of Ulster (they were less worried by the Protestant Belfast workers), who were asked to submit to Catholic majority rule. It was a class price, and a racial price. For the Tories it seemed an intolerable price. On the other hand, if they let stability slip, who would be the ultimate beneficiary of chaos? In a sense, Britain on the eve of the 1914 War was a microcosm of the whole of Europe: if one army moved, others would follow. It would be – the phrase recurs throughout the history of the English – 'a leap in the dark'.

The Tory Party never finally made up its mind. But it came close to, indeed made plans for, rebellion and treason. In *The World Crisis,* Churchill was later to write: 'It is greatly to be hoped that British political leaders will never again allow themselves to be goaded and spurred and driven by each other or by their followers into the excesses of partisanship which on both sides disgraced the year 1914.' This even-handed apportionment of blame ignores the truth. The Liberals were legitimately enforcing the law of the constitution; the Tories contemplated resisting it. It was one thing for Edward Carson to organise the Ulster Covenant in 1912 to defeat Home Rule. The Ulster Irish could scarcely be denied the use of physical action which had been freely applied by the majority. But it was a different matter for English statesmen actively to assist them, not only to ignore the verdict of the imperial parliament but to seek to frustrate the legal acts of the executive. Bonar Law came close to the brink when he said: 'I can imagine no length of resistance to which Ulster will go, which I shall not be ready to support.' But 'support' was a vague word; it might mean anything, including treason; or it might mean nothing at all. After all, the Tory leadership had climbed down over the Parliament Bill, leaving Lord Halsbury and his fellow 'ditchers' high, dry and impotent. Less vague, however, was the kind of language which began to circulate, in secret letters and memoranda, when Alfred Milner took a hand.

Lord Milner was, in 1913, an unemployed proconsul with a long record of violent activism in South Africa. There he had recruited and trained a crew of able and ambitious young men who believed, in varying degrees, in his concept of a race-empire. Milner did not understand the English, and in particular he neither understood, nor liked, English democracy. He hated the Liberals because they had made his employment of Chinese indentured labour in the Rand an emotive and damaging issue in the 1906 election. He had been humiliated in the House of Lords in consequence. He had, too, a personal grudge against Asquith who had married Margot Tennant, whom Milner desired. In Ulster, Milner saw a noble cause: the rescue of a white settler colony of superior British stock from submersion in a sea of inferior Celts. On 9 December 1913 he wrote to Carson (marked 'very confidential'):

... there must very soon ... be a 'rebellion' in Ulster. It would be a disaster of the first magnitude if that 'rebellion' ... of loyalty to the Empire and the Flag – were to fail! But it must fail unless we can *paralyse the arm* which might be raised to strike you. How are we to do it? That requires forethought and organisation *over here* ...

'Paralyse the arm' could only mean the sabotage or suborning of the law-enforcement agencies of the Crown, in the last resort the army. Milner believed in peacetime conscription, and through the National Service League he had developed contacts with many army officers, including the Director of Military Operations at the War Office, General Henry Wilson. Wilson was a plausible conspirator, a fanatical 'Ulsterman' and the centre of many devious webs, as we shall see. He was game for anything. Others were less anxious to stick their necks out and risk prosecution. In February 1914 Milner secured effective control of the English-based Union Defence League, with the object of organising an 'English Covenant'. The League included many sonorous establishment names, including Lord Roberts, Kipling, Dr Warren, the President of Magdalen, and the constitutional lawyer A. V. Dicey. It was also heavily financed by the plutocrats and the landed millionaires. From secret code-lists in Milner's papers, we know that Astor promised it £30,000, Rothschild, Iveagh and the Duke of Bedford £10,000 apiece. Various plans were kicked around, some legal, some not. Dicey wanted the King simply to dismiss Asquith and dissolve Parliament. But supposing the Tories failed to win the election? Another plan was for the Lords to veto the annual Army Bill. But if the armed forces ceased to exist, who was to hold down the industrial rabble? There were schemes to induce army officers to resign, rather than coerce Ulster, pay them compensation, and guarantee reinstatement when the Tories returned to power. But was not this incitement to mutiny? Some of Milner's proposals were evidently so wild that they provoked a letter from Dicey hinting plainly that he was exposing himself to the Treason Felony Act. What Milner had in mind was revealed, in some detail, by a secret memorandum found in his papers. Under a spurious cover of pseudo-constitutionalism, it provided for a straightforward *coup d'état* in Ulster:

The difficulty in taking immediate action is to make sure that the steps taken will not give too severe or sudden a shock to the British instinct for legality.... From this point of view it is essential, in the present conflict between the rights of the people and a Tyrannous Parliamentary Executive, acting in the name of the Crown ... to keep in mind those fundamental principles of our constitutional law which make the responsibility for the maintenance of the King's peace rest not upon the Crown and its ministers and officials, but upon the local magistrates, and in which respect of the duty of assisting the magistracy make no distinction between officials, military or civil, and ordinary citizens.

In short, said Milner, the Crown should be by-passed. The lord-lieutenants, deputy-lieutenants and magistrates of Ulster should meet, elect a provisional committee, and enrol 'special constables'. They would then form a Provisional government and send out letters (drafts of which were attached to the memo), calling on the military and police authorities to provide assistance, and to carry out no instructions save those of the Provisionals. All this, of course, was treason. Moreover, an Ulster *coup* made no sense, even in the short run, unless physical retribution from England was to be averted. And how was this possible, unless the imperial government itself were overthrown, by force or mutiny? The logic of Milner's plan must have involved the destruction of English democracy, just as in 1958 the Algiers military conspirators risked ruin unless they captured the Paris machine.

Were the Tory leaders privy to Milner's scheme? Would they have supported it? Walter Long, boss of the Tory agricultural interest, was president of the UDL and discussed with Milner, for instance, plans to print currency for the rebels. But Bonar Law developed acute cold feet in the spring of 1914. He discarded the Army Bill proposal, and was under growing pressure from Tory constitutionalists to keep well inside the law. Many ordinary Tories were shocked by the Curragh resignations, by Ulster gun-running and by Fenian attempts to follow suit. One cannot see a wily bird like Balfour, for instance, dipping his feathers in treasonable waters. The English ruling class is not easily persuaded to risk a stand-up fight in the open, especially when their opponents have their hands on the formal levers of power. Asquith, another wily bird, was feeling his way towards a compromise in July 1914, and if the crisis had been allowed to develop the likelihood is that the English would have muddled through and hailed the result as a miracle of constitutional good sense, etc. Milner might have found himself a very lonely conspirator indeed. As it was, the Ulstermen fell on the European crisis, which meant the indefinite suspension of Home Rule, with joyful relief.* Without exception, the plotters were the most rabid

* But so did Asquith. 'This,' he said to Lady Ottoline Morrell on 25 July, 'will take the attention away from Ulster, which is a good thing.' On August 3 he told his Cabinet colleague, J.A. Pease: 'The one bright spot in this hateful war, upon which we were about to enter, was the settlement of Irish strife and the cordial union of forces in Ireland in aiding the Government to maintain our supreme National interests.' See Cameron Hazlehurst: *Politicians at War, July 1914–May 1915* (London, 1971), Chapter 1.

and active supporters of immediate intervention. Thus Ireland played a notable part in the English nemesis.

The World War of 1914–18 was the greatest moral, spiritual and physical catastrophe in the entire history of the English people – a catastrophe whose consequences, all wholly evil, are still with us. How did it happen? Why did the English hurl themselves bodily into a European convulsion, whose origins were not their concern, and whose outcome – whichever side won – could not conceivably benefit English interests? The more one examines this question, the more certain one becomes that there is no answer which makes sense. For once the political genius of the English failed them. This was a war they should not have fought. It went against the whole grain of English history. The English had always sought to reinforce and emphasise that detachment from the Continental land-mass which the Channel provides. They had struggled successfully to exclude themselves from all Continental systems, political, military, religious and economic. They had completed this process in the sixteenth century, thereby unleashing the process of dynamic growth on their island, on the basis of which they had created the modern world. They had henceforth intervened on the Continent merely to redress developments which threatened their separate status. This is the true meaning of the word isolation: the attitude of mind of a people who live on an island, and wish to keep the sea as their frontier. It does not preclude contacts, exchanges, cooperation: but it inhibits the systematic involvement with the land-mass which diminishes, and in the end destroys, the island privilege. Isolation, thus properly construed, is the most consistent single thread running through the tapestry of English history. In 1914 it snapped. England slipped back into the Continental system, and has stuck there ever since. How?

The question is all the more agonising because this calamitous involvement took the form of a test to destruction between the English and German peoples. From the perspective of two World Wars, this does not now seem so extraordinary. From the perspective of the nineteenth century, it would have seemed unbelievable. The English and the northern Germans had cultural and racial affinities which went to the roots of their societies. England herself had been created by Germanic settlers looking for a freer and a better life – as the English had later created America. In the eighth century, in a spirit of racial piety, the English had converted Germany to Christianity and civilisation. Germany had received the flame of religious freedom via Huss and Luther from Wyclif, inspired by the oldest intellectual traditions of the offshore island. The English were xenophobic, often racist: but they had never hated the Hanseatic traders and seamen in the way they hated the French and Italians. The Germans, indeed, had always occupied a privileged place in the English consciousness of the world. Modern Germany had been created not so much by Bismarck as by Castlereagh, in the aftermath of the Napoleonic wars, as a defence against, and counterpoise to, the social anarchy of France and the communal barbarism of Russia. In the nineteenth century the Germans emerged as model citizens in a world which seemed to be shaping itself on English lines: industrious, law-abiding, hating domestic violence, devoted to education, obsessed not by logic but by the practical arts of science and technology, wholly reconciled to constitutional order and stability, ripe to be influenced still further in the direction the English were travelling. Even the worst elements who had a hand in German policy in the period 1870–1914 were no more, or less, typical of the Germans than, say, the members of the Percival and Liverpool governments were typical of the English, during the equivalent stage of the growth of political enlightenment in England. The very aspects of Germany which the English found most irritating – their drive for industrial exports, their love of the sea and naval power, their desire for colonies to bring order and discipline to the benighted heathen of Africa and the Pacific – echoed precisely the urges, ideals and objectives of the English.

The paradox of the English and the Germans at each other's throats was heightened by the fact that England embarked on the conflict in the company of the French and the Russians, two peoples notorious for their record of political expansion beyond their cultural frontiers. From the seventeenth century, the French had consistently sought to advance north and east into non-French-speaking territories, and the English had spent much treasure, and even some blood, in trying to prevent them. Napoleon III's chief war-aim in 1870 was the acquisition of German-speaking Luxembourg; even as late as the 1950s, French proconsuls in the Saar were trying to persuade its wholly German inhabitants to accept permanent inclusion in the French State. It was a matter of opinion whether France or Germany had a better title to Lorraine, but there could be no argument that Alsace was predominantly German in culture. Certainly France in 1914 gave no currency to the belief that the problems of Europe could only be settled permanently and peacefully on the basis of self-determination.

Russia had for centuries pursued an even more ambitious and consistent policy of acquisition by virtue solely of her State power. No country in the world had, for generations, inspired more horror and repulsion in the English people. Predatory in Europe, predatory in Asia, Muscovite Russia had engulfed vast territories and entire peoples in one gigantic tyranny, a Eurasian empire which sought not to elevate its colonial subjects, but to depress them, not to guide them towards self-government but to imprison them for ever in a racist autocracy. Tsarist Russia was regarded by the English as the most brutal and reactionary power on earth. Many of them thought the anti-Jewish persecutions of the 1890s, which stopped only just short of Hitler's final solution, the most dreadful human calamity of the nineteenth century, practised moreover by a Christian power which claimed civilised status. In his immensely popular book, *Democracy and Liberty* (1896), surveying the social achievement of mankind, W. E. H. Lecky had written:

> Nowhere, indeed, in modern Europe have such pictures of human suffering and human cruelty been witnessed as in that gloomy Northern Empire, where the silence of an iron despotism is seldom broken except by the wailings of the famine-stricken, the plague-stricken, and the persecuted.

After the turn of the century, the pogroms were renewed, providing the first great impetus for the Zionist settlement in Palestine. Russia stood in startling contrast to Germany, where the wealthiest and most civilised Jewish community on earth flourished. Nor were the Jews the only victims. Tsardom persecuted Polish Catholics and Lithuanian Lutherans, the members of scores of dissenting sects, Ukrainian and Georgian nationalists, Tartars and Balts, Letts and Finns, and the many tribal and nomadic peoples of Central Asia. In so far as Russia was changing – and she was, at remarkable speed – she was moving in directions which perceptive English observers found ominous. The Russians were the last in Europe to industrialise, but by the 1890s they were doing so rapidly, in huge State-financed and -controlled units. Tsardom in fact was laying the foundations of Stalinism, and by the time war broke out Russia already possessed all the most frightening characteristics of the Soviet slave-state: a ubiquitous secret police, rural communes and a collectivised agricultural sector, organised anti-Semitism, the systematic persecution of national minorities, the cult of the army, territorial aims in all directions, State management of the economy, a growing State sector in manufacturing, State-run unions, and a high growth-rate, especially in heavy war-orientated industry.*

If there was to be a clash between these two peoples, and the Germany which Britain had helped to appoint to the role of peacekeeper in

* See Lionel Lochan: *The Making of Modern Russia* (London, 1962).

central Europe – what John Morley called 'the high-minded, benignant and virile guardian of the European peace' – why should Britain become involved at all? Here we come to the heart of the matter: it was the imperialist hubris of the English which provoked their nemesis in Europe. The assumption was made by Salisbury and many of his colleagues, an assumption maintained by most historians since, that imperialism and isolation were compatible and indeed complementary. In fact they were in inevitable conflict. The true isolationist was the 15th Earl of Derby, who deplored alike colonial expansion and intervention in Europe, seeing the connection; this was why he left Disraeli and joined the Gladstonian Liberals, whom he served as a cautious and far-sighted Foreign Secretary. Disraeli's re-entry into Europe, with Britain as a leading negotiating power, coincided with the propagation of the imperialist idea: the two policies reacted on and reinforced each other.* In a practical sense, the English imperialist spasm was an attempt to escape from Britain's economic difficulties, an easy alternative to tariffs and, still more, to the distasteful business of becoming an efficient manufacturing nation. But other powers had their difficulties. Others sought an easy escape from them. Seeing Britain avoid the competition in efficiency, they began to compete with her in the acquisition of space. Britain's escape merely provoked a European stampede. As Britain was first at the colonial carrion, her successful greed aroused the resentment of all the late-comers; and this, in turn, a defensive snapping of the lion's jaws. In his last professional comment on international politics (1908), Lord Sanderson, Derby's devoted secretary at the Foreign Office, observed: 'It has sometimes seemed to me that to a foreigner reading our press the British Empire must appear in the light of some huge giant sprawling over the globe, with gouty fingers and toes stretched in every direction, which cannot be approached without eliciting a scream.' In his first dispatch as French ambassador in London (1898), Paul Cambon noted with alarm the nervous and antagonistic postures the English imperialists were liable to adopt at the least alarm – a reversal, he rightly considered, of the pacific internationalism the English had pursued for 80 years. The Boer War found Britain not so much in isolation, which implies lack of contact, but in potential hostile involvement with her imperialist

* This point was eagerly grasped by Léon Gambetta, the ideologist of France's policy of revenge against Germany. France should congratulate herself, he wrote in July 1878: 'England has broken away from what I would term her insular policy to renew her tradition of a Continental policy... England has made a brilliant re-entry into the European conflict.' He foreshadowed a British-French-Russian alliance, to surround Germany and smash her. See J.P.T. Bury: 'Gambetta and England', in *Studies in Anglo-French History*, edited by Alfred Colville and Harold Temperley (Cambridge, 1935).

competitors. It threatened to be a recapitulation of the events which, in the early 1780s, ended Britain's first Empire. The European powers, watching the progress of the Boer War, were torn between envy of Britain's possessions, and contempt for her incompetence in defending them. Britain's traditional strength and safety lay in the fact that she had no common frontiers with European powers to provoke conflict. This was still true in 1870. But by 1914 she had common frontiers with her fellow imperialists all over the globe. Imperialism thus forced her into contact with Europe, in the quest of settlements, in the demarcation of spheres of influence, ultimately in the search for allies. Put another way, imperialism made it certain that another European conflict would be global, and very probable that Britain would be dragged into it.

Imperialism also ensured that such a war would be radically different to any other, fought by the antagonists in a spirit of total and destructive commitment. For imperialism implied racial superiority. It was a deadly game, played by self-appointed master-races. Increasingly, in the speech and literature of the pre-war epoch, we find the terms of a debased Darwinism employed in political discussion. Weird theories of international determinism were evolved on the basis of racial origins. Pro-German Englishmen spoke of the French as 'Celts'; the pro-French referred to the Germans as 'Huns'. A future war, wrote the Kaiser, would be 'the last battle between Teutons and Slavs', and he feared it would find 'the Anglo-Saxons on the side of the Slavs and Gauls'. The imperialist doctrines of white racial superiority over the dark-skinned peoples thus produced a calamitous debasement of spirit among the whites themselves. They began to see each other not as sophisticated human beings but as rival herds of highly bred animals, doomed to slaughter each other like beasts in desperate encounters for the survival of the fittest breed. Inflamed by its urge to impose 'civilised' white-racial rule on African primitives, Europe itself became the theatre of tribal conflicts, fought by warriors armed with all the resources of modern military technology. Such a war, springing from the racist roots of imperialism, would involve not just courts and governments and professional armies, but entire peoples, seeking to survive and to exterminate, using the genocidal white fire-power which had made imperialism possible.

This was the prospect which English hubris invoked. The English thus made a salient moral contribution towards the age of total war. But they might still have escaped involvement in the catastrophe they helped to create. They were dragged in as a result of a malfunction in their political system, a narrow but crucial failure in the democratic process. How this came about is worth examining in a little detail. The story is part diplomatic and political, part military. The two aspects are connected, but distinct: the first determined whether the English would be committed to war or not, the second determined the type of war they would fight. Let us look at them in turn.

In December 1905, Sir Edward Grey became the British Foreign Secretary. The reasons for his appointment were entirely connected with the internal power-struggles within the Liberal Party, and in particular with the manoeuvrings which surrounded the formation of the Government. With Haldane and Asquith, the other two leading Liberal imperialists, he had signed the 'Relugas Compact' to keep Campbell-Bannerman, the 'pro-Boer' and their titular leader, out of effective power when the Liberal moment should come. The Compact broke down (thanks largely to that decided woman, Lady Campbell-Bannerman), but, in submitting, the imperialists got themselves key jobs: Asquith the Exchequer, Haldane the War Office, and Grey the FO. He had once served there as under-secretary (where he had reflected the predominant anti-French mood of the day). Otherwise he was wholly unqualified for his post. He had spent a lazy boyhood and youth, and had been sent down from Oxford for 'incorrigible idleness'.* He read little, and spoke no foreign language. He came from the Northumbrian gentry, and his interests in life were entirely confined to the countryside, where he spent his leisure hours in the paradoxical pursuit of protecting birds and killing fish. Apart from one visit to the West Indies, he never set foot outside the British Isles, and he had a particular dislike of the Continent. In the Commons, Grey was respected, especially by the Tories; it was thought that, for a Liberal, he had 'the right tone'. Suspicious of democracy, he held himself aloof from the radical tide which swept into the Commons with the spectacular victory of January 1906. Unlike Gladstone, he did not believe it probable that the mob possessed collective wisdom. He was thus the perfect instrument for the English catastrophe. Indeed as a man he was remarkably accident-prone (more so than any other contemporary statesman, except the luckless Emperor of Austria). His own nemesis was pretty impressive. His first wife was killed in a carriage-smash. His second wife died suddenly. One of his brothers was trampled to death by a buffalo. Another was torn in pieces by a lion. Both his houses were burnt to the ground. He went blind, and died childless, and his peerage

* In the 1920s he was nevertheless elected Chancellor of Oxford University, thus providing a further illustration of Oxford's preference for 'character' over education.

became extinct. Everything he touched, including British foreign policy, seemed fated to disaster.*

Grey inherited from his predecessor, Lord Lansdowne, a political agreement with the French, under which the two countries had successfully reconciled their conflicting interests in various parts of the world, notably in Morocco and Egypt. It was part of a general attempt to smooth down the ragged edges produced by the imperialist scramble. One had been signed with Japan. It was hoped to make others with Germany and Russia. But the French *entente* was unusual. In the first place, it had acquired unexpected relevance at the end of 1905, against the background of a dramatic Franco-German confrontation in Morocco. Secondly, to many influential people in England, it meant a great deal more than a mere understanding. As long ago as 1878, the King, then Prince of Wales, had told Gambetta that he wanted an Anglo-French alliance. Edward had a German accent, and of course a German father, but he was by taste and temperament a passionate francophile. He loathed the men who ran Germany, especially his nephew, the Kaiser. The short, occasional minutes he wrote on the diplomatic dispatches shown to him testify to an invariable suspicion of German motives: 'As absurd as it is false' (a German complaint); 'A case of bullying as usual' (German treatment of Spain); 'In plain English – Germany ousts France from Morocco and puts herself in her place'; 'Germany is certain to act against us – behind our back.' Edward was not alone in his sentiments. A growing number of Englishmen were alarmed by Germany's naval programme, which threatened the maintenance of England's traditional two-power standard. This, in itself, might not have mattered. Far more important was the emergence of a dominant anti-German group in the Foreign Office.

When Grey took office, the FO was on the eve of a historic transformation. Lord Sanderson, its chief permanent official for many years, was about to retire. The son of a Tory MP, he had been appointed by patronage. He ante-dated the age of competitive examinations and the emergence of a thrusting bureaucratic *élite*. He maintained the old traditions of the Office derived from Palmerston (whose portrait adorned

* Grey seems to have suffered from a death-wish. In his early forties, he noticed with pleasure that his hair was growing grey, and wished he was thirty years older. In 1912 he wrote to a friend that he looked with horror on the 'hideous cities' and 'ghastly competition' of the modern world; if God shared his view, 'then the great industrial countries will perish in catastrophe, because they have made the country hideous and life impossible'. During the last weeks of peace, in July 1914, Grey's numb imperviousness to disaster was coloured by desperate attempts to give up smoking, which he had been told, no doubt wrongly, was responsible for his failing eyesight. For more details about this odd fish (and a more generous estimate of his career), see Keith Robbins: *Sir Edward Grey* (London, 1971).

his sanctum). He thought the clerks should wear tall hats in the season, and cultivate good handwriting for copying dispatches. He revered the memory of his old chief, Derby, whose speeches he had edited and who had left him a legacy of £10,000. Like Derby, he thought there was no such thing as 'good' and 'bad' powers, towards which Britain should adopt permanent attitudes: she should, on the contrary, strive to maintain friendly relations with all states, on a basis of give-and-take. Above all, policy should be made by the Secretary of State and the Cabinet, and openly defended by them in Parliament. The duty of the Office was simply to assist. Sanderson had big, thick glasses and was known as 'Lamps'. He played the flute.

Sanderson had been 47 years in the FO when Grey's reign began, and early in 1906 he retired. Sir Charles Hardinge, his successor, was a different type altogether: a thrusting, highly organised imperialist, a future Viceroy (who was nearly killed when a bomb was thrown into his howdah), and an efficiency expert. He carried through the revolution 'Lamps' had long delayed, and reorganised the Office as a high-powered, policy-making machine. The age of the mandarin had come. Senior officials were relieved of routine duties and encouraged to 'think out' policy positions and express them in lengthy memoranda. Files were re-designed so that dispatches could be impressively minuted by the hierarchy before they reached the minister. An internal Office consensus began to emerge and operate powerfully at the political, policy-making level. From the start there was never the smallest doubt what the consensus would be. 'Thinking out' encouraged activism, and activism meant selecting 'friends', identifying 'enemies' and working against them, forming alliances, moving into the European system. The new mandarins were all francophiles, most of them imperialists and Unionists. There was Sir Francis Bertie, a large aristocrat known as 'the Bull', later sent to Paris as ambassador, where (wrote Grey) he pursued the Anglo-French cause 'in the most efficient and wholesome manner'. There was Louis Mallet, Grey's secretary; George Spicer, the Assistant Clerk; Arthur Nicolson, passionately pro-French, who presided at the Office on the eve of Armageddon and was (said his son) strongly pro-Ulster; above all, there was the Chief Clerk, Eyre Crowe.*

Next to General Wilson, Crowe was the great spider-figure of British involvement. He was the son of a diplomat and art-historian, and of a German woman called Asta Von Barby. He had been born in Leipzig, and educated in Dusseldorf and Berlin; he spoke German perfectly,

* For further information about the Foreign Office at this time, see G.W. Monger: *The End of Isolation: British Foreign Policy, 1900–1907* (Edinburgh, 1963), and Zara S. Steiner: *The Foreign Office and Foreign Policy, 1898–1914* (Cambridge, 1969).

English and French with an accent. Though he married a German, and despite his background, he loathed the Kaiser's Reich. He thought it the enemy of libertarian democracy. But oddly enough, he was himself almost a caricature of an eccentric, somewhat aggressive German professor. He had sandy, crinkly hair, wore odd suits (his hats had incensed old Sanderson), travelled by the tube (or 'the unterground' as he called it), and spent many industrious hours delving into diplomatic and military history, amassing evidence for his theories. He was a captain of volunteers, and had even tried to fight in South Africa. The others called him 'the Bird'. Under Hardinge, he was the organiser of the mandarin take-over, and the man who fed ideas into the new machine. The Office was his entire life: any outsiders, especially politicians, who ventured to hold views on foreign policy he dismissed as 'meddlesome busybodies': diplomacy was a matter for the experts. It was all very similar to what was going on in Berlin.

In 1905, there were a number of people in London anxious to transform the French *entente* into a *de facto* military alliance, using the Moroccan business as a pretext: Lord Esher, Edward's friend and a francophile busybody-behind-the-scenes; Sir George Clerk, Secretary to the Committee of Imperial Defence, Wilson of course, and Paul Cambon. The last had been instructed by his government to get military talks going early in 1905, when the Tories were still in power, but had made no progress. There seemed a better chance with the new, ignorant and gullible Grey; and on 29 December 1905, Grey received a letter from Colonel Repington, Military Correspondent of *The Times*. Repington was a former officer who had resigned his commission following a scandal with a married woman, a born gossip and intriguer, and a propagandist for the French alliance: Wilson writ small, in fact. He asked Grey, on behalf of Major Huguet, the French Military Attaché, whether it would be proper for questions to be put to the French staff about their requirements. Grey agreed. Cambon followed this up in January with a personal visit, in which he proposed that the French naval and military attachés should hold talks with the Admiralty and War Office. Again, Grey had no objection.

From those fatal encounters all else flowed, and the English moved forward to Armageddon. One says the English: but, of course, the English knew nothing about the encounters. The most fateful decision in their history was taken secretly, in an atmosphere which smacks of conspiracy. In 1906 no English Cabinet, certainly not a Liberal one, would conceivably have approved a military alliance with France, or any other great power; it was still less likely that such an alliance would have received the endorsement of Parliament, and least of all of the electorate. It is important to understand Grey's position. He may have been led to suppose that military talks had taken place in the past, under the previous Government, and that he was being asked to authorise continuity rather than innovation. But in that case he should have called for the relevant papers; and in any event he should have sought the authority of the Cabinet. But it seems more probable that he knew what he was doing, and was already under the influence of the Office 'consensus'. This being so, he lied to the House of Commons when he told it on 19 February: 'They [our relations with France] remain exactly as they were' – a plain misstatement of fact. Furthermore, from this point, Grey's own actions became increasingly furtive. The Prime Minister was, indeed, informed of the talks, by a circuitous route, but was dissuaded by Grey from calling a Cabinet meeting to discuss the matter. 'Certain ministers,' said Grey, 'would be astonished at the opening of such talks.' This was an understatement: the Cabinet would have vetoed the whole enterprise. Campbell-Bannerman's behaviour was also irresponsible.* Obviously the talks would lead, were in fact designed to lead, to joint contingency planning, with the inevitable result that the French would believe Britain had accepted a moral commitment to aid her in the event of war with Germany. 'C–B' was doubtless unaware of the pressure within the Office to transform the arrangement into a *de facto* alliance. But he saw the central point. 'It comes close to an honourable arrangement,' he wrote, 'but let us hope for the best.' His eyes were further opened when he met Clemenceau and discussed the joint talks. 'C–B' told him it was most unlikely the Cabinet and Parliament would agree to send troops to the Continent. Then what, asked Clemenceau in astonishment, was the point of the talks? There was no mistaking which way things were moving, as the Foreign Office put on the pressure. In the summer of 1906 Grey tried to prevent Haldane (who of course was privy to the secret) from attending the German army manoeuvres, on the grounds that this would annoy the French; after much fuss, Haldane went; but the following October, by threatening to resign, Grey successfully prevented the band of the Coldstream Guards from paying a courtesy visit to Germany, as 'our foreign policy will not stand any more'. Knollys, the King's secretary, thought it would seem extraordinary that

* Especially since he was the most widely travelled, and in some ways the best-educated, of the Liberal ministers. He spoke several languages fluently, and knew Europe well. Alone of the Liberal *élite*, 'C–B' had something approaching contempt for the 'ancient universities'; he thought the education they provided was purely social, and that even in the subjects which they claimed to teach better than anywhere else, the classics and theology, they were in fact incompetent. But 'C–B' was lazy, and by 1906 inclined to avoid points of controversy.

'the sovereign of this country, supported by the Secretary of State for War, cannot even send a military band abroad without the approval of the Foreign Office'. The truth was that, unknown to the Cabinet, Parliament or the public, Britain was already becoming the political puppet of France.

At the beginning of 1907, Crowe emerged from his think-tank and produced a vast memorandum, written on pale green paper, about British foreign policy. It was partly a one-sided and elaborate review of Anglo-German relations, and partly a call for systematic sternness in resisting German pretensions. He argued that the 'new spirit' Britain had shown over Morocco (that is, pro-French puppeting) had shaken German aggressiveness and made her think twice:

> In this attitude she will be encouraged if she meets on England's part with unvarying courtesy and consideration in all matters of common concern, but also with a prompt and firm refusal to enter into any one-sided bargain or arrangements, and the most unbending determination to uphold Britain's rights and interests in every part of the globe. There will be no surer way to win the respect of the German Government and of the German nation.

This, of course, was a certain formula for disaster. The memo, circulated to the Cabinet, made no reference whatsoever to the Anglo-French staff talks, a 'one-sided bargain' if ever there was one. It failed to urge that the process of settling outstanding disputes which had been applied to France in the form of the *entente* could equally well be applied to Germany. On the contrary, it was an elaborate scenario, based on selective evidence, allotting hero and villain roles to great European Powers, something which went contrary to the whole spirit of Britain's traditional Continental policy. Old Sanderson, in retirement, was kindly sent a copy. To the consternation of the FO, he replied with a strongly dissenting memo of his own, replete with mature wisdom and notably free from the highly emotional bees buzzing about 'the Bird's' brain. It evoked a condescending comment from Grey, and was eventually circulated (a whole year later), with an elaborate refutation from Crowe.

This was as close as the Cabinet got to being informed of British foreign policy, for which they were responsible to Parliament and the nation, until 1911. Grey was now wholly in the hands of the mandarins, and quite determined to by-pass the constitution. He resented the fact that MPs were allowed to question him. As he wrote to Nicolson (3 October 1906): 'The Members have now acquired the art of asking questions and raising debates and there is so much in foreign affairs which attracts attention and had much better be left alone.' Haldane created a general staff and a 'striking force' for dispatch to France. On Repington's advice, the name was changed to 'expeditionary force' so as not to alarm the English public. When the creation of this force, an entirely new departure in British military policy and one specifically planned to dovetail into the French alliance, was announced, Haldane told the Commons that it was needed because 'we have to protect the distant shores of the Empire from the attack of the invader'. This came close to a deliberate lie.* Why the suspicions of other ministers were not aroused is difficult to understand. Of course there was a tradition that foreign policy was primarily a matter for the PM and the Foreign Secretary: in Gladstone's last Cabinet he had emphasised the point by sitting with Rosebery at a separate table. But there was no precedent for this deliberate deception. Old 'C–B' went to his grave without informing the Cabinet of the military talks, which proceeded remorselessly, deepening Britain's commitment and imposing a tightening grip on her freedom of action. What is even more sinister is that Asquith, who succeeded as Prime Minister in April 1908, had to wait three years until Grey told him about the military talks. He mulled this over for several months in his ponderous way and then told Grey (September 1911): 'The French ought not to be encouraged, in present circumstances, to make their plans on any assumption of this kind' (i.e., possible British assistance). Grey replied: 'It would create consternation if we forbade our military experts to converse with the French. No doubt these conversations and our speeches have given an expectation of support. I do not see how that can be helped.' Finally, at the insistence of John Morley, who had got wind of the business, the Cabinet was at last informed in November, and held two long discussions. It laid down that there was to be absolutely no commitment to the French, and that no further talks were to be held without specific Cabinet approval. But even at this stage the Cabinet was not told the whole truth (indeed, throughout the period Grey withheld a great many compromising Foreign Office documents from Cabinet scrutiny). When, a year later, they discovered that the talks went back to 1905, they instructed Grey to write, and themselves amended in draft, a formal letter to Cambon emphasising the absence of a British commitment. This was the letter which Grey read to the House of Commons (omitting a crucial and damaging passage) on 3 August 1914 in the notorious speech which made a British declaration of

* Haldane, a loud, self-confident lawyer, with a high opinion of his own powers of perception and judgment, underestimated the deviousness of the British brasshats: 'The dear generals are angels, no other name is good enough for these simple, honourable souls.' See Peter Rowland: *The Last Liberal Governments, the Promised Land 1905–10* (London, 1968).

war certain. But as far as the Foreign Office, Grey, the military staffs and the French Government were concerned, it remained for all practical purposes a dead letter. They acted throughout on the assumption that, in the event of a German attack on France, Britain had an inescapable moral commitment to come to France's aid with all her resources.

The rest of the political and diplomatic story is soon told. When the balloon went up at Sarajevo, Grey did his limited best to avert a European conflict. But this was impossible. We now know that the entire German ruling establishment, civil, military and industrial, were determined to fight a preventive war against France and Russia on an issue which made Austrian support certain, and that they leapt at the Serbian pretext.* Grey's only choice was whether or not to keep Britain out. But his mind had been made up (or, rather, it had been made up for him) that, if war did break out, Britain must fly to France's side and thus make possible the destruction of German militarism. To the British war party, the question of Belgium was irrelevant; or, rather, Belgium was merely the instrument by which Grey could sell the war to his colleagues. The Foreign Office had long assumed, with Grey's approval, that the British must, if necessary, acquiesce in a unilateral French violation of Belgian neutrality.† The Anglo-French staffs were quite prepared to violate it jointly; the Belgian invitation was merely a bonus. Of course, Asquith pounced on the Belgian issue as the one way to hold his Cabinet and hence his party together – his chief consideration throughout. To help him, Grey obligingly misled the Cabinet on both 31 July and 2 August about his talks with the German ambassador, Lichnowsky. The trick worked: only Morley and Burns resigned. (No one thought to ask: if we are going to war for Belgium, how do we propose to save her? A good question: there were, in fact, no plans to help Belgium.) Grey got his war. After it was all over, Sir Arthur Nicolson's son, Harold, wrote: 'Sir Edward Grey shares with Bethmann Hollweg the honour of being, alone of pre-War statesmen, morally unassailable.' Alas, the evidence now shows that both were as guilty as any. Grey, indeed, has the sombre distinction of having

* Fritz Fischer: *Germany's Aims in the First World War* (English edition, 1967), especially the evidence of Count Tisza, Count Czernin ('Germany demanded that the ultimatum to Serbia should be drawn up in those sharp terms'), Josef Baernreither, Otto Noetzsch, Arthur Von Gwinner, Admiral Georg Von Muller, Kurt Riezler, pp. 88–92.

† On 15 November 1908 Crowe presented a memo on Belgian neutrality. Hardinge minuted: 'Supposing that France violated Belgian neutrality in a war against Germany, it is, under present circumstances, doubtful whether England or Russia would move a finger to maintain Belgian neutrality...' But if Germany did so, 'it is probable that the converse would be the case'. Grey thought Hardinge's comment 'to the point'. See *British Documents on the Origins of the War*, Vol. IX, (London, 1926), pp. 375–8.

inflicted more damage on the English people than any other man in their history.

The English entered the great Continental war as a French political puppet; worse, from the outset, they became a French military puppet. Grey's policy of deception, of a secret, unwritten alliance, brought Britain the worst of all possible worlds. As Grey could give no formal commitment, he could demand no right of reciprocation: in particular, he could not ask for (and he never discovered) the terms of France's alliance with Russia, the key to the whole mechanism of the war. Britain thus committed herself to abide by an unknown chain of events ultimately set in motion by the Tsarist autocracy. There was a still more bizarre feature of the *entente*. The Anglo-French staff talks committed the British army, such as it was, to help the French. But they did not give the British staff the right to know what the French war plans were: that was conditional on a formal alliance. So the expeditionary force became the junior partner in an enterprise the objects of which were concealed from it. As Churchill had noted in 1912: 'We have the obligations of an alliance without its advantages and above all without its precise definitions.' The French claimed they reciprocated in the naval strategy by taking over responsibility in the Mediterranean, thus allowing Britain to concentrate her battle-fleet in the North Sea. On the eve of war, they used this as moral blackmail against Britain by claiming that they had deliberately exposed their Channel coast. 'Are you going to let Cherbourg and Brest be bombarded when it is by your advice and by your consent, and to serve your interests as well as our own, that we have concentrated all our ships far away?' shouted Cambon at Grey on 1 August. In point of fact the French had no alternative but to deploy their entire fleet in the Mediterranean: it was essential to provide cover against the Austrian navy while they ferried their North African army to France.

As for the army talks, the politicians did not feel it was their business to inquire. As Grey told Asquith: 'What [the general staff] settled I never knew – the position being that the Government was quite free, but the military people knew what to do, if the word was given.' From the very first moment the staff talks started, the fate of a million British soldiers was sealed. When, early in 1906, a delighted Major Huguet crossed to Paris with permission to ask the French staff how Britain could help, he found the experts of the Deuxième Bureau drawing up a plan for the invasion of England! But they quickly snapped up Britain's offer. Naturally, the British troops should simply form a unit of the French army! (*'L'armée britannique ... devra ... être liée à celle de l'armée française, c'est-à-dire, être placée sous la même direction.'*) The French stuck relentlessly, and successfully, to this course, although they did not get formal supreme control until 1918. The idiocy of the British commitment was later described by Sir William Robertson, CIGS: '... we, on our side, were not able to insist upon our right to examine the French plan in return for our cooperation. When the crisis arose there was no time to examine it, and consequently our military policy was for long wholly subordinate to the French policy, of which we knew very little.' This, in fact, was an understatement: in practice, Britain's military effort was subordinate to French policy until the last shot was fired: and French policy was to commit both nations to fighting a war of destruction on the Western Front, with Britain supplying her full share of the cannon-fodder.

To involve Britain in a great Continental land-war was also the object of General Wilson, and indeed of most of the leading British army commanders. It was a simple matter of *déformation professionelle.* Wilson noted in his diary that Stamfordham, George V's secretary, 'said among other things that I was more responsible for England joining the war than any other man. I think this is true.' Wilson's Gallic strategy triumphed for two reasons. The politicians thought their responsibility ended with the decision to go to war. The rest was up to the commanders. The idea that in a pitiless and total struggle between entire peoples some degree of political control was necessary simply did not occur to them. But the English were also betrayed by the incompetence of their admirals. As professional men, the admirals of course had no desire to concede the leading role to the army; but they took no steps to prevent it from becoming inevitable. During the Agadir crisis, in August 1911, Asquith called a meeting of the CID to consider: 'Action to be taken in the event of Intervention in a European War.' This was the only time, until after war was declared, that the civil and military powers met to discuss grand strategy, and one young officer, Captain Hankey of the Marines, realised it was a crucial occasion. A week before the meeting, he sent a long letter to Reginald McKenna, First Lord of the Admiralty, urging that the navy take active steps to present a viable alternative strategy:

> It is of course notorious that the DMO, General Wilson, who has brought this question [the expeditionary force] to the front has a perfect obsession for military operations on the Continent.... He holds the view, not only that military action is indispensable in order to preserve the balance of power in Europe, but that we require a conscript army for the purpose. If he can get a decision at this juncture in favour of military action he will endeavour to commit us up to the hilt.

He advised that the Admiralty should either decline to give a date for transportation of the troops, or take the attitude that the expeditionary force idea was 'altogether a wrong one'. Not only was this very shrewd advice ignored: the meeting was a fiasco for the admirals (and a catastrophe for the English people). The Admiralty had declined to create a general staff; it had done no homework. When Sir Arthur Wilson, the First Sea Lord, proposed, as an alternative to the army plan, a scheme to land troops in Germany, he was held up to ridicule. That, said Sir William Nicholson, the CIGS, was a formula for disaster: hadn't the Admiralty studied the maps of Germany's strategic railway system, and noted the speed with which it could concentrate overwhelming force on the north-west coast? Of course not, said Admiral Wilson: it was not their business to have such maps. Nicholson (known to his naval enemies as 'Old Nick'): 'I beg your pardon, if you meddle in military problems you are bound not only to have them, but to have studied them.' After the meeting, Hankey sadly confessed that he was 'driven to admit that the Senior Service on this occasion have sustained a severe defeat'.*

So the English went to war. There was still a minute chance that they might avoid the Flanders holocaust. The day after the declaration, on 5 August, Asquith summoned a grand war council, what Henry Wilson termed 'a historic meeting of men, mostly entirely ignorant of their subjects'. Present were four politicians, one admiral, two field-marshals (including Old Roberts, who was 82), seven generals and two colonels. The gathering spent some time, said Wilson, 'discussing strategy like idiots'. There was now no dissent to the view that the army must cross the Channel. The argument, according to Grey's account, 'related solely to the moment at which the British Forces could be used best and most effectively to help the French Army'. The options were narrowing to vanishing point. Oddly enough, it was only General French, who was to command the force, who sought a muddled escape from the ineluctable mincing-machine. He, said Wilson, 'plumped for going at once and deciding later where to go – but then he dragged in the ridiculous proposal of going to Antwerp'. This, indeed, made a bit of sense in terms of British interests. Antwerp, after all, was what the war was supposed to be about, from Britain's point of view: she had created an independent Belgium, she had signed the Belgian guarantee of 1839 precisely to 'unload the pistol pointing at England's heart'. But the idea was brushed

* In November Hankey pointed out: 'If the army has been committed to the centre of the campaign at the outset of war . . . the great advantage of seapower is to a great extent thrown away.' Stephen Roskill: *Hankey: Man of Secrets*, Vol. I, 1877–1918 (London, 1970).

aside contemptuously by the War Office experts: all the staff-work had been done, the loading schedules drawn up, the French railway timetables compiled: the expeditionary force had to form the left flank of the French army.* So the doom of a generation was sealed. For all practical purposes, the British political system might not have existed. The 'frocks' had abdicated. The entire nation had been consigned, bound hand and foot, into the custody of the generals. The English people were given the role of stage-extras, to wave flags now, to suffer and be slaughtered later. For all the protection their democracy afforded them, they might have been better off living in imperial Germany: there, too, the generals were in command, but at least they knew their business.

One of the few men who foresaw the magnitude of the catastrophe overtaking England was John Morley. He had sought in vain to avoid the commitment. Now he resigned, not in the belief his going would have any influence on events, but because, like Bright in 1882, he wanted to protest, in the only way he could, against something he knew to be utterly wrong and contrary to reason and humanity. He said the war would destroy European civilisation. Even if Germany were defeated, the ultimate result would be the establishment of Russian tyranny in Eastern and Central Europe: 'We are only playing Russia's game.'† He thought it intolerable that the great Liberal Party, shaped by Gladstone and generations of enlightened men of affairs and intellectuals to advance peaceful prosperity and individual freedom, should submit to 'wholesale identification with a Cabinet committed to intervention in arms by sea and land in Central Europe and all the meshes of the Continental system'. The war, he realised, would destroy the Liberal Party: the first step came in 1915 in the shape of coalition with the Tories who 'counted Liberalism, old and new, for dangerous and deluding moonshine'. Morley defined as his chief reason for writing his *Recollections* the desire to 'keep bright for younger readers with their lives before them the lamp of loyalty to Reason'. Two worlds

* Lord Kitchener, in his formal orders as Secretary of State for War, commanded Sir John French: 'The special motive of the force under your control is to support and cooperate with the French army against our common enemies.' The victor of Omdurman added a pathetic word of warning: 'Officers may well be reminded that in this, their first experience of European warfare, a greater measure of caution must be employed than under former conditions of hostilities against an untrained adversary.' The Earl of Ypres: *1914* (London, 1919), pp. 14–15.

† Morley expressed his terror at 'the half-barbarous Russian swarms'. He thought that 'the Slav peoples are the most instinctively and phrenetically communistic in their aspirations'; allied with French revolutionary ideas, through the Franco-Russian alliance, the Russians would do fatal damage to Europe. See D.A. Hamer: *John Morley, Liberal Intellectual in Politics* (Oxford, 1968).

disappeared in 1914 and the events it set in motion: the world of the European territorial aristocracy, and the world of liberal idealism. The collapse of the first brought rejoicings among the educated; only slowly was it realised that the destruction of the second was even more important, above all for the English.

The war had a third consequence, the nature of which has not yet been fully grasped. To understand it, we must adopt a long perspective, and analyse the half-century and more which has followed 1914 not as two distinct wars, followed by two distinct periods of peace, but as a continuous unity. One might call it the re-entry of the English into the Continental system, from which they had broken away at the time of the Reformation. The breach with Rome was, above all else, an affirmation of English sovereignty, a unilateral declaration of independence. When Henry VIII declared 'This realm is an Empire' he meant simply that England acknowledged no other earthly authority, and was wholly in control of its political, legal, religious and economic arrangements. By a semantic paradox, the eventual erosion of the sovereignty thus acquired began in the late nineteenth century when 'empire' had assumed a quite different significance. Imperialism brought the English not more freedom but less. It drove them to form relationships of mutual interest or antipathy with the other Great Powers of Europe which inevitably involved a loss of choice. Sovereignty was surrendered by the English in 1914, and has never been fully recovered: even as I write this, large British forces squat behind their weapons in the heart of Continental Europe, and their removal (it is thought) would entail such a woeful chain of consequences, to Europe and the world, that no British Government dare make such a move, however desirable it might be from Britain's viewpoint. The year 1914 began a play which has no ending and Britain, having accepted a leading role in it then, is still a conscript actor on the stage, mouthing lines not of her choosing.

Of course, in 1914 Grey and the English war party believed that Britain had no alternative but to throw her weight into the scales of battle. France, without British aid, would be destroyed; her destruction would leave Germany dominant in Europe, and would ultimately entail the dismemberment of the British Empire. The view was wholly fallacious. Certainly, Germany was anxious to create a supranational economic structure in Central Europe, something which her industrial dynamism made inevitable, in any case, and which indeed has now come into being. France would have to find her place in such a structure (and has now done so). Britain, too, would have to come to terms with it, as she is attempting to do today. But there is no evidence that Germany in 1914 had any intention, or even desire, to damage essential

British interests. Even the movement for a big German navy was on the wane.* A victorious Germany would have provided problems for Britain, but problems far less acute than those created by British participation in the war.

However, it is most unlikely that Germany would have won the war in any decisive sense. France saved herself at the Marne, a battle whose outcome owed nothing whatever to the presence of the British expeditionary force. Once the German master-plan was frustrated, the Germans would have faced the problem which baffled the Allied Commanders: in an age when military technology placed all the advantages with the defence, how to defeat any enemy determined to remain on the defensive? As it was, the Germans fought a defensive war on the Western Front; once they switched to the offensive in spring 1918, the ratio of casualties moved immediately and heavily against them, and brought about their military collapse. Evidently, the Germans had no better answer to the problem than Haig or Joffre or Nivelle. Had Britain stayed out of the conflict, France would have fought a wholly defensive war, the Germans would have attempted the grand offensive, found the casualties unacceptable, and chosen a political compromise as an alternative to a costly military stalemate. An uncommitted Britain would, of course, have assisted in ensuring that the compromise was fair and workable. The great Continental powers would have learned a salutary lesson, and the outcome would have been a more stable, and chastened, Europe.

Instead, British involvement ensured that all the Great Powers would be tested to destruction. The addition of the industrial, and above all the manpower, reserves of the British Empire to the forces of the *entente* powers made possible their adoption of an offensive strategy. The figures showed an enormous manpower preponderance against Germany: and the figures could not lie! Thus, the 'big push' became the war-winning weapon, and when that failed the still more costly 'war of attrition'. In the event, this failed too, in any decisive sense. In the autumn of 1918 Haig commanded the only army still theoretically capable of major offensive operations. He was the last man on earth to shrink at casualties: but even he was by then unwilling to engage in the huge

* Partly because a growing number of Germans realised that it not only provoked English hostility but diverted resources from Germany's army. In a lengthy letter to Edward Marsh (Private Secretary to Winston Churchill, First Lord), Hugh Watson, of the British Embassy, Berlin, commented (12 March 1913) on 'a widespread feeling in Germany that Naval competition with England is hopeless, and that Germany must stick to her proper arm of defence, the army. Indeed it is true to say that Germans are realising at present that the army is the Nation's life, and the Navy a subsidiary, if not a luxury.' See Randolph S. Churchill: *Winston S. Churchill* (London, 1969), Companion Vol. II, Part 3, 1911–14, pp. 1716–19.

battles still necessary to complete the destruction of the German forces. The armistice was aptly named: the war was inconclusive. The Germans rightly claimed that the Versailles Treaty did not accurately reflect the military balance in November 1918. As Mussolini remarked (it is his only remark worth recording), the Allies should either have taken Berlin or imposed a far less onerous settlement. The first was beyond their strength, the second beyond their wisdom. Hence the resumption of the conflict was inevitable. The Second World War took place not so much because no one won the First, but because Versailles did not acknowledge this truth.

Of the combatants, Britain was in many respects the chief sufferer, as the United States was the only beneficiary. Of course it was a paradox that a Liberal Government should lead the English into a great Continental war: but the paradox took more than two years to resolve. The essence of English liberalism was the moral and practical virtue of freedom: free the individual from artificial restraints, and he will realise the full potentialities of mind, body and soul. But how do you fight a total war on these principles? The tragedy is that Asquith and his colleagues tried to do it. English individualism died the hard way, in the Flanders mud. The English had always taken it for granted that men should be hired to fight wars (especially abroad), as they were hired to sail ships, drive coaches, build roads. It was a job for the poor or the professional, to perform for wages. The idea of military service as a function of citizenship had never worked in England, and when it reappeared at the end of the nineteenth century it was seen as a foreign institution, advocated by the militarists and the extreme Right; it was anathema to the Liberals and the budding Labour Party. If more men were needed, they would of course volunteer: that was the true Liberal way to fight a war, if indeed there was a Liberal way to fight a war.

The result was a liberal expenditure of the more eager, the more conscientious, the more responsible-minded elements in Britain's youth: a massacre of the adventurous *élite* of all classes. The old professional British army had been virtually wiped out by the autumn of 1915; conscription did not begin to fill the depleted trenches until 1918: in the years between, the years of the 'big push' and the theory of attrition, the overwhelming majority of those killed were volunteers, the simple patriots and idealists who had been brought up against the emotional background of the imperialist hubris. In other European countries, conscription ensured a rough and ready equality of sacrifice: the butchery took its toll of the manhood as a whole, dealing an impartial justice to the reluctant and the brave, to those whose instincts were to lead, and those who followed. In Britain it was concentrated on the best. Of British adult males under 45, 10 per cent were killed, over 20 per cent wounded, many very seriously. Those who survived had no doubt whatsoever that the finest of their generation had gone, a feeling impossible to quantify or prove, now perhaps dismissed as mere sentiment, but an overwhelming conviction which reflected a terrible truth. Such losses had an inevitable effect on the subsequent quality of British society, at all levels and in all spheres. It brought about a perceptible lowering of public and private spirit, of the kind which Archbishop Wulfstan of York had analysed and lamented in his famous sermon more than nine centuries before.

The material losses were in some respects heavier and more decisive. Britain entered the war the richest nation on earth, the greatest creditor in the history of international accountancy. It emerged with the national debt increased from £650 million to £7,435 million; overseas assets had been ruthlessly sold off to pay for war-imports, and in addition a huge debt had been contracted with the United States, balanced only by loans to Britain's enfeebled European allies which could not be repaid. The City of London had, in effect, financed the war against Germany, as it had financed the war against Napoleonic France, and British industry was left to pick up the bill. But the economy which had been rapidly expanding in the early decades of the 19th century was now stagnant. After a brief recovery in 1919–20, recession set in, continuing for 20 years, until the war against Hitler restored growth. The crisis of 1931–2 was merely the bottom of the low trough which marked the entire inter-war period. In Britain, unemployment, which had averaged 6 per cent in the last decade before 1914, averaged 14·6 per cent in the years 1920–38. Though industrial production and real wages eventually rose, the rise was almost entirely confined to the second half of the 1930s, under the stimulus of rearmament; and the probability is that the second German war alone prevented a further down-turn of the economy.

With a depleted manhood, and an enfeebled financial base, the English were thus left as reluctant custodians of an unrealistic European peace. There was no one else to do the job. The United States, which had gained most from the war, and lost least, repudiated the European responsibilities she had begun to assume, and retired into isolation. France was the chief architect of the peace, which she had shaped with one end in mind: her own physical security against Germany, which she sought to reinforce by paper alliances with the small nation-states created from the ashes of the eastern empires. But after 1929 the French economy ceased to expand, indeed sharply contracted. France was incapable of exerting what military strength she possessed beyond her frontiers, indeed incapable, as it proved, even of defending them. The collapse of the old great-power structure of Europe had been noted, at the time of Versailles, as a source of future conflict. The League of Nations had been created with this in mind, to supply a collective management in place of the diplomatic balance which had gone. But the League could not be greater than the sum of its parts: it had no armed forces, and only a small and diminishing stock of moral authority. And its effective parts, in practice, amounted to an uncertain Britain and an increasingly reluctant France. Europe was policed by the walking-wounded.

Hence, once a nationalist regime was again in the German saddle, it was inevitable that the Germans would seek to exercise the authority and influence in Central Europe which their economic strength and dynamism made possible. And it was only a matter of time before such a regime came to power. Hitler and his gang were the mere beneficiaries of a natural process which would have taken place anyway. He inherited the essential aims of the Kaiser's Reich; he pursued them with reckless speed and ruthlessness, but any German Government of the 1930s would have moved in the same direction. And who was to stop the process? Russia had been one of the principal victims of the peace-making: she was bound to assist any process of revision which might restore some of her lost territories and widen the *cordon sanitaire* against Germany – to work, in fact, towards the kind of Europe she got in 1945. France was allied to the phoenix-states of Versailles. But she could not, and in any event would not, fight for them. Her army was geared to purely defensive warfare. The idea that there was some point of no return, when a Germany which had recovered her nationalistic spirit could have been curbed by a threat, a gesture or even a limited police-action, is an illusion. If France had acted to prevent the remilitarisation of the Rhineland, the Second World War would merely have occurred three years earlier. Again, the Czechs provided a pretext for war, not a means to avert it. In 1938 the Czechs had 36 divisions; the Germans had 21, with 14 forming; the French had 80. If the Czechs had fought, the French were bound to assist them. But the Czechs would not defend their freedom, either in 1938 or 30 years later. The Poles, indeed, were willing to fight: but not in the one set of circumstances which made sense, in alliance with Russia. Britain and France were willing to give them a paper guarantee, which they knew they could not honour. It was the story of Belgium all over again. Britain could not physically prevent Germany from acting the role of acquisitive aggressor in central Europe. All she could do was declare war, resume the struggle suspended in 1918, and hope that in the resulting trial of strength she would survive.

This is what in fact happened. The muddles, inconsistencies and lack of logic of the 1930s merely reflected the restraints imposed on Britain by the loss of her sovereignty in 1914. But this time the roles of Britain and France were reversed: Britain was now the senior partner, and France her reluctant puppet – and, as it turned out, a wholly unreliable one. Britain accepted a role beyond her means, which she had never entertained in all of her history: as the protector of Europe. Of course it was an absurdity. Fortunately, America and Russia were brought in to relieve her of an impossible burden. But the result was a European structure which bore no relation to the one Grey thought he was trying to preserve in 1914. Indeed, it largely fulfilled John Morley's prediction. From Britain's point of view, the great wars against Germany were exercises in futility.

They were also an inconclusive exercise in self-discovery. What were the English? Were they Europeans or not? Did the frontiers and arrangements of Central Europe have such a crucial bearing on their own well-being and destiny that Englishmen must fight to maintain them? Before 1914 the question was not put to the English people as a whole, or even their elected representatives. It was answered, in the affirmative, in an atmosphere of furtive secrecy by a few highly placed men. There was nothing democratic about it. Had the democratic machinery been employed, the answer must have been different. Britain must have repudiated the Continent, and kept a watching-brief in its war. But after 1918 the English had no such wide choice. They were enmeshed in a great variety of commitments. The question was no longer: should they exist? It was: how should they be discharged?

It was at this precise point that the English people, as a whole, were given the right to determine the answer. It was not merely that the suffrage was extended: in 1918 and, again, in 1928 to embrace the entire adult population.* It was that the ruling *élite* actually began to practise democracy, instead of seeking to circumvent it. The key Englishman between the wars was Baldwin. In 1923 he was made Prime Minister and leader of the Conservative Party, in preference to the élitist Curzon, precisely because he was a commoner who transcended the traditional class barriers. The choice was symbolic; it was also important in real terms. Baldwin did exactly what was expected of him. He preached the doctrine of social reconciliation. He actually

* In 1911 about 60 per cent of adult males had had the vote. But many of the poorest were not registered. Moreover, 500,000 wealthier electors had two or more votes: in January 1910 two brothers had 35 votes between them.

believed in democracy – something towards which even Gladstone had only fumbled near the end of his life. Baldwin accepted the sovereignty of the people, not just in theory but in practice. He wanted to make it work. He brooded over the record of the Putney debates. He was fond of quoting Colonel Rainborough: 'Really, I think the poorest he that is in England hath a life to live as the richest he.' He reflected, too, on the Athenian experiments in democracy. 'Our colossal task,' he said, 'is to take over these principles of personal participation in government, of cooperative discussion, of active consent, which were effective in these tiny groups of citizens and believers, and apply them to the immense populations of modern states and empires.' Of course he was not a socialist: he thought socialism was a negation of true democracy, a degradation of man. Man progressed from status to contract: socialism was a reversal to status. Man should participate in public life as a responsible individual, not as a member of a class or group. And what was more important than that these millions of responsible individuals should decide the vital issue of peace or war, and that their elected leaders should accept their decision?

In the 1920s the world seemed at peace, and no decision was called for. Germany was beaten, Russia impotent, France, Japan and America allies or partners. The League settled minor disputes: there were no major ones. There was no decision to disarm; it was taken for granted. Spending on defence, in real terms, fell to well below the pre-war level. The famous ten-year rule was framed, and observed until 1932. Every element in society – including, above all, Churchill, whom Baldwin installed as an enthusiastic Gladstonian Chancellor of the Exchequer – sought to limit armaments, unilaterally or by agreement. Britain abandoned the two-power standard and her role as a world naval policeman. With general approval, she ceased to be the world's greatest air power. She maintained what was officially called a 'limited liability' army. Her Empire had actually increased as a result of the War, and she ran more than a quarter of the globe with armed forces which, even by mid-Victorian standards, appeared sketchy. All of this may have been unwise: it was undoubtedly democratic; it was the English choice.

Of course, once the nationalists came to power in Germany, the entire picture changed overnight. The era of stability ended: not just Germany but other powers sought to profit from the new mobility. What were the English to do? The choice was up to them. Baldwin would carry out what they wished. But the English did not know what the answer was. They lacked the materials with which to conduct a fruitful analysis of how the 1914 War came about. True, the old belligerents had produced massive volumes of 'secret' documents. But they obscured rather than clarified the issue. The truth about the real war-aims of Germany, the manoeuvring of Grey, only emerged in the 1950s. Lord Sankey observed to Clifford Allen: 'I should think there were very few people who would not now admit the war of 1914 was a tragic mistake.' This was the general view; but it was based on a fallacy. The 'mistake' lay in British participation. Nothing could have stopped the war, on which Germany was determined. But this point was not grasped. The watershed years 1928–32 brought a torrent of anti-war novels, poems, autobiographies, plays, essays by the embittered participants, such as Sassoon, Graves, Aldington, Blunden, Sherriff, Remarque. They piled on top of an edifice of war-memoirs, from Grey, Churchill, Lloyd George, Wilson, Repington, Ludendorf and countless others. In all this welter of analysis and explanation, the truth was somehow hidden. War was attributed to muddle, to the failure to resolve legitimate grievances, to the high level of armaments, to the capitalist system, to the greed of the armaments kings. No one in England insisted that German aggression was the culprit. To do so would be to admit that a second war was inevitable (though Repington provided a sinister clue by entitling his memoirs, published in 1920, *The First World War*). Keynes's analysis of the Versailles treaty as an unjust and Carthaginian peace was the accepted orthodoxy. No one would fight for Versailles, or rearm to preserve it. But if Versailles was a dead letter, what was the rule of law in Europe?

The English people were not merely muddled and misinformed about the origins of the 1914 War; they were divided on how to prevent a recurrence. The chasm in British politics which the events of 1931 opened ended any hope of a party consensus on foreign affairs. The official opposition was no longer given confidential briefings: indeed it no longer wished to have them. After Hitler came to power, Baldwin and many other Tories moved slowly towards rearmament, but he did not wish to advance more quickly than he thought the nation would follow. After Germany pulled out of the Disarmament Conference in October 1933 and, in effect, reasserted her full sovereignty as a Great Power, the Tories immediately suffered a disastrous by-election reverse in Fulham. The swing was not surprising after the 1931 landslide: the defeat was almost certainly caused by the means-test, and had little to do with foreign affairs. But Baldwin interpreted it as a massive national verdict against the arms-race. Very well; democracy had spoken, he would have to wait until the public mood had changed as, by 1935, he reckoned that it had. Then he fought the 1935 election on a programme of modest rearmament, and won it. He was later pilloried for trying to practise, no doubt ineptly, the theory of democratic control in foreign policy.*

The truth is, no section of English opinion had a solution to the problem of what to do with Germany. Some thought rearmament would lead to war: they were right. Others thought the lack of armaments would invite war: they were right too. War was inevitable. The Labour leaders, and other well-meaning people, thought the League would provide collective security, and thus an alternative to rearmament; in fact, if its controlling members lacked adequate arms, it was merely a system of collective insecurity. International leagues can only work, like democratic parliaments, if they are based on the rule of law: otherwise they are themselves liable to become the instruments of injustice, as the United Nations has often shown. But the rule of law must be enforced: and no one proposed the League should have an army of its own, least of all its most passionate supporters. The Left thought you could not be a true supporter of the League without opposing rearmament; the Right thought you could not oppose pacifism without opposing the League. The Right at least had its own hereditary and instinctive reflex that arms would come in handy. Most of the Left was committed to sheer nonsense. Attlee said: 'We stand for collective security through the League of Nations. We reject the use of force as an instrument of policy . . . Our policy is not one of seeking security through rearmament but through disarmament.' One can read this statement forwards, backwards and sideways, and indeed upside down; it is just words without meaning. British democracy could not cope with the problems of foreign affairs; perhaps nothing could.

The English dilemma was resolved in a characteristically English

* The treatment of Baldwin is an excellent example of the instant English myth. What Baldwin actually said, in his 'appalling frankness' speech, was that in 1933 and 1934, the mood of the country was such that no government could have appealed to the electorate on a programme of rearmament and got a mandate; by 1935 the mood had changed and 'we got from the country – with a large majority – a mandate for doing a thing that no one, 12 months before, would have believed possible'. This was twisted to mean that Baldwin fought the 1935 election on an anti-rearmament policy because he could not have won it any other way. The myth first appeared, in its mature form, in an article in the *Sunday Express* (3 September 1939) by Peter Howard, later head of Moral Rearmament: 'He said he had not told the electors the truth about rearmament at the 1935 general election because he believed that if he had done so they would not have voted for him. Lord Baldwin did more damage to democracy than any other Premier in Britain, and certainly more damage than any other man except Cromwell [*sic*].' Cassandra, in the *Daily Mirror*, put it another way: 'Here was an old and stupid politician who had tricked the nation into complacency about rearmament for fear of losing an election.' It was even believed, by some, that Baldwin had publicly advocated 'unilateral disarmament'. All Baldwin had done, in fact, was to practise democracy and tell the truth – a sombre warning for future politicians. See Keith Middlemass and John Barnes: *Baldwin* (London, 1969), pp. 970–2.

fashion. Ultimately, everything depended on the attitude adopted to Hitler. The English Right did not greatly object to what Hitler was doing inside Germany. In some ways it met with their approval. They certainly did not regard the nature of the Third Reich as justification for an anti-German posture. Some of them had wanted an ideological crusade against Soviet Russia in the early 1920s, but those days were gone: there could be no question of seeking to influence the internal affairs of a right-wing regime. On the other hand, they began to object strongly to what Hitler was seeking to do outside Germany: this was the kind of old-fashioned aggression they recognised and resented. With the Left it was the other way round: they deplored the Hitler regime, but they thought it wrong to deny Germany the right to seek to modify an unjust peace-settlement. The two wings of the English political spectrum were thus at cross-purposes. They came together, and achieved a remarkable degree of unity, because of two factors which were irrelevant to the main issue: the Jews and Spain. As news of Hitler's systematic destruction of the immensely rich and influential German Jewish community penetrated the Tory consciousness – and it took some time – the attitude of the Right towards the Nazi regime changed fundamentally. What was once seen as a bulwark of social stability was now revealed as its enemy: Hitler, in Tory eyes, ceased to be a conservative and was recognised as a dangerous radical. With the Left, a similar process of conversion about Hitler's external aims took place over the Spanish Civil War. Hitler and his acolyte Mussolini were already domestic tyrants: now they became the leaders of an international crusade to destroy democracy and socialism. Thus by roundabout ways, and for emotional reasons which were diverse or even conflicting, the English nation reunited round the proposition that a war against Hitler had to be fought. The unity, when it came, was impressive and sustained, but it had little to do with logic, and still less with the formal machinery of democracy that the hapless Baldwin had tried so earnestly to employ.

Curiously enough, the pattern and tempo of British rearmament, though determined purely by accident, could not have been more nicely judged. It was as though some residual English deity was pulling invisible strings. The life cycles of modern weapons pose delicate problems of timing, which are only imperfectly understood now, and were largely ignored in the 1930s. If the English had rearmed rapidly from the moment Hitler came to power, they would not have averted war, but they would have fought it with vast quantities of already obsolete equipment. The French, the Italians, the Russians, even to some extent the Germans themselves, armed their forces with tanks and aircraft which were already out of date by 1940. In the case of France and Italy, the handicap was conclusive. The weapons produced by the very limited British rearmament of 1935–7 were largely useless. One product was the Singapore base, originally proposed in the 1920s but delayed for economy reasons. It was unfortuate it was ever built at all: it could be defended only in the context of decisive sea and air superiority (which was lost), and the net result was the surrender of the two extra divisions we invested in it. Similarly, a large British army would merely have swelled the total of British prisoners and equipment captured by the Germans in 1940. One consequence of the delay in British rearmament was a decisive shift in British resources from offensive to defensive air power. In 1934 Baldwin was told that there was no way to defeat the bomber; the RAF was therefore a deterrent, designed to meet terror with counter-terror. If Britain had then rearmed rapidly, she would merely have built more bombers, whose limited range and primitive navigational and bomb-aiming equipment made an ineffective strategic weapon. As it was, fear of the German bomber led to a highly productive investment in the research and development of radar, and, as progress was achieved, in the production of fast fighter-aircraft and the training of pilots in sophisticated defensive techniques. By a bizarre paradox, the delay in rearming made possible a British victory in the air-battle of 1940. Moreover, thanks to the delay Britain adjusted to a war economy in a more systematic and efficient manner than the Germans. In March 1938 the Government allowed rearmament to interfere with the normal patterns of trade, and the TUC agreed to drop craft restrictions on output; in February 1939 aircraft production 'to the limit' of available resources was authorised. A mass of planning regulations were introduced. Thus Britain possessed the elements of a wartime system of priorities long before the Germans or any other belligerent. In September 1939 she was making more tanks and aircraft than Germany, and spending a higher proportion of her Gross National Product on defence. These measures, made more effective by the long and cautious run-up to full war production, were adequate to ensure Britain's survival. It was beyond her capacity to do much more. Germany could only be defeated by the new super-powers, and only their continued presence in Europe after 1945 could prevent a German resurgence. As it was, Britain's contribution to the war effort had been excessive in relation to her resources. It was enough to buy her a ticket to the top table, but as a spectator rather than a participant. She had the glory, such as it was, but little of the substance of victory. She had accumulated a mass of overseas debts, sold virtually all her remaining overseas liquid assets, lost two-thirds of her gold reserves, accumulated embarrassing sterling balances in London, and run her industrial plant practically into the ground. She was old, tired, sick and unhopeful. Forty years of acting the part of a Continental Power had destroyed the English hubris. The nemesis was still in progress.

The way in which the English disposed of their Empire was almost as confused and unsystematic as the way in which they acquired it. Seen from the perspective of the 1970s, it can be presented as a quarter-century of English statesmanship, devoted to the planned transfer of power to elective and responsible governments, within a flexible framework of Commonwealth association – the fulfilment of the vision J. S. Mill had outlined in 1861. It is true that over 500 million former British subjects acquired their freedom, for the most part without violence, and that the new states thus formed, with the exception of Burma, chose to retain constitutional links with Britain. As with almost any other English institution, it is possible to trace back the Commonwealth concept a very long way, and impose a framework of historical development on a series of events which were haphazard and often unconnected. But in fact it was Rosebery, a leading imperialist, who first used the term British Commonwealth; and what he understood by it was essentially a group of white, English-speaking states, united by innumerable similarities in their institutions and way of life. The multiracial Commonwealth of today evolved largely by accident, as a result of Britain's growing weakness in world affairs: there is little evidence that anyone planned it. Like Magna Carta it was an old English muddle. The decisive moment came in 1946–7, when the state of the British economy forced the Attlee Government to speed up the progress of India towards self-government by the rapid withdrawal of British troops and administrators: this meant not merely a partition on religious lines, which was opposed to the whole drift of enlightened British policy in the past, but the creation of a theocratic Moslem State in two halves separated by a thousand miles of Indian territory: a formula for future disaster. It also meant the evacuation of Burma, where little progress had been made in training *élites* for responsible self-government. The liberation of the Indian subcontinent was an act of deliberate statesmanship in the sense that Britain took a voluntary decision which otherwise would have been forced upon her by events, and so avoided fruitless bloodshed and expense. But the legacy she left behind did, and does, her little credit.

With India and Pakistan as members of the Commonwealth, the precedent had been set for all the former colonies and dependencies to attain self-government within the same framework. But surprisingly little effort and thought seems to have been devoted to making this process orderly and systematic. Moreover, the free movement of labour and colonisation in the past had left complex plural societies in many territories which made it difficult to transfer power on a basis of stability and natural justice. Malays and Chinese in Singapore, Malaya and Borneo, Negroes and Indians in Guyana, Turks and Greeks in Cyprus, Jews and Arabs in Palestine, Europeans, Indians and Africans in Kenya, Europeans and Africans in Rhodesia – to give freedom to some meant its denial to others. There were other complications. In parts of Africa, particularly on the West Coast and in East Central Africa, the old colonial boundaries had been drawn up in a hasty and arbitrary fashion, which made little economic sense and conflicted with the tribal geography; the largest and most important of them all, Nigeria, was divided by religious as well as ethnic factors. Here again, any scheme of independence must involve grave injustice to minorities, or highly sophisticated, perhaps unworkable, federal constitutions. Above all, there was an unresolved doubt in the minds of successive British governments as to the role Britain herself was to play in the emerging Commonwealth and in the world. Was she to be a titular dignitary, concentrating on her own insular affairs? Or the centre of a global system of finance, trade and mutual defence? Was the Commonwealth to be the coda of an old theme, or the germ of a new one? Was Britain to be an Atlantic power, or a European power, or a world power? A medium-sized power or a great power? Was she to conduct her policies through the Commonwealth, the Atlantic alliance or the United Nations – or all three simultaneously?

Certainly, British national interests would have been best served if the English had drawn the logical conclusion from the liberation of India, and had evacuated all their overseas territories as rapidly as purely practical arrangements could be made. But the English found it difficult to make up their minds about any of the questions which their tottering Empire raised. They were tugged by the conflicting forces of greed, pride, responsibility and exhaustion. Withdrawal from India removed the keystone, indeed the whole *raison d'être,* of the East of Suez Empire, with its network of military bases. But the supposed need for direct control of Malayan rubber and tin, and still more of Middle Eastern oil, supplied a specious justification for maintaining it. Britain worked through a succession of expedients, often ill-planned, mutually contradictory, or abandoned long before fulfilment. She knocked together federations in the West Indies, Nigeria, East Africa, the Rhodesias, Malaysia and Aden: then saw them disintegrate, or survive only through bloodshed and hideous injustice. Sometimes she simply abandoned her responsibilities and scuttled, as in Palestine, where she believed the extermination of the Jews would win her the friendship of the Arabs.* Sometimes she hung on desperately, as in Cyprus, in the belief that evacuation would precipitate racial and international war. In East Africa she abandoned the whites and the Asians (and indeed many minority tribes) to an uncertain fate. In Southern Rhodesia she abandoned the Africans to white supremacy. In the Sudan she abandoned the Nilotic population to racial and religious persecution by the Arabs. She did not raise a finger to protect the black and coloured peoples of South Africa, or the tribesmen of South-West Africa, or the Christian Biafrans; on the other hand she sent troops and police to protect the morals of tiny Anguilla. Sometimes she fought, sometimes she surrendered, sometimes she did both in turn. She scuttled from the Canal Zone in 1954, returned by force two years later (in the company of the French and the Israelis – an astonishing reversal of traditional British policy towards the Arabs), and left as abruptly as she came. She defended the Jordanians and the Kuwaitis from subversion, then decided to scrap all her military commitments in the area. In Malaysia, she fought a successful guerrilla war, imposed self-government, constructed a multi-racial confederation which promptly collapsed, defended its components against Indonesia, then discovered that financial reasons made her presence 'unnecessary'. She arrested nationalist leaders, imprisoned them, released them, honoured them. She planned majestic systems of international defence, invested hundreds of millions in military bases – in Cyprus, East Africa, Aden, the Persian Gulf, the Indian Ocean, South-East Asia – then left them in a hurry, often before they were finished.

By the end of the 1960s, the game was up. All the viable territories had attained independence. The English had largely lost their interest in the Commonwealth. The Tories had never shown much enthusiasm for one with a multi-racial composition; and now the move towards Europe led to the rapid devaluation of the Commonwealth ideal. The Labour Party, with ostensible reluctance, and much breast-beating, followed in the Tory wake. The fact that, in 1971, Britain no longer presided *ex officio* at the Commonwealth Conference, and that it was held in Singapore, not London, was a sure sign that the English were pulling out, in all but name. Indeed, they had now acknowledged that Britain was no longer a world power. As recently as 1965 Harold Wilson had used wild words about Britain's frontier being still on the

* On 17 June 1948 Harold Nicolson recorded in his diary a conversation with Ernest Bevin at the Persian Embassy: 'Nobody is going to tell him that in principle it does not pay better to remain friends with 200 million Moslems than with 200 thousand Jews, "to say nothing of the oil".'

Himalayas. But in 1968 the decision was finally taken to scrap the East of Suez policy (what remained of it) and in 1971 the Tory Government tacitly admitted this could not be reversed. Nemesis was complete. The English reverted to their role in the sixteenth century: a medium-sized, relatively prosperous power, perched off the coast of Continental Europe, uncertain once again about their true relationship to the land dimly seen from the cliffs of Dover.

To a remarkable degree, the experience of acquiring the status of a great imperial power, and then losing it – an experience which spanned almost exactly a hundred years – has left few traces on the English people. For perhaps two decades they allowed themselves to be enthused by the imperialist afflatus. But the mood subsided during the Boer War; it had vanished entirely by 1918. Between the wars, the English governed a quarter of the earth more from habit and a residual sense of duty than from a deliberate sense of mission. After 1945 the public consciousness played little part in the disposal of the Empire. The English relinquished their vast overseas possessions with indifference, tinged occasionally with relief. Ministers, officials, governors, military men took the decisions, good, bad or morally neutral, almost wholly uninfluenced by popular pressures, one way or another. To fight or withdraw? To go or to stay? The English did not really take much interest. The killing of British soldiers by terrorists aroused the occasional flicker of anger; there was a spasm of disgust when the terrible story of the Hola Camp massacre in Kenya came to light. But such episodes were quickly forgotten: if anything, righteous anger and self-criticism militated alike in favour of more speedy retreat. For a few weeks the Suez adventure produced genuine drama, and deep divisions, at least among the upper and middle classes. But it, too, soon passed into oblivion, and left no discernible mark on the course of British politics. The end of Empire did not become an issue at any general election, or even at a by-election. Follies went unpunished, achievements unrewarded. No one was blamed for the tragic mess Britain left in Palestine – although the consequences are with the world to this day. No one was praised for the brilliant operations in Borneo, one of the most successful campaigns in British military history. Indeed, it was often difficult to know who were the heroes and who the villains. One moment the English public were told that Jomo Kenyatta was 'the prince of darkness and death', the next that he was a benign elder statesman, the guardian and toast of Kenyan civilisation. Nkrumah changed from a rebel and a convict, to a liberator and an ally, to a tyrant and oppressor, to an exile, all for no reason discernible to the mass of the English people. Archbishop Makarios was transformed almost overnight from a scheming Levantine prelate, the puppet-master of murder, into a pillar of the Commonwealth and the honoured guest of the Queen. The scenes changed rapidly; the cast swopped roles; the familiar lines were repeated, the English audience remained silent, ultimately bored. In 1968, the Government were able to announce, with justifiable pride, that in the previous 12 months, for the first time in memory, perhaps since records were kept, no British soldier had died on active service anywhere in the world. The news was welcome: but the only discernible reaction was a drop in recruiting. Only in one respect did the English take a keen and baleful interest in the affairs of the Commonwealth: public pressure compelled the politicians to end the unrestricted entry to Britain of Commonwealth citizens. It was as close as Britain ever came to eliciting a popular mandate for the end of Empire.

The English, indeed, have remained strikingly impervious to the events of the external world during this terrible century. They paid a heavy price, in human and material terms, for their intervention in the affairs of Europe: but two world wars, the acquisition and loss of vast territories, played little part in the continued growth of their institutions, the slow, almost imperceptible evolution of their attitudes and way of life. During this last hundred years, Austria and Turkey have disintegrated, Russia, Germany, France and China have endured invasions and revolutions, fundamental assaults on the whole structure of their established societies, changes in constitution and law and status; Japan has been the victim of nuclear assault, prolonged occupation, political experiment; entire nations have been born or engulfed; even the United States, quarantined by the world's two greatest oceans, has found her domestic peace shattered by the pressure of her involvement in the world beyond. But the English have kept their own affairs rigorously apart from their dealings with other peoples. They have contrived to develop on strictly indigenous and traditional lines. Britain remains the most stable country in the world, absorbing change within the rigorous framework of continuity.

The most stable country in the world: but there is a more sombre side to the story. With rare consistency, the English throughout their history have chosen to buy stability: but they have always had to pay a heavy price for it, in the suppression of energy, the denial of natural genius, the frustration of dynamic forces within their society, in hopes deferred and opportunities missed. In the half century since the 1914–1918 War shattered the old mould of life, the English have picked their way prudently and safely through the tumults of change: but they have drawn the penalties which must befall those who fail to invest in adventure, experiment and risk.

The First World War raised unimaginable spectres but also dreams of justice. In the ephemeral literature of the time, we can detect an almost universal expectation that Britain was on the eve of a swift and majestic transformation, which no force within her would be able to resist or even delay. There was a flavour of the 1640s, of 'living long in a little time', an eagerness to embrace the future. 'We are living at a time when days and weeks have the fullness and significance of years and decades,' wrote one observer in 1916. 'That horrible ogre, Tradition, lies in the dust,' thought another. 'There are certain great historic events, like the Protestant Reformation and the French Revolution, which have altered mankind for good. The war was one of these far-reaching forces.' This was a comment from 1920.

Yet when it came to the point, the English refused, as so often before, to take that famous 'leap in the dark'. In 1916, in desperation, the English had abandoned liberalism and invoked all the resources of State management in a bid for sheer survival. The experiment was rewarded: the war-economy Lloyd George created sustained Britain during those crucial months of effort in 1918 which finally broke the will and discipline even of the people and armed forces of Germany. But the peace was surrendered without a fight. 'The war,' reported the Cabinet, 'has brought a transformation of the social and administrative structure of the State, much of which is bound to be permanent.' The question remained: how much? The mistake was made of opting for too little. The men who had won the war by their endurance and sacrifice had little choice in the matter: only one in four of them, in practice, got the chance to vote in 1918. The Parliament then returned reflected not so much the earnest desire to start afresh as an overwhelming emotional resolve to restore the vanished pre-war world – something which, across the distorting perspective of Armageddon, seemed like a lost paradise.

Of course, such a world could not be restored. What emerged instead was a confused and unsystematic amalgam of liberalism and State meddling. The brief post-war boom brought an orgy of de-control, a clearing of the decks for unrestricted free enterprise. The chance to nationalise the mines, the largest, weakest, most strife-torn sector of British industry, was missed. Consultation through State bodies between both sides of industry broke down. The economic arrangements of the Versailles Treaty inhibited any lasting revival of world trade, which faltered, then stabilised well below its pre-war level. Prices had more than tripled since 1914; wages now began to fall from their high wartime averages. The war had stimulated, even created, new industries in petrol-engines, electrics, chemicals: government failed to reinforce the development, concentrating instead on tinkering efforts to revive the old staples, now in irreversible decline. Their financial and economic policies made matters worse. They deflated by vicious cuts in public expenditure, and high interest rates. This raised the unit-costs of British exports, and so handicapped industry's efforts to retain a falling share of a shrinking world market. One pillar of the old international economy, the free movement of goods, had collapsed in the tariff wars which preceded the age of imperialism. Another, the free movement of labour, had gone in 1914. Britain's economic and financial establishment sought blindly to uphold the third and last, the free movement of money, by returning to the gold standard. The net effect, an upward revaluation of the British currency, against all the evidence of her industrial performance, merely raised still further the price of British exports. The only course which seemed open to mine-owners and manufacturers was to slash wages and increase the hours of work, thus provoking a despairing defensive reaction from the proletariat.

British governments were dominated by parallels from the aftermath of the Napoleonic wars. It was the only historical guide they had; they studied it eagerly; but of course it was false. They did not know how to reduce unemployment, or curb inflation, or revive industry. The only fixed point in a changing world, it seemed to them, was gold: this they worshipped, in the hope of miracles. In December 1914 J. M. Keynes had rejoiced in the blow he saw had been inflicted by war on the uncontrolled international monetary system: 'If it prove one of the after-effects of the present struggle, that gold is at last deposed from its despotic control over us and reduced to a constitutional monarch, a new chapter in history will be opened. Man will have made another step forward in the attainment of self-government, in the power to control his fortunes according to his own wishes.' But nothing had been learned, nothing forgotten: gold, like the Bourbons, was hoisted back on to the throne. Indeed, it was not until 1936 that Keynes was able to produce his general formula for an escape from disaster, and by that time gold had collapsed under its own weight, in the shambles of 1931. John Galsworthy, updating his *Forsyte Saga* in 1929, had observed: 'Everything being now relative, there is no longer absolute dependence to be placed on God, Free Trade, Marriages, Consols, Coal and Caste.' The trouble was, such ancient shibboleths, and many others, survived with just sufficient strength to prevent any fundamental re-thinking.

Like Britain herself, they were walking wounded, still on their feet, still trying to direct events.

As in post-Napoleonic times, ministers mistook the consequences of their own economic policies for a growing desire on the part of the workers to promote class war and destroy society. It was happening abroad: it could happen in Britain. Their behaviour strikingly echoed the aggressive panic of Sidmouth, Castlereagh and Liverpool. Once again, the English working class was made the victim of guilt by foreign association. As in 1815, they wanted merely to hang on to their wartime gains, meagre as they were. As in 1815, they were all allotted a leading role in an international conspiracy against property. Asquith's government had successfully coped with very similar industrial convulsions in the pre-war period, using the absolute minimum of force. But men who had presided over the wartime carnage had lost their faith in liberal methods, their faith indeed in the extraordinary restraints and good sense of the English people. They thought almost entirely in terms of violence and weapons: it was they, not the workers, who were planning class war. Some of the conversations between Ministers, recorded in the Cabinet papers and Tom Jones's diaries, make hair-raising reading: they reflect a hysterical fatalism which, happily, was entirely absent in the workers' camp. Here, for instance, is Lloyd George:

> We cannot take risks with labour. If we did, we should at once create an enemy within our own borders, and one which would be better provided with dangerous weapons than Germany. We had in this country millions of men who had been trained to arms, and there were plenty of guns and ammunition available.

If Lloyd George could not keep his head, it was unlikely anyone else would. Churchill: '. . . there would have to be a conflict in order to clear the air . . . By going gently at first we should get the support we wanted from the nation, and then troops could be used more effectively.' In January 1919, tanks, troops and machine-guns were concentrated in Glasgow, for no apparent reason. Tanks were also deployed in Liverpool: a battleship anchored in the Mersey, with steam up. 'Loyal ex-servicemen' were recruited to form Defence Units, 70,000-strong.* Brass-hats were called to the Cabinet; Ministers trembled as they moved divisions across the table, as if they were re-fighting the Battle of the Somme. Austen Chamberlain: 'We are in front of a situation here which may require all our forces. I am all for holding the British coalfields

* Under the supervision of the minister Tom Jones called 'Sir Hindenberg Geddes'. Public money was also used to bribe newspapers, particularly in Scotland: Thomas Jones: *Whitehall Diary, i, 1916–25* (London, 1969), p. 139.

rather than the Silesian ones.' Lord Birkenhead: 'We should decide without delay around which force loyalists can gather. We ought not to be shot without a fight anyway.' This peer was the head of the law, the official keeper of the King's Conscience. He shouted to hecklers at a public meeting: 'Howl on, you wolves of Moscow! We shall slit your soft white throats for you!'

Of course when the General Strike finally came, it was an old-fashioned English anti-climax. As might have been expected, as even a superficial glance at English history would have shown, neither the workers nor their leaders had the slightest intention of launching a revolution. They did not want to take over the State. They wanted to keep their present wages, or express solidarity with those facing cuts. The Communist, Willy Gallacher, afterwards commented ruefully: 'We were carrying on a strike when we ought to have been making a revolution.' But this was to miss the point. There was no possibility of 'making a revolution'. The strikers' news-sheet, the *British Worker*, put the position with absolute truth: 'The General Council *does not* challenge the constitution.' It was 'engaged solely in an industrial dispute ... there is no constitutional crisis'. The apparatus of government repression was not required. Strikers and policemen played football: there was nothing else to do. Middle-class armies of enthusiasts, recruited by both sides, found themselves in the ranks of the unemployed.* As during the Peasants' Revolt, the English workers were seeking some ancient, idealised norm of social justice: they demonstrated their solidarity, then dispersed, leaving it to the authorities to redress manifest inequities. Ramsay Macdonald commented: 'If the wonderful unity in the strike ... would be shown in politics, Labour could solve the mining and similar difficulties through the ballot-box.' This was rather Baldwin's view; after the strike was over he was anxious to de-escalate the social struggle, to encourage Labour to take power peacefully, though of course 'not in my time, O Lord'.

But could the Labour Party become the instrument of an economic and social revolution carried through by parliamentary processes? Could it propel the nation across another watershed, as in 1832? Labour was the residual legatee of the Liberal Party, a casualty of the war

* Evelyn Waugh, the most precise fictional chronicler of the years 1920–50, catches the mood as usual: 'We dined that night at the Café Royal. There things were a little more warlike, for the café was full of undergraduates who had come down for "National Service". One group, from Cambridge, had that afternoon signed on to run messages for Transport House, and their table backed on another group's, who were enrolled as special constables. Now and then one or other party would shout provocatively over the shoulder, but it is hard to come into serious conflict back to back, and the affair ended with their giving each other tall glasses of lager beer.' – *Brideshead Revisited*, Book One, Chapter VIII.

which had destroyed liberalism as a dynamic force. But in all essentials Labour was a characteristic English muddle, unlikely to achieve anything except by accident. Part of it was social democratic, vaguely committed to a watered-down version of Continental socialist theory. Part of it was working-class Liberal, reflecting the deferential history of the Lib-Labs who had been allowed a servant's place at the Asquithian feast. The only part which mattered was the trade union element: and even this was torn between non-party syndicalism, and the use of Parliament to promote working-class objectives. In 1918 the Labour Party adopted a socialist constitution, in which it did not believe and never sought to implement, but which, equally, it was treason to amend or challenge. It rejected revolution emphatically; but it also rejected alliance with the more liberal capitalist elements in society. It thus had no chance of achieving the electoral consensus which had brought about majority Liberal governments. It was condemned either to opposition, or rule through a parliamentary minority. It lacked both the votes to implement socialist policies, and (because of its exclusiveness) the talent to manage a capitalist economy. To take office was therefore to accept direction by the civil service, or invite disasters: in 1931 Labour even contrived to choose both. Its leading spirits included those who shunned respectability and those who sought it. It took elaborate, even savage, measures to dissociate itself from communism, but broadcast its weird illusions about the Soviet Union and so exposed itself to the forces of xenophobic panic. For a party which made a fetish of internal discussion, and spent weary hours elaborating its programme, its record of legislative achievement between the wars is remarkably thin: for all practical purposes it consists of John Wheatley's Housing Act of 1924. There was, indeed, no agreement as to what Labour was supposed to do in office. Educate itself for power? Or actually exercise it? When the first Labour Government was formed in 1924, David Kirkwood rejoiced: 'Bishops, financiers, lawyers, and all the polite spongers upon the working classes know that this is the beginning of the end.' But Macdonald asked his colleagues to purchase or hire court dress and wore it himself on every possible occasion. This was only one example of a total confusion of purpose. Labour attributed its failures to conspiracies and 'bankers' ramps'; in fact it was the victim of its own ignorance, irresolution, division and cowardice. It reached its nadir not in its calamitous defeat at the hands of the National Government in November 1931, but in August, three months earlier, when Labour ministers did not know, and feebly failed to discover, that they could take Britain off the gold standard. Leadership of this quality could invite only one verdict. Nor was there much improvement in the

1930s: Labour's contradictory and unrealistic attitude to Germany was matched by a failure to evolve a policy to cure unemployment above the level of mere slogans. Labour lost the 1935 election decisively: it stood little chance of doing significantly better in 1940. The lost decades of the 1920s and 1930s were the responsibility not merely of a Conservative governing class but of a working-class opposition which proved itself equally unfit to rule. Nor is this surprising. A people which allows the best of its youth to be murdered cannot expect miracles from the survivors.

Yet the English eventually made progress of a sort, at their own exasperating pace. During the 1930s some lessons were learned. British industry became markedly more efficient, as free trade was belatedly abandoned, small and medium-sized firms concentrated into larger units, and science and technology occupied a larger place in British education, business and government. Slowly the policies of the 1920s were put into reverse. Slowly, too, as a new generation moved into position, a new consensus – mainly middle-class, detached rather than springing from party – began to form around a policy of State action to promote social justice. In July 1935, for instance, the Liberty and Democratic Leadership Group produced a five-year plan for implementation in a single parliament. It provided for public ownership of transport, electricity, the Bank of England, mining royalties and armaments, for food subsidies and cheap milk for children, for compulsory schooling to 16, part-time education to 18, and a steeply progressive tax structure. Two years later it added a national health service – first mooted in England in the 1640s. The old dichotomy, it claimed, between 'a wholly competitive capitalist system and one of State ownership, regulation and control' was 'wholly beside the mark . . . For it is clear that our actual system will in any case be a mixed one for many years to come.'

The movement received a powerful impetus not so much from the onset of the war against Hitler as from the desperate predicament of the nation in 1940–1. The unity the English then found inspired not merely the Churchillian resistance but an immense yearning for social justice, and a feverish search for the practical means to secure it. It was as though the Government signed a social contract with the people: the creation of an equitable society must proceed *pari passu* with the quest for victory, yielding in priority only to the absolute necessities of the war effort. The machinery required to win was equally the instrument of welfare economics. The budget of 1941, whose philosophy was based on Keynes's pamphlet, *How to Pay for the War* (1940), for the first time dealt with the national income as a whole, organising the use of total resources to meet priorities fixed by deliberate political decisions. The White Paper, *Employment Policy* (1943–4), accepted that 'a high and stable level of employment' must be maintained even if this involved risking traditional objects of government policy, such as the parity of sterling. Rationing and war-taxation ensured equality of consumption, national service and the direction of labour equality of sacrifice. The Emergency Hospital Service adumbrated the public control of health and medicine. At the end of 1942 Sir William Beveridge produced proposals for a national insurance scheme which went well beyond the brief he had been given by the Government, and called for basic reforms not merely in insurance but in health, education, housing and the provision of jobs. It was, in effect, a blueprint for the Welfare State; and its enthusiastic reception by the public, its endorsement by the Labour Party, and the suspicion, even hostility, it aroused in many Conservative quarters, were pointers to Labour's electoral victory in 1945. The public rightly judged that Labour was more likely to fulfil the social contract than its opponents. But the ideas behind the Welfare State, the outline and even the details of the legislative programme which embodied it, were essentially the product of a national consensus, of which the Labour Party was the political beneficiary. Left to its own resources, Labour has never shown itself to be an efficient instrument of radical social change, or even of self-sustaining reformism.

Indeed, even provided with a programme, and the public impetus to carry it through, Labour showed some characteristic weaknesses as a governing party. The universalist principles of social welfare, though accepted in theory, were abandoned in practice, leaving holes in the Welfare State which were eagerly widened by succeeding Tory Governments. Labour ministers lacked the will to impose a salaried service on the medical profession, or even to reform its structure. Private practice, private beds in hospitals, private health insurance were permitted. A two-tier service thus emerged, and the discrepancy between the treatment available to the rich and the poor widened with the failure to build more hospitals or train enough doctors. It is an astonishing fact that the doctors, fearing over-competition in their profession, were able to persuade Labour to cut the number of medical students by 10 per cent, so that those taking medical degrees actually fell from 14,147 in 1949–50 to 12,937 in 1956–7. In 1951 Labour itself surrendered the principle of non-contribution, and failed to re-establish it when it returned to office 13 years later.

Over a much broader front, Labour retreated from the brink of structural change. War-taxation had redistributed income: but this was to tackle the effect, not the cause, of inequality. No steps were taken to tax wealth, or even capital gains, until the great bonanza of the 1950s and early 1960s was already over. Indeed, the net effect of Labour's policies was actually to promote inequality. During the war, the party had expected to have to cope with unemployment rising to 8 per cent: in fact full employment was maintained, for reasons no one fully understood, and instead Labour faced the baffling problem of inflation, for which it had no remedy, theoretical or actual. Labour activists attempted to defend inflation as a form of income-redistribution, from which the organised working class must benefit. In reality, the victims were the poor, and the gainers were the rich: Labour's policy of dividend restraint accelerated the process by adding enormously to share-values, which of course were wholly untaxed; and the party proved incapable of devising any effective system of taxing the rise in the price of land which inflation promoted and sustained. Thus, it is a curious fact that the entire post-war period has shown a steady, in some respects spectacular, consolidation of the power and wealth of the capital-owning class. The wage-earners, if they were lucky, kept pace with inflation, sometimes ran ahead of it; but it was the owners of property who added substantially to their wealth and real incomes. Labour nationalised the loss-making sectors of the economy, over-compensating their shareholders; it left the 'commanding heights' and the wealth-generating sectors virtually untouched. All that the working classes won from a quarter-century of boom conditions was the illusion of affluence in the form of consumer-durables, and even this was denied to perhaps one-fifth of the population. Actual wealth remained in roughly the same hands.* To some extent Labour perceived this, as it campaigned for office in 1964: its leaders spoke of policies to reward 'those who earn money as opposed to those who make it'. But it proved incapable of translating the distinction into realistic policies. Indeed, in Labour's second post-war term of office, the gap between the rich and

* There are various ways of computing the position, but all produce much the same story. In 1911–13, 5 per cent of the population owned 87 per cent of the country's wealth; the bottom 90 per cent owned a mere 8 per cent of the wealth. In 1936–8, the top 5 per cent still owned 79 per cent. Despite the impact of war and post-war taxation, the top 5 per cent still owned 75 per cent of all wealth in 1960. Moreover, 58 per cent of all investment income was received by a mere 1 per cent of the population. In terms of ordinary shares, the concentration was still higher: the richest 1 per cent held 81 per cent of company stocks and shares; the richest 5 per cent held 96 per cent. Estate duty, the only form of tax on wealth, has totally failed to effect any redistribution of property. In the period 1965–9, it brought in a mere 3·5 per cent of central government revenue from taxation; a century before, 1868–77, it had brought in 7·9 per cent. Expressed as a proportion of national income, it was a mere 1 per cent, the lowest figure since the period 1916–19. A century of trade union activity, and four Labour Governments, have made no discernible change in the ownership of wealth in this country.

the poor, which had widened significantly during the 13-year Tory interregnum, grew at an accelerated rate.

One crucial reason for Labour's powerlessness to effect radical change in the rewards of society was the low priority it accorded to educational reform. Education, as we have seen, was the great progressive failure of nineteenth-century England. The pattern has been repeated in the twentieth century. Brougham had advocated a State comprehensive system as long ago as 1810: the English have still to create one. By 1900 primary education was available to all, but only a tiny proportion of the working class got secondary education, and virtually none went to universities. By 1914, only 200,000 children were at secondary schools, three-quarters of them fee-paying. Between 1890 and 1910, six new universities were created in England and Wales: none at all in the 1920s, and only three university colleges in the 1930s. Secondary education for all was delayed first by the cuts of the early 1920s, then by the 1931 crisis, then by the Second World War: it was not made a reality until the late 1940s. Even so, the Newsom Report of 1963, *Half Our Future*, told a dismal tale of the national failure to cultivate the potential gifts of children between 13 and 16 years who possessed average or less-than-average ability. In higher education the story is much the same: in 1962 the total of places in all establishments was not much over 200,000, and a huge programme of capital expenditure will merely double them by the mid-1970s. Education in the twentieth century provides a typical example of the English time-scale of reform. The pioneers make a proposal; a quarter of a century later it is generally accepted by enlightened opinion; chance and accident, financial cuts and economic crises, the churches, the Lords and other obstacles to progress delay it for another quarter-century; implementation takes 10 years or so. By then the reform is universally accepted as obvious common sense, and pious regret is expressed that it was not accomplished sooner. Meanwhile, the rest of the world has moved on, usually faster.

As in the nineteenth century, the slow spread of educational opportunities not merely failed to undermine the class structure but actually reinforced it. It is significant that education was the one aspect of welfare on which the Tory-dominated wartime coalition was prepared to legislate before victory was won. The 1944 Act at last conceded the principle, generally accepted by reformers more than 20 years earlier, that education is a continuous process, moving through its primary, secondary and higher stages. But it was also based on the proposition that 'it is just as important to achieve diversity as to ensure equality of educational opportunity'. In fact diversity and equality are mutually destructive: if diversity is promoted, equality must be sacrificed. Labour's failure to grasp this point, and to amend the 1944 Act accordingly, illustrates its misunderstanding of the realities of power and the meaning of social justice. Not only was the private sector in secondary education preserved, but a two-tier form of educational apartheid was introduced in the public sector. The English thus provided three distinct grades of secondary education, all ultimately financed by the public, either directly or through tax-reliefs: one overwhelmingly upper-class, the second overwhelmingly middle-class, the third overwhelmingly working-class. The divisions naturally ensured that higher education, too, was provided (or not) on a class basis, the State again footing the bill. Thus, by a new variety of the old English vanishing trick, the many found themselves perpetuating, and indeed financing, the privileges of the few. The consequences of the 1944 Act were overwhelmingly obvious by the late-1950s: yet in six years of office, a new Labour government made no move to introduce an alternative system. Studying the educational record of four Labour regimes, spanning more than half a century, the historian can only conclude that the Labour movement, as a collective force, does not believe education has any important role to play in social improvement and the promotion of human happiness. Yet the whole of history, especially of English history, shows that it is the most important single factor. The educational revolution in sixteenth-century England set in motion the chain of events which produced the modern world; and, conversely, England's educational failures lie at the very root of her decline as a dynamic society.

This melancholy truth leads us to a related, and still more dismal, aspect of the English problem. A high rate of economic growth cannot be sustained unless there is a correspondingly high rate of investment in the education of the people – not merely in technical skills but in social responsibility through the liberal arts. The English ruling establishment, in which I include the Labour leadership, have fobbed off the working class with a minimum education; and the country has in consequence received a minimum growth-rate. The relative industrial decline of Britain since the 1870s was produced initially by the inability of the entrepreneurial class to adjust to new technologies, itself a reflection of the kind of education they chose to give themselves. It has since been dramatically reinforced by both the power and the limitations of the British trade unions.

In a society which is politically free but socially unjust, the workers will naturally combine together to influence events. The kind of influence they exert will depend very largely on the kind of education they receive. If British trade unions are traditionally-minded, obscurantist, incompetent, anti-progressive, even reactionary – and of course they are – the fault lies with society as a whole. The English have chosen to provide the minimum of education for the mass of the people: and they have therefore produced a blind and obstinate giant, with a marked tendency to delinquency. Seen in the long perspective of the last 100 years, the influence of the British trade union movement has been almost wholly destructive. This is not an accident. The British trade unionist has failed to define his objectives, and has therefore exercised his power without knowing what its ultimate purpose is supposed to be. Why is a trade union in business? Is it to raise wages absolutely, or to raise them relative to other categories of workers, or relative to other classes of society? Is it to maintain differentials? To prevent unemployment? To increase employment? To improve the conditions of work? To raise productivity, or to lower it? To force the pace of change, or to restrain it? Does it have wider aims – to promote equality, to raise the social status of its members, or the entire working class, to reform society, to influence foreign policy? Or does it merely seek to protect the particular influence-group it represents? If so, should it protect the industry as a whole, or simply the work-force in it? Trade union leaders in this country have never asked themselves these questions, let alone asked their members. And, lacking a sense of direction, the unions have inevitably become a profoundly defensive and conservative force in English society, a restraint on change perhaps more effective than any other, even in a country where such restraints are ubiquitous. Their most obvious and pervasive effect has been to impede the growth-rate of the British economy, particularly since 1945, during a sustained period of global expansion which offered unparalleled opportunities to the British people to raise their standard of life and the quality of their public services. By inhibiting the use of technological change in raising productivity, and by promoting wage-inflation, they have damaged the interests of most of their members, helped to widen the gap between wage-earners and property-owners, and inflicted particular injury on the lower-paid workers and the poor.* Their record in denying

* Relative gains by strongly unionised groups of workers (in terms of real wages) during the century 1870–1970 have almost invariably been determined by the factors of full employment and technical change, rather than by union action. The group which has made the biggest gains, relative to 1870, is composed of members of the armed forces, who are not unionised at all; here of course, full employment and increased specialisation have been decisive in raising earnings. Needless to say, the influence (or non-influence) of trades unions on real wages is a matter of acute controversy; see Guy Routh: *Occupation and Pay in Great Britain, 1906 1960* (Cambridge 1965), pages 150 ff; and R. Ozanne: 'Impact of Unions on Wage Levels and Income Distribution', *Quarterly Journal of Economics*, l. 73, No. 2, May 1959.

equality of opportunity to women, to immigrants and to the unskilled is lamentable. But in a more fundamental sense their conservatism has crippled the energies of the progressive forces in Britain by denying freedom of action to the Labour Party, the political instrument they finance and dominate. The woeful failure of the Labour Government in the 1960s, though attributable in part to the same causes which operated in 1931, was ultimately the responsibility of the unions, which forced ministers to resort to the reactionary device of repeated doses of deflation to balance Britain's external account. When ministers, in desperation, turned on the unions in 1969 and mounted a frontal assault on union privileges, the unions merely tightened their grip round the party's throat and compelled a humiliating surrender.* This made a Conservative return to office inevitable, and, by a characteristic English paradox, opened up the way to reform (resisted, of course, by the Labour Party, which had reverted, after its brief dash for freedom, to its status as the trade unions' poodle).

But reform, in this as in so many other spheres, has come very late in the day. By buying stability at the cost of change, the English risk forcing desperate remedies on themselves. A century ago, by choosing imperialism as the easy alternative to industrial efficiency, they became the prisoner of hubris and invoked the nemesis of two world wars. Now, having wasted a quarter-century of peace and economic buoyancy, they find themselves pushed unwillingly into a Continental system from which it has been their historic mission to escape. And, by a paradox which the English should find heart-breaking, the 'Europe' with which they contemplate merging their identity is in all essentials the historic concept of German expansionism – the Mitteleuropa of the Kaiser and Hitler, cleansed, to be sure, of their barbarism, invested with the Napoleonic trappings of French legalism, but nevertheless enshrining the proposition that Europe must coalesce against the outer darkness of the world beyond. It was in 1890 that the German academic Gustav Schmoller had first advocated this 'regrouping of civilised Europe':

> He who is perceptive enough to realise that the course of world history in the twentieth century will be determined by the competition between the Russian, English, American and perhaps the Chinese world empires, and by their aspiration to reduce all the other, smaller, states to dependence on them, will also see in a central European customs federation the nucleus of

* The Chief Whip reported to the Cabinet that a majority of the Parliamentary Labour Party, under union pressure, would oppose legislation based on the White Paper *In Place of Strife*. At that point, in the words of the Prime Minister (to the author), 'the cabinet turned yellow'.

something which may save from destruction not only the political independence of those states, but Europe's higher, ancient culture itself.

It was another such academic, Hans Delbrück, who saw this 'Europe' as an alternative to the 'cultural monopoly of the Anglo-Saxons' and the 'Russo-Muscovite world'. And it was the Kaiser who seized on the phrase 'the United States of Europe' as something to flourish in the faces of the Americans and the Russians. In 1961, half a century and two world wars later, Edward Heath, an ardent English advocate of Europe, unconsciously echoed the old Emperor's misty visions: 'We now see opposite to us on the mainland of Europe a large group comparable in size only to the United States and the Soviet Union, and as its economic power increases, so will its political influence.' The Kaiser, it should be added, at times wished to include the English in his community, but feared that England's involvement with her Empire might prove an insuperable barrier. He could not have foreseen the day when the English, having lost not merely their Empire but their faith in themselves, would appear on Europe's doorstep as a despised and persistent suppliant, petitioning entry. 'When a people loses its self-confidence,' wrote Walter Lippman, 'it welcomes manacles to prevent its hands shaking.' Thus the story of the offshore islanders threatens to come full circle. The Roman world made the medieval church its residuary legatee; the heritage, after many vicissitudes, has passed in turn to the new Carolingians of western Europe; and the English, who won their independence from the first, and defied the second, now seek a humble refuge in the third. But perhaps the spirit of Pelagius is not yet dead.

Epilogue

THE position in which the English find themselves as they approach the last quarter of the twentieth century is a dangerous one. They feel they have lost their greatness; they fear they are losing their self-respect. In point of fact, the English predicament is not as serious as many of them suppose – are, indeed, taught to suppose by their harassed and nervous leaders. The 'greatness' the English have relinquished – their dominion over a quarter of the earth, their function as a world power – was more a source of weakness than of strength; it inhibited rather than liberated. The notion that the English, having given birth to the modern world, are now in the true sense effete, is misconceived. The English have not stepped down from a throne: they have left a prison. They are now more free than at any time in the last hundred years: free to decide upon the direction in which they wish to go, without regard to the wishes and interests of imperial partners and subordinates. Their responsibilities to others have been handed over gracefully, or snatched from them: they can now make their own unqualified choices, as a self-governing, independent island people. They have eased off the burden of a bankrupt estate, and must now make their own way in the world. This is not the stuff of tragedy:

> Nothing is here for tears, nothing to wail
> Or knock the breast; no weakness, no contempt,
> Dispraise or blame; nothing but well and fair,
> And what may quiet us in a death so noble.

The death of empire should be the rebirth of a people. But this, alas, is not the way the English see it. And herein lies the danger. The English suffer bitterly from a sense of loss. They feel they have forfeited caste and status. They resent a world in which their high, authoritative tones are no longer heeded, or even heard. They watch, in bewilderment and some anger, humanity ordering its affairs without their supervision, often in the teeth of their advice. The loss is felt as keenly on the Left, as on the Right. If the longing of the English to rule is frustrated, so is their equally eager desire to do good. English philanthropy, no less than English imperialism, has a huge, unsatisfied appetite. There is still, among the English, a hunger to be significant in the world. They wish to count as they once did; and the knowledge that they no longer do so breeds a certain despair, which in turn provokes a feverish quest for remedies which may prove to be desperate. Ideally, perhaps, the English ought to sit and be quiet for a time, to invoke a national mood not, indeed, of repose, but of concentration and introspection. But the English are activists: they suspect the process of thought unrelated to immediate and practical decisions. They must deploy their energy: the risk is that they will deploy it in the wrong direction, recklessly pursuing false solutions to non-problems. For the English to lose an empire is no great matter: to lose their judgment is serious.

Yet this offshore island people, for all their self-conceit, are no strangers to despair. They have worried themselves through history. The 'groans of the Britons' are their first independent record of an island consciousness. The note of lamentation, of impending catastrophe, recurs monotonously through the centuries. The island paradise is always under threat of imminent extinction. Myth and historic fact reflect the same image of perpetual twilight. Arthur is a figure of tragedy, Alfred an overburdened and often bewildered statesman, nobly exhausted by what must have seemed insuperable difficulties. The caterwauling of Wulfstan adumbrates the gratifying gloom of Dunkirk. With the Confessor, both the twelfth-century plaster effigy and the eleventh-century reality (dim though it is) breathe resignation and decay. *The Anglo-Saxon Chronicle* is a uniquely consistent narrative of distress and woeful prognostication: wailings at the Danes, at the Normans, at the depredations under Stephen, sustain the theme almost down to the last folio – 'God Almighty have mercy on that wretched place! . . . May Christ establish counsel for his wretched people! . . . God Almighty destroy all wicked plans! . . . Christ take counsel for the wretched monks of Peterborough and for the wretched place!' The words are those of a well-fed clerical historian, but the tone is authentically English. The Middle Ages are punctuated by these national crises of confidence, in which the dissolution of the realm and the general destruction of ordered society is confidently predicted and (by divine intervention) narrowly averted. Magna Carta was drawn up by men who believed themselves not on, but over, the brink of catastrophe; the commons rose in 1381 to save, as they believed, a kingdom from servitude and impoverishment; Parliament itself was forged essentially as an instrument to avert disaster: in 1386 it announced 'imminent ruin . . . the kingdom is impoverished, the magnates saddened and the whole people is weakened'. The common people in 1450 lamented that the King 'has lost his law, his merchandise is lost, his common people is destroyed, the sea is lost, France is lost, the King himself is so beset that he may not pay for his meat and drink'. For nearly half a century Margaret

Paston's letters catalogue the seeming disintegration of what she termed 'this troublous world' of England, where 'a man's death is little set by nowadays'. One might have thought that a staunch reformer like Thomas Brecon would have rejoiced at the splendour of the Reformation, but he reported, as Henry VIII neared his end, that 'the state of England was never so miserable as it is at present'. Worry, worry, worry was the theme of Elizabethan times, with the Queen's cousin, Lord Hunsdon, foreseeing 'the utter ruin of the whole country'; half a century later the position was seen to be infinitely worse, with Parliament noting 'the pressing miseries and calamities, the various distempers and disorders which had not only assaulted, but even overwhelmed and extinguished the liberty, peace and prosperity of this kingdom. . . .' Ten years later William Oughtred confessed himself 'daunted and broken with these disastrous times'. 'Never was there, my Lord,' wrote Pepys in 1659, 'so universal fear and despair as now;' but seven years later he notes: 'Our losse both of reputation and ships having been greater than is thought hath ever been suffered in all ages put together before . . . every day things look worse and worse;' and a few months later: 'I do fear so much that the whole kingdom is undone.' 'Good God!' whined the Duke of Buckingham in 1714:

> . . . how has this poor Nation been governed in my time! During the reign of King Charles the Second we were governed by a parcel of French whores; in King James the Second's time by a parcel of Popish priests; in King William's time by a parcel of Dutch Footmen; and now we are governed by a dirty chambermaid, a Welsh attorney, and a profligate wretch that has neither honour nor honesty.

At intervals in the decades following the 1740s, Horace Walpole bewailed England's decline, his shrill voice rising *fortissimo* as the American colonies were lost, and final doom sealed with the sale of his father's collection of paintings to the Russians. Ten years later it was no better: from the Right, James Woodforde moaned: 'Pray God, however, prevent all bad designs against old England!' From the Left, Dr Currie thought 'the nation was never in such a dangerous crisis'. Forty years later, Croker saw 'the King enslaved, the House of Lords degraded, the Bill passed, the Revolution, I may say, consummated . . . Depend upon it, our Revolution is in a sure, and not slow, progress... we are now become a fire-ship, which will spread the conflagration.' The gallery of woeful prophets, always assured of an attentive, and usually of a sympathetic, audience, stretches on to the English crack of doom.

What distinguishes the present chorus of self-doubt and criticism is not the fear of internal chaos, or of external perils, two predominant themes in the past, but a nagging anxiety about Britain's performance in the international league-tables of material prosperity. The English, who invented modern competitive sport, are obsessed by the statistical evidence of their decline in the world championship, and the impending verdict of relegation to some lesser category of breeds. This touches their pride, and with reason. From the very dawn of recorded history, their island, or at least the lowland zones which constitute the heartlands, have provided a high standard of life for those fortunate enough to inhabit them. The offshore islanders, in this respect, have never felt themselves to suffer by comparison with any race with whom they came into contact: they have been good providers for themselves, making industrious and profitable use of the modest but adequate resources nature placed at their disposal. Now, for more than a century, Britain has been in relative (and recently in pronounced) decline by comparison not only with traditional rivals but even with audacious newcomers. That Britain should be overtaken by the United States was bearable, had indeed been predicted by a multitude of English observers as long ago as the 1840s and 1850s. More galling has been the astonishing recovery of a truncated Germany from her terrible adversities. France, for centuries the object of English hostility or condescension, now enjoys a higher standard of living; so do Switzerland, once the mere holiday-home of the English upper middle class, Holland, a former economic satellite, and Belgium, a Foreign Office creation. Italy, on present trends, will surpass the English in the next decade, and so will Japan. Yet only 109 years ago, Lord Palmerston, airily justifying the destruction by the Royal Navy of a Japanese port, commented: 'I am inclined to think that our Relations with Japan are going through the usual and unavoidable stages of the Intercourse of strong and Civilised nations with weaker and less civilised ones.' Now Japan is the world's third largest industrial state, and is rapidly turning the British dominion of Australia into an appendage for her raw materials.

Why should the English worry? Are they not better off than ever before in their history? The answer is irrelevant. The English are, and have always been, a highly competitive nation, constantly measuring themselves by external yardsticks. To them, honestly acquired wealth is the reflection and reward of moral probity: 'Virtue is now by much the best bargain.' To admit failure in the race to affluence is to confess a collapse of national character. The English have always striven for the paradox of motion within a framework of stability: the stability remains, the motion falters. Thomas Hobbes, who was (except in his rigour) the most characteristically English of philosophers, generalised from English attitudes to propound a Galilean theory of politics:

So that in the first place, I put for a generall inclination of all mankind, a perpetuall and restlesse desire of Power after power, that ceaseth onely in Death. And the cause of this, is not always that a man hopes for a more intensive delight, than he has already attained to; or that he cannot be content with a moderate power: but because he cannot assure the power and means to live well, which he hath at present, without the acquisition of more. (*Leviathan* (1651), Chapter II)

Substitute 'wealth' or even 'standard of living' for power, and we have an accurate observation both on the acquisitive world today, and of the fear of the English for their place in it. The instincts which Hobbes observed in individuals appear still more strongly in nations, and transcend forms of political organisation. Communist states study their tables as eagerly as capitalist ones; the statistics-race is superimposed on the arms-race; and 'peaceful competition' joins (but does not replace) war in global strategies. Angry and bewildered, the English suffer from an acute reinfection of the disease they have transmitted to the world.

In their search for a cure they are tempted by some weird and drastic specifics. The Suez Affair of 1956–7 left the English ruling class dazed and rattled; and in its aftermath one could observe the first tentative gathering of an establishment consensus in favour of a Continental solution. The movement acquired momentum among the high-minded, or at any rate the highly-placed, throughout the 1960s. It had no popular roots: but neither did the entry into the military system of the Continent in the decade 1905–14. It is eloquent testimony to the panic of the English *élite* that such a solution could even be considered. Being unable or unwilling to reconstruct the national economy in the 1880s, the English had collapsed into empire. Now the Empire had gone: the economic solution still eluded them; so there was to be a collapse into the Continent.

Why this should bring about a restoration of English dynamism could not be, or in any event was not, explained. The creation of a European entity, initially commercial, ultimately political and military, was a German concept: its geographical centre of gravity must inevitably be the territories bordering the Rhine. The French accepted the concept provided they wrote, enforced, and in the last resort rewrote, the rules, in strict accordance with their national interests. The Germans agreed, confident that their energy would in the long run transcend any Gallic bias built into the structure. The French feared this too, but trusted to their diplomatic and legal skill, and their well-exercised powers of blackmail. It was not to be supposed that they would permit a new and substantial entrant to upset the balance of a carefully judged treaty drawn up to safeguard their interests. On the contrary, the French had to ensure that the workings of the rules would impose, and continue to impose, a burden on the British economy sufficient to prevent Britain from displacing the political fulcrum of the Community, which rested in Paris. Otherwise the whole object of French strategy would be defeated.

It is not clear whether the English *élite* have ever grasped this point: if the conditions of British entry were likely to stimulate rapid growth in the British economy, then the French could not conceivably permit it to take place. But then the English enthusiasts have not been notable either for the clarity or the consistency of their reasoning. The prime object of entry was at times said to be economic, at times political. The advantages on the second point could not, obviously, be quantified; they must presumably depend, in the last resort, on the advantages secured by the first. But these, too, it seemed, could not be quantified; or, if so, were seen to be non-existent, very likely negative. As the economic argument for entry collapsed, the political one was stressed. But if membership would weaken Britain economically, how could it strengthen her politically? The enthusiasts could provide no answer; or, rather, took refuge in obscurantism. 'The inescapable need,' wrote one of them (*Sunday Times*, 31 January 1971), 'is for an act of faith ... Such faith is rarely if ever created by pouring over investment tables or boxing with shadows or even counting parliamentary majorities.' What was needed were 'twentieth-century evangelists'.

Such robust disdain for actual calculation of profit and loss, of advantage and disadvantage, would doubtless have appealed strongly to that sixteenth-century evangelist, Thomas More. The debate in which he participated revolved around similar issues. The English had discovered, by bitter experience, that the balance of national advantage did not lie with their continued adherence to the old treaty of Rome, and wished to renounce it. They rejected an elaborate system of international law which imposed a one-sided economic burden on them and inhibited their sovereignty. In resisting majority opinion, More did not seek to advance practical arguments on specific issues: he could scarcely do so. Still less was he prepared to abide by the verdicts of parliamentary majorities. 'I am not bounded, my Lord, to conform my conscience to the Council of one realm against the General Council of Christendom.' For him, it was simply a matter of faith. He was, naturally, overruled, by less elevated and more sensible spirits. We admire his courage and rejoice in the defeat of his cause. If faith, that is the blind and unquestioning acceptance of established authority, had prevailed in sixteenth-century England, there would be no modern world, including no Common Market, to argue about.

But the marketeers are not the only twentieth-century evangelists who seek to persuade the English to substitute faith for reason. Nor, for that matter, are the English alone in their exposure to the ceaseless voices which preach the abdication of mind. The United States of America, that strange enlargement, mutant and perversion of the English social genius, is now the theatre of wild experiments in politico-religious thought, some of them clumsily re-enacted here. In the intellectual desolation of this century, a new generation of witch-doctors peddle their cures and find fanatical disciples, especially among the young. In the middle decades of the nineteenth century, educated and enlightened people, most notably in Britain but to some extent in all countries dominated by western European ideas, believed strongly that an age of reason and tolerance was dawning. They believed mankind would progress steadily, if slowly, towards a style of life in which each individual would obtain, as of inalienable right, not merely a rising standard of material comfort, but intellectual, cultural and spiritual fulfilment. This was the liberal ideal: and there was rising confidence that it could eventually be realised. Such optimism did not imply hubris. J. S. Mill, for instance, wrote modestly not of 'progress', or even of 'reform', but of 'improvement'. There would, in the decades and centuries to come, be countless marginal 'improvements' in all aspects of life, which in aggregate would bring about a true, but gradual and peaceful, revolution in the human condition. This was essentially an English concept, reflecting the empirical optimism which the English experience seemed to justify. It rejected as unrealistic, and possibly dangerous, the millenarian philosophies which sought to short-circuit and accelerate the process of history by gigantic solutions based on abstract models of how human beings were supposed to act. Of course, in the England of 1880–1914 a certain note of scepticism was heard: a tendency to appreciate the complexities of modern civilisation, and to insert shades of grey in the simple silhouettes of moral and material advance. But this was merely to impose a necessary corrective on a general optimism which was not itself challenged.

The terrible events of two World Wars, of the painful and unsuccessful convalescence which followed the first, and of the twilight peace – overshadowed by the threat of still greater catastrophe and punctuated by savage outbreaks of violence and persecution all over the world – which followed the second, ought to have demonstrated, beyond all possible argument, that there was no alternative to the slow, pragmatic liberal solution. The breakdown of 1914 occurred not because liberalism was strong, but because it was weak; not because it was tried and found wanting, but because powerful men rejected it. The horrors of our time are the responsibility not of liberals, but of their avowed and dedicated enemies. Nazi Germany, Soviet Russia, Communist China, the three major tyrannies of our century, together with the multitude of minor regimes they have spawned, inspired, sustained, cosseted or imposed – in Spain and Greece, in Eastern Europe, Latin America, Africa and Asia – have all been consciously based on the deliberate and systematic denial of liberal assumptions. They have sought to subject, not to exalt, the individual; to suppress, not to reveal, truth; to regiment, not to tolerate; to impose verdicts rather than offer choices; to use violence as opposed to argument; to torture, not to heal. Whatever their declared political complexion, they have one thing in common: an unqualified detestation of liberal ideas and the relentless pursuit of those who seek to express them.

The tragedy is not merely that such regimes flourish, but that their tenets and practices infect societies where liberalism still survives. From 1914 onwards, the tolerance to many forms of evil – especially of violence – has been progressively raised even in the liberal democracies. At first in self-defence, initially with repugnance, later with case-hardened enthusiasm and with growing proficiency, these liberal-based states have adopted some of the methods and attitudes of their enemies. They made nuclear weapons, and used them; destroyed whole cities, more recently whole countrysides; and armed and protected monstrous petty satrapies with the supposed object of preserving their own liberal institutions. The agony of the United States – the stage on which the crisis of the twentieth-century liberal conscience is played – is the agony of a society corrupted and discoloured from within by the very process of repelling corruption from without. But is this a failure of liberalism? On the contrary, it is a failure of societies insufficiently liberal. It demonstrates not the limitations of liberal ideas, but the lack of courage, or numbers, of those who hold them, the sheer inadequacy of the liberal impress on Western civilisation. The conclusion to be drawn is not to reject liberalism, and seek alternatives – there are, indeed, no rational alternatives – but to reinforce the liberal elements which survive, and build new ones. Liberalism is weak not because it is itself inadequate, but because it is starved of enthusiasm, conviction, intellectual nourishment, above all of followers.

Why should it be necessary to restate such a proposition, so obvious as to be self-evident? Yet the ugly fact remains that millions of young people, born after even the twilight of liberal certitudes, not merely ignore the proposition but, if challenged, specifically and emphatically reject it. Liberalism is caricatured not as the opponent and victim of the totalitarian horrors, but as (in some mysterious way) their ultimate fount of origin. The just are held responsible for the deeds of the malefactors, and the rule of law for its breakdown. What is most disturbing about this thesis is not its perversity but its consequences. When intellectuals reject liberalism, they do not take refuge in scepticism (which is itself an essential element in the liberal method). They do not believe in nothing. On the contrary: they believe in anything. The rational physician is dismissed. But the pain and fear remain, and the intellectual witch-doctor inherits the land. In the 1870s, the age of imperialism was born in the bizarre imaginings of overweening academics. Now a new generation of evangelical professors have set up their pulpits. A sensible man trembles when a distinguished scholar deserts his discipline and begins to propound theories of the universe. But the young rejoice at the prospect of exciting and radical solutions to problems which have baffled generations of dull empiricists. Academic imperialism coincided with – was made significant by – the first great expansion of higher education. The academic adventurism of today flourishes in the explosion of learning which has brought millions, tens of millions, and will soon bring hundreds of millions into the universities. There, reacting against conventional disciplines which were already crumbling, they have formed eager armies, militant in body, docile in spirit, for the deranged and the ambitious to command. Magic herbs have been plucked from the overgrown ruins of traditional theories, stirred into the pot, and a powerful, aromatic brew distilled, and then poured into young minds contemptuously purged of the liberal ethic. In one way or another, the new creeds are all variations, sometimes exalted, sometimes debased, of political nostrums long since tried, and found wanting – or indeed wholly destructive. The only novelty lies in the permutations of falsehood and unreason which academic ingenuity has contrived to concoct. Some cults preach varieties of Marxism, some proletarianised refurbishings of Hitlerism, some a combination of both. Unknowingly, the ancient and dusty battles of nineteenth-century revolutionaries are re-fought in new and fashionable trappings. With each successful assault on the customary standards of deduction and logic, of proof and plausibility, the threshold of reason is lowered. Sciences like psychology, pseudo-sciences like sociology, are wheeled in to supply destructive ammunition. Murderous and deluded gods are hoisted into the new pantheon and worshipped: Marx and Lenin, Sorel and Trotsky, Mao Tse-tung, who presides serenely over the human sacrifices of his State religion,* Che Guevara, the pathetic would-be caudillo from Argentina, Ho Chi Minh, the hero and architect of a modern thirty-years' war. National and racial deformations are eagerly embraced: regimentation from China, intellectual bullying from Germany, nihilism from France, petty-Caesarism from Latin America. As the intellectual structure collapses, theories degenerate into slogans and arguments are met with abuse and fists. Thus, inevitably, monsters re-emerge from the graves in which (it was thought) liberalism had buried them. Inverted racism is acclaimed. Black Power, a hideous blend of African tribalism, Arab anti-Semitism, and the ferocious Moslem cult of male supremacy, is made to appear a force for liberation and justice. Real crimes, crying to heaven for vengeance, in Czechoslovakia, in the Sudan, in Iraq, in East Pakistan, are ignored or even justified unless they fall within the narrow categories of approved grievances. There seems no point at which the betrayal of mind is to stop. Some preach that truth is only to be found in hallucination, that drugs are the panacea for the disjunction of the times. Not opium of the people: opium for the people. Others rake over the debris of collapsed religions: primitive Christianity, varieties of Hinduism and Buddhism. Some take the ultimate step, and argue that only the mad are sane.

Above this seething cauldron of unreason brood those malevolent spirits of the twentieth century, intolerance and violence. Critics are dismissed as (to quote the *Guardian,* 1 October 1970) the 'imaginative and conceptual slaves to the Liberal Encyclopaedia'. And to abuse is added the threat of suppression. One notable *guru,* Herbert Marcuse, openly asserts:

> The whole post-fascist period is one of clear and present danger. Consequently true pacification requires the withdrawal of tolerance before the deed, at the stage of communication in word, print and picture. Such extreme suspension of the right of free speech and assembly is indeed justified only if the whole of society is in extreme danger. I maintain that our society is in such an emergency situation, and that it has become the normal state of affairs.

Indeed, Marcuse appears to believe that the young people to whom he preaches welcome such extremities, that they are 'a new type of person

* Until the Chinese revolution, Christian missionaries had circulated among the Chinese copies (in various versions) of the penny catechism, in red covers. The ingenious Mao produced his own catechism, also in red, called *The Sayings of Chairman Mao,* and known as 'The Little Red Book'. The idea was appropriated by the Danes, who wrote a modern catechismal guide for schoolchildren, *The Little Red Schoolbook,* which soon found its way, in a translated and modified form, to Britain. Thus our children got back their catechism, embodying, of course, a new ethical orthodoxy. The ways of providence are mysterious.

with another instinct for reality, life and happiness; they have a feeling for a freedom which has nothing to do with, and wants nothing to do with, the freedom practised in senile society'.* This traditional (that is, real) freedom is opposed for many specious reasons, but chiefly because it demands the right to pursue truth in an ideological vacuum, itself a cardinal sin among the new totalitarians. Four revolutionary sociologists from France make no bones about it: 'The hypocrisy of objectivity, of apoliticism, of the innocence of study, is much more flagrant in the social sciences than elsewhere, and must be exposed.'†

Do they mean exposed or, rather, suppressed? If intolerance is preached in a framework of violence, it must take an active form. And the systematic justification and use of violence is the one common denominator of all the new political nostrums; indeed, violence is often seen not so much as a necessary evil but as a creative element, what one writer, nodding approvingly at Sorel, terms 'the existential, therapeutic, and even virile virtues of revolutionary violence'.

Here, I think, is one convincing reason why the alternative evangelism of the campus is unlikely to make much headway among the English, young no less than old. In 1968, the year of student revolt, a juvenile, unsophisticated and abortive echo of 1848, some of us hoped that the English might make a particular and characteristic contribution to this international phenomenon, adding an element of constructive and empirical reformism to a movement whose vehemence was merely driving itself into the sand. In fact the English made no significant gesture, merely staging (as in 1848) a feeble imitation of the Continental convulsion, decorated with some campus agitprop devices transmitted by American students. As the students were unable to put forward an articulate, let alone plausible, theory to explain and justify their movement, or even a set of realisable objectives, it was interpreted merely as a physical assault on the university system itself. To the English public, traditionally starved of higher education, beginning at last to get it in increasing quantity, and aware also of a belated improvement in its quality and relevance to their needs, such an assault

* Quoted in *Student Power: Problems, Diagnosis, Action,* edited by Alexander Cockburn and Robin Blackburn (London, 1969), p. 372. The delusion that a 'new' kind of freedom, or liberty, can evolve from totalitarian violence was also current in the 1930s, before and during the Stalinist terror. In 1937, the climax of the great purges, G.D.H. Cole wrote: 'Already, I believe, the Soviet Union is feeling its way towards the restoration of many of those liberties which had to be curtailed. . . . We shall find they are not merely putting back the liberties they have restricted, but establishing a new and higher kind of liberty, hitherto unknown in the world – a liberty extending to every section of the people, and women equally with men!' – *Proceedings of the Second National Congress of Peace and Friendship with the USSR.*

† Quoted in *Student Power*, p. 378.

seemed misguided, even disgraceful – a view eventually shared by the great majority of students themselves: 'The expense of spirit in a waste of shame.'

The new radicalism indeed evokes the powerful hostility of deep English reflexes, above all the suspicion of Continental ideas, associated in the English mind with the theory and practice of violence as a political form. To this is added the growing belief that American society, too, is now acclimatised to violence – that it is an ineradicable element in American political processes – and that American ideas could, and unless resisted will, spread the infection here. It was grimly noted by the English that the only tangible result of the student movement in Britain was the inauguration of a further cycle of violence in Ireland, the traditional theatre of extremist politics within the British orbit. The latest phase of Ulster politics, indeed, provides for the English a lurid, textbook demonstration of the futility of violence.

It is, of course, a sermon preached to the thoroughly converted. Indeed, the radical movements on the Continent and in the United States reinforce the insular elements in English life. The English regard – have always regarded – violence as the supreme political cancer. One could fairly say that the entire history of the English is the story of the conquest of violence. The English settlers of the fifth and sixth centuries were refugees from instability and violence as much as from hunger. From the very earliest times, the concept of the King's peace, of a carefully demarcated area radiating from the presence of the Sovereign, from which violence was banned and argument replaced physical action, has governed political development. The King's house, from which violence was quarantined, became the King's court; the court his council, the council his parliament, the parliament the political nation, the political nation the entire nation: thus the area of tranquillity was progressively expanded to embrace the whole realm, as every element in society was enabled to forsake the violent search of redress in favour of argument, the politics of words. A strong central authority, exercising a monopoly of violence, has always seemed to the English not only desirable but essential: the argument has revolved around its composition, not its powers. The English rightly perceived, from an early date, that safety, and prosperity, and ultimately liberty, spring not from political organisation but from the absolute rule of law: only on the basis of the rule of law can effective political organisations be constructed. In the *Leviathan,* Hobbes deduced the English instinct from first principles, and formulated it into a theory of absolute sovereignty: the only way for men to avoid death by violence, to pursue wealth, and develop their individualities in security, was to acknowledge a perpetual sovereign power, against which each of them would be powerless. This was not (as used to be supposed) a formula for the totalitarian State: on the contrary, it was a formula for the night-watchman State. Once the individual agreed to relinquish power to a sovereign body, the composition of such a body, and the way that composition was determined, could be settled by experiment: it might be, as Hobbes admitted, 'any Assembly of men whatsoever'. In practice, it was on the basis of Hobbes's reasoning, and through its popularisation by Locke, that modern parliamentary sovereignty emerged. Parliament, once elected, is absolutely sovereign: it is bound neither by its previous decisions, nor by a written constitution. It is, if it chooses to be, the valid and responsible instrument of revolution. Indeed, one might say that the English have made careful provision, within their political system, for the practice of revolution within a non-violent, indeed perfectly legal, framework. The English hatred of violence springs from the conviction that it is unnecessary. And if it is unnecessary, to exercise it is a crime against nature, at any rate against English nature. The English system can accommodate the revolutionary: it repels those – it will always repel those – who do not believe revolution can be achieved without force. This is so not because the English are docile: on the contrary, they are capable, and know themselves to be capable, of great violence. Their saving grace is their self-knowledge, and the use they make of it: the evolution of the English public system is a long, and ultimately triumphant, exercise in self-control. There is a paradox here, as Sir William Temple noted in his *Essay upon the Origin and Nature of Government* (published 1751): 'Nor do I know if men are like Sheep, why they need any government: Or if they are like wolves, how can they suffer it.' The truth is, the English are wolves who have accepted the discipline (and rewards) of the sheepfold: they will turn with wolfish ferocity, and sheep-like unanimity, on anyone who seeks to break down the walls with which they curb their appetites.

The solution of violence, then, no less than the solution of surrender and Continental collapse, offers no prospect of a cure for the English. Those who proffer it doom themselves to futility and frustration. For the English, and ultimately, I believe, for the world, there is no alternative to liberal empiricism. There is no cause for despair, and no case for desperate remedies. There is nothing that is wrong in English society, or in world society, that cannot be adjusted for the better, provided we keep our heads and exercise our reason. Everything worthwhile the English have achieved, for themselves and others, has been built upon the great tripod of the liberal ethic: the rejection of violence, the reaching of public decisions through free argument and voluntary compromise, and the slow evolution of moral principles tested by experience and stamped with the consensus. All English history teaches that these are the only methods which, in the end, produce constructive and permanent results. We must, in short, resume the quest for improvements.

But 'in the end', of course, we are all dead, as Keynes observed. If there is no place for violence, there is certainly a place for impatience in English public life. It is not enough to restate the inevitability of the liberal processes: we must seek for ways to speed them up. The disenchantment with liberalism springs not so much from its premises as from its performance. The theoretical assault on its aims and methods is little more than an angry attempt to rationalise the disappointment, even disgust, which radicals rightly feel at its failure to maintain, let alone increase, the pace of reform. All societies are mobile, in the sense that they have an inherent tendency to produce inequalities and social injustice unless political correctives are constantly applied: thus action is required merely to counter the vicious thrust of uncontrolled human greed and ambition. Liberalism must move even to keep society from receding into the darkness. To propel it towards the light requires a steady acceleration of effort.

Measured by this necessity, the performance of the current political repository of liberal action in this country, the Labour Party, has been so woeful as to question its constitutional capacity to discharge the role. By comparison, the achievements of its predecessor, the Liberal Party, during the years 1868–1914, were very striking. Labour has now held the liberal mandate almost as long, for nearly half a century. The only time it has shown itself able to discharge it was in the immediate post-1945 period, when it inherited both a radical programme and a national consensus for reform: even so, it was notable that Labour's dynamism declined sharply between 1945 and 1951, when it wearily relinquished office. The 1964–70 government was by any standards a fiasco: Labour proved unable even to hold the social drift. The most it could do was to provide a favourable parliamentary climate, in which backbenchers removed specific barbarism and disabilities – capital punishment, and the anomalies of the homosexual, divorce and abortion laws – leaving the general social and economic structure intact, if not reinforced. The pre-war defence that Labour had never held real power is no longer an answer to the accusation of failure. It is true that in 50 years as one of the two major parties, Labour has held office only for 15. But its weakness is reflected not merely in its capacity to lose elections but in its inability to achieve anything when it wins them.

Labour is a prisoner of its origins. It suffers from the English mortmain. It is, by historical definition, both a theoretical socialist party and a class one. These two dead hands from the past grip its arms, and dangle it, like Pinoccio, over the political stage. But it evidently does not believe in social ownership. It has never sought to implement the specific socialist aims of its constitution. Nor, in practice, does it believe in class warfare. Thus it gets the worst of all possible worlds. Its theoretical socialist commitments prevent it from appealing, as the Liberal Party was able to do, to the progressive elements within the *bourgeoisie;* but, in office, it nevertheless attempts to work the capitalist economy without their support and expertise. At the same time, as the political expression of the trade union movement, it is tied to one of the most conservative and tenacious elements in society whose whole philosophy is defensive and sectional. Middle-class elements within the Labour Party must accept the status of second-class members; but Labour has never been able to win the support of the whole working class even at the polling stations, still less in the operation of its policies.

The experience of English history seems to suggest strongly that little in this country can be achieved through the politics of division, and that social improvements are brought about by the creation of a national mood, expressed through instruments which have genuine national qualifications. Unless Labour can transform itself into a national movement, embodying the enlightened elements in all classes and occupations, free of historical commitments, free of disabling alliances, devoted solely to the practical pursuit of social justice, its record in the future will be no better than in the past. It must abandon the fallacy that any particular class, by virtue of its condition, has a monopoly of moral rectitude, and work towards the grand coalition of the liberal-minded. It must seek to evolve an *élite,* classless because drawn from all classes, providing, in Gladstone's words, 'a natural leadership, based on free assent'; otherwise, the 'supremacy of political power', theoretically vested in the majority, will remain, as he observed, 'in danger of slipping through their fingers'. This is the real lesson of our times.

But of course it is in the nature of the English to make such a programme difficult. By a strange paradox, they long for a national unity, exercised by a strong central authority, while assiduously creating obstacles to it. They are a nation not merely of classes, but of clubs. They create miniature sovereign societies, loyal to themselves, exclusive, jealously preserving privileges acquired over many generations. The English army, still a social rather than a military institution, is a series of clubs. English education is a series of clubs. Each public school, each grammar school, increasingly each comprehensive school, is a club. An Oxbridge college is a club. The Anglican Church is a club, dignified by the name of Establishment, but with increasingly free and easy rules. The medical profession is dominated by two clubs, the colleges of physicians and surgeons, with a joint club committee in the GMC. The legal profession is two clubs, the Bar Council and the Law Society, with real power vested in the socially dominant branch. The City establishment is a club, with its functional branches in the Bank of England, the Capital Issues Committee, the Takeover Panel. The civil service is a club, with its own commissioners. Racing is run by a club. The trade unions are clubs, linked together in a club federation under a general council which is the epitome of a club committee.

To a greater or lesser extent all of these clubs exercise a measure of sovereignty over their members, decide recruitment and expulsion, enjoy legal immunities and privileges, negotiate separately with government, and above all maintain their own courts outside the jurisdiction of the common law. The justice dispensed in a court-martial, in a diocesan court, by the disciplinary committee of the General Medical Council, the Bar Council and the Stewards of the Jockey Club, or by a kangaroo court in a factory, may vary greatly in quality: but it usually has a Star Chamber flavour. Moreover, all these clubs maintain – are indeed designed to maintain – systems of restrictive practices, the victims of which are the public as a whole, though they are invariably defended as operating in the general interest: 'The restrictive practices of the legal profession,' claimed the Council of the Law Society (April 1971), 'are a form of self-discipline, its members imposing on themselves standards of competence and conduct designed solely for the benefit of the public.' This is the invariable refrain, from trade unionists and other privileged groups alike. Every professional body believes itself the best judge in its own cause, not least the politicians, who constitute the supreme English club, and defend their club privileges with unrivalled ferocity.

The fact remains that there is a general and historic tendency in English society for powerful, self-organised groups to opt out of the law and create their own franchises. The result is not merely national division, not merely even injustice (from which, at one time or another, we all suffer), but a system which operates at all levels, and over virtually every branch of the nation's life, as a powerful obstacle to reform. All clubs are conservative, all are selfish, all are, in the true sense, anti-social. By eroding the operations of the State, they limit the public power to effect those improvements which it is the object of liberal politics to secure. Thus, the true reformer will appear in the guise of Henry II, breaking down the castles and stockades of the private franchises; or, like Edward I, he will ask *Quo Warranto,* by what warrant, right or authority any group claims and exercises privileges above the common folk. This is a question to be put not merely to trade unionists, but to those who provide and enjoy private education, and the other inequalities which spring from the class and professional structure. The levelling of institutional inequalities is the key to the gospel of improvement.

Indeed, without institutional reform, we shall not make much progress towards a just society. The club-principle is the enemy of the social-principle. So long as men are permitted to band themselves together in fierce groups, to protect and advance their sectional privileges, society will remain a jungle, and its rewards will be apportioned according to strength, organisation and the relentless pursuit of self-interest. The law confines itself to protecting life and property; it should protect and advance merit, industry and usefulness. What a man or woman is paid is not everything; but it is the most important single element in his or her material well-being. It is monstrous and uncivilised that it should be determined by the brute force of organisation. As long ago as 1947, a government White Paper admitted:

> The last hundred years have seen the growth of certain traditional or customary relationships between personal incomes – including wages and salaries – in different occupations. These have no necessary relevance to modern conditions. The relation which different personal incomes bear to one another must no longer be determined by this historical development of the past, but by the urgent needs of the present.

What is more urgent than social justice? And how can this be brought about unless the 'historical development of the past' is taken to mean the self-perpetuating, self-expanding accumulations of property as well as the warfare which trades and professions wage to secure their slices of the national income? But the White Paper declined to face the problem of past injustices and inequality. It even declined the role of arbiter in the jungle:

> It is not desirable for the government to interfere directly with the income of individuals otherwise than by taxation. To go further would mean that the government would be forced itself to assess and regulate all personal incomes according to some scale which would have to be determined. This would be an incursion by the government into what has hitherto been regarded as a field of free contract between individuals and organisations.

Here was a radical government, speaking at the height of its power and self-confidence! Yet social justice is impossible until rewards are socially determined. And who is to make that determination, if not the public speaking through the organs of government? It cannot be left to gang warfare as at present. Ultimately society must accept the public responsibility to settle incomes according to social worth. The private franchises will be destroyed, and the writ of the people will arbitrate economic relationships.

The separatist or club-principle in society promotes cultural as well as material evils. If some groups of men are permitted to set up sovereign enclaves to promote their sectional interests, others will imitate them. Once society relinquishes its right to interfere in one sphere of human professionalism, the demand for immunities will multiply. If you leave incomes to be settled by trade unionists and industrialists, if you surrender religion to the clergymen, and medicine to the doctors, and jurisprudence to lawyers – if you make the professionals the arbiters of their professions, and the experts of their expertise – you must end by relinquishing art to the artists. Culture ceases to be a common heritage and becomes the private empire of professional creators. Art itself begins to operate behind club doors, firmly barred to the public; it becomes incestuous, or the property of narrow factions, ultimately the victim of fashions, set, maintained, applauded, picked up and cast down by tiny minorities of insiders. When time gives historians the perspective to judge our age, they may epitomise its barbarism not in our violence, intolerance and irrationality, but in the suicide of western art. What happened to these people, they may ask? Why did they suddenly abandon the traditions of more than two millennia, throw away the accumulated standards of successive civilisations? They may date the process from the 1880s, or thereabouts, when the creators withdrew from the public the right of arbitration and formulated the doctrine of art for art's sake. The creators claimed they sought release from an increasingly materialistic world; in fact, they were asserting hieratic privileges, and so invoking the atrophy which afflicts any priestly caste which denies participation to its congregation. The ivory tower became not a refuge but a prison, and confinement produced madness. Just as, in politics, the retreat from liberal certitudes opened the way to the academic witch-doctors and the charlatans of race and violence, so the cultural bastions were toppled one by one, each falling more easily than the last. Painting became a matter of mere pigments hurled at, or trampled into, the canvas; music dissolved into discordant or random sounds; poetry sought meaning in the meaningless, and ceased to be read or learned or recited except by its practitioners. Hideous buildings arose, to complete the sense of oppression induced by cultural abdication. From the universities, the prevailing fashions sent forth field-grey regiments of professional experts, critics, evaluators: the Non-Commissioned Officers of Bedlam. But if the public could be drilled into submission – even be made to pay for the anti-culture – they could not be made to love it. Sullen and resentful, they watched the carnival from outside, as they had, through most of history, watched the processes of government. But they had at least been taught to read: music, painting, even poetry, might be successfully presented to them as unfathomable mysteries, requiring professional interpreters, the priestly caste. But they knew that prose should mean something; and here, significantly, the anti-culture made least progress. There were, even, ominous signs of a revolt.

Perhaps for this reason, the guardians of cultural anarchy invoked the crude instrument of sex: a new opium of the people. Novels, plays, films had shown an obstinate reluctance wholly to exclude traditional cultural standards: sex was unleashed like a ferret to hunt them out. In the nineteenth century, the responsible creators had struggled against the prevailing mood to give sexuality, in their work, its appropriate place as an aspect of human experience. The best of them had coped with this problem skilfully, had even transmuted censoriousness into a useful artistic discipline. The Victorian attitude to sex had caused great human misery (we assume) but surprisingly little damage to art. Now the position was turned upside down, and sex engulfed all. A Gresham's Law began to operate, as the base currency of sexual exploitation forced art to race for cover. Those who had sneered at the Victorians enthusiastically adopted cultural postures which were far more grotesque. Ancient obscenities became the emblem of free speech, the hallmarks of articulation; the barrack-room was invoked to supply the conversational deficiencies of the salon. The sexual act, the sexual organs themselves, upstaged the imagination, with every variety of perversion, sadism and brutality employed to evoke fading responses from a dazed but subservient public.

Obviously, the cult of sexuality, like the wider barbarism of which it forms part, damages art; in a sense replaces art. There is no agreement on whether it damages people. But in so far as it unleashes violence, it produces inequity, and punishes the weak. Nor is it surprising: when democratic control of culture is abandoned, we must expect the oligarchy to observe the club-rules of the jungle. A characteristic of the new cult is the off-hand, even contemptuous, degradation of women, one more aspect of the de-humanisation which has marked the century of total war. Women become mere objects for the satisfaction of masculine appetites which they may not share; the context in which the claims of sexual 'freedom' are advanced is essentially a male one. Thus the beneficiaries of the sexual revolution are men, and women form its principal victims; indeed, for them, it has all the salient features of a counter-revolution.

The quality of a civilisation is always reflected in its treatment of women. It is an infallible yardstick. 'Who – Whom?' The Marxist question is even more effective in a social and sexual context than in a political and economic one. It is of the essence of the new barbarism that it breeds an aggressive and hypocritical male-supremacy. Women form the true proletariat of the anti-liberal world: the sexual revolution imperceptibly shades into counter-revolution, with women its chief victims. They are also, in England, the great excluded from the club-principle. The private franchises exist largely without them, often against them. There are no female baronies or enclaves, above the law. There is no institution, or custom, or attitude in England which evokes in the male that deep spasm of righteous but impotent anger which all women feel, from time to time, when they contemplate the crude injustice of a society made by men – made, that is, by a section of society from which they are, by definition, excluded. It is the same anger once felt by the colonial Briton, by the medieval peasant, by the religious dissenter, by the eighteenth-century middle class, by the nineteenth-century working-man. The anger can only be assuaged by the systematic application of liberal principles: for it is the function of liberalism to redress the weakness of nature by reason. This is what civilisation is about. It is what England is about. Is it not strange that England, the birthplace of liberalism, should have left this huge repository of human injustice virtually intact? Would it not be both logical, and historically apt, that the English should make equality for women their next great contribution to the happiness of mankind?

It is not simply a question of justice. Progress consists in the unleashing of human energy. Societies advance when they adapt their institutions to tap the hidden resources of submerged classes. In the sixteenth century, by breaking away from the static world of Roman Christianity, the English ended the clerical control of intellectual inquiry, and unleashed the layman. In the seventeenth and eighteenth centuries they unleashed the merchant and the manufacturer. So the modern world came into existence. More and more individuals found it possible to realise their potentialities and make their full contribution to society. As the catchment area of liberation widened, there was a sharp acceleration in material progress. Yet even today, half of humanity is, for all practical purposes, still submerged. Generation after generation, vast human resources are wantonly squandered by our failure to permit women to fulfil themselves. We are not merely denying opportunities to women. We are denying our own future. The first society, the first nation to unleash its women will inaugurate the next great phase in human development, as momentous as the Reformation or the industrial revolution. Will it be England, as before?

There is another, and related, opportunity facing us. 'The poor ye have always.' Do we? Need we? One of the depressing constants of English history, observable from the very earliest recorded times right up to today, is the huge anonymous presence of the poor. They are always there, faintly echoed in charters and chronicles, in sermons and prayers, in works of economics and political philosophy, in parliamentary reports, in ministry statistics. Their numbers are variously computed according to the science of the day: sometimes they are thought to constitute a quarter of the nation, sometimes a fifth, or a tenth, or a twentieth, depending on definition and the standards of measurement. They arouse pity, resentment, fear, usually a combination of all three. They are the people who, whatever the rules of society, seemingly cannot make a living within them, and must receive some form of external assistance to raise them to the level of comfort currently regarded as bearable. Over the centuries, England's rulers have (by the standards of other countries) devoted an unusual amount of thought to the problem: they have not been content to regard poverty as an inescapable fact of life. They have nagged away at it, trying private, public, institutional charity, penal laws, moral pressure and exhortation, doles and subsidies, emigration. Yet the poor remain. They reproduce and perpetuate themselves. If the records existed, most of them could doubtless compile a genealogy of poverty reaching back to the Dark Ages, just as Queen Elizabeth II can trace herself back to Ine and beyond. The dogged persistence of English poverty, its causes and characteristics, have successfully defied the last hundred years of 'improvement'. Much of what was written about poverty in the 1890s is still valid today. There is very little to add to an analysis of its structure published in 1913.

The truth is, all governments have asked: 'What can we do for the poor?' Sometimes they have asked: 'What can the poor do for themselves?' They have never asked: 'What can the poor do for us?' Yet this is the true liberal question. The real tragedy of poverty is not suffering, or degradation, but waste. Over the centuries, tens of millions in this country alone have failed to realise their potentialities and make, each one of them, his or her unique contribution to the wealth and happiness of all. We calculate the negative cost of poverty, in subsidies and welfare payments, in crime and the relief of sickness: but we ignore the huge positive loss of our failure to integrate perhaps one-fifth of our people in the productive and creative nation. As with women, we leave unutilised a huge segment of the nation's resources. We argue about palliatives, with varying degrees of humanity, while ignoring the hard-headed alternative of a cure. No form of investment is more fruitful, by any system of calculation, than that which breaks the reproductive cycle of poverty. Here is another great task for English ingenuity.

It is no accident, of course, that the great reservoirs of English poverty are to be found outside the south-east heartlands. No student of history can do other than acknowledge a profound respect for the constant and remorseless force of geography. Wealth and power in England, in Britain indeed, has always gravitated towards the south-east. Beyond lay the 'brute and beastly shires', the 'bare Scotch firs', and 'blind Wales'. For a brief period, during the first and second phases of the industrial revolution, it seemed possible that the English fulcrum was shifting to the north and west: but even while the hand of Manchester was felt at Westminster, there was a notable and increasing tendency for the wealth generated by northern resources and skills to flow down into the lowlands. By the turn of the present century, the historic pattern was re-establishing itself: the 'distressed areas' were coming into being. For more than 50 years, successive governments have tried to revitalise a lost industrial supremacy in these parts – a supremacy which can now be seen as a deluding aberration. Huge and increasing sums of public money have been devoted, by a variety of subsidies, grants and tax devices, to the forlorn task of refurbishing the great nineteenth-century centres of industry. Like the Roman attempt to urbanise Britain, it has all been a colossal exercise in waste and futility. As with poverty – indeed the two problems are very largely one – ever more expensive palliatives have diverted attention from a possible cure. Wise government does not resist nature: it reinforces, and so harnesses, its thrust. Seen in terms of geography and environment, the English industrial revolution was a crime against nature, and in the end an unsuccessful one. It devastated a wilderness made for enjoyment, not profit. As the axis of progress shifts back into its accustomed place in the English lowlands – bringing, despite every effort to soften its ravages, a steady erosion of the natural landscape—it seems folly to resist nature's attempt to recover the lands it lost in the north and west. Ought we not, rather, to assist the transfer of industry and populations to the south-east, to rationalise a natural *fait accompli,* and at the same time to bring back the wild places for the recreation and solace of all? All over the world, sensible and sensitive men and women are awakening to the dark side of material progress, and are struggling to find means to limit its poisons. They are discovering that humanity needs nature just as much as the wealth which destroys it. But so far the radical solution – the division of the earth into territories which people inhabit and exploit, and those which they enjoy – has seemed too drastic. Yet at one time it seemed too drastic to challenge religious authority, to assert that slavery was *contra naturam,* to permit people to speak and write freely, to grant political rights on the basis of personality, not property. In all these matters, the English, in their empirical and pertinacious way, taught themselves, and so the world, that the drastic solution was right, and in the end inevitable. It is not beyond their powers – it will be, perhaps, their privilege – to codify, for the first time, the rights of nature.

Do I make, in this book, too large a claim for the English? Writing during a period of despondency in English affairs, I would answer that this, if a fault, is forgivable. Certainly I have not sought to gloss over the vices and weaknesses to which the English seem peculiarly prone, or to ignore the tragedies they have manufactured for themselves and others. They have been, they still are, one of the most active races in the history of the world, with an enormous capacity for good, and evil. On balance, they have performed useful services to humanity. Now, in their maturity, having lost their hubris, having survived (one trusts) their nemesis, they are poised for a fresh experience, 'a nation not slow and dull, but of quick, ingenious and piercing spirit; acute to invent, subtile and sinewy to discourse, not beneath the reach of any point that human capacity can soar to'. The words are those of John Milton, the great English poet of renewal and recovery.

APPENDIX I

The Danes and the English

THE ninth- and tenth-century Danish settlements were often military in character, and place-names ending in -by and -thorpe indicate the names of individual commanders. Five Danish forts became towns: Lincoln, Nottingham, Derby, Leicester and Stamford. The settlements had something of the character of a mass-migration, with the Danish element varying from one-third to two-thirds of the population; Yorkshire and Lincolnshire were the areas of most intensive settlement. There are over 600 place-names ending in -by (meaning place or town, hence by-law); some 300 ending in -throp and the same number in -thwait; and about 100 ending in -toft – at least 1,500 Danish place-names in all. In some parts of Lincolnshire and Yorkshire, place-names of Scandinavian origin make up 75 per cent of the total. Danish Christian-names persisted in large numbers until well into the twelfth century, indicating that assimilation was very slow. Danish patronymics, especially those ending in -son (Jackson, Johnson), became a permanent feature of English nomenclature. The influence of Danish on the English language was profound, especially after the Danes became assimilated. Old English borrowed few Danish words, and these mostly associated with hostile contacts. Later, as the Danes adopted English, Danish contributed at least 900 words designating common, everyday things and fundamental concepts, and arguably about 1,000 more. Among them, taken at random, are: *band, bank, birth, brink, creek, dirt, fellow, gap, kid, leg, link, race, rift, root, score, scrap, seat, sister, skin, skirt, sky, steak, thrift, trust, want* and *window*. Danish even contributed pronouns, prepositions, adverbs and – most remarkable of all – part of the word to be (*are*). *They, their* and *them* are Scandinavian; the expression '*they are*' is thus wholly imported. Danish influence contributed greatly to the simplification and streamlining of English grammar and syntax which marked the transition from Old to Middle English, and hence to the structure of the modern English language. Locally, in dialect, it was even more pervasive. Alternative English and Danish variants of common words continued to jostle each other for superiority until the end of the Middle Ages, with the Danish form occasionally winning, as in *eggs*. We know this from a famous and revealing passage in Caxton's preface to his paraphrase of the *Aeneid* (1470):-

For we englysshe men ben borne under the domynacyon of the mone, whiche is never stedfaste, but ever waverynge, waxynge one season, waneth and dyscreaseth another season. And that comyn englysshe that is spoken in one shyre varyeth from a nother. In so moche that in my dayes happened that certayn marchauntes were in a shippe in tamyse, for to have sayled over the see into zelande, and for lack of wynde, thei taryed atte forlond, and wente to lande for to refreshe them. And one of theym named Sheffelde, a mercer, cam in-to an hows and axed for mete; and specyally he axyd after eggys. And the goode wyf answerde, that she coude speke no frenche. And the marchaunt was angry, for he also could speke no frenshe, but wolde have hadde egges, and she understode hym not. And thenne at laste a nother sayd that he wolde have eyren. Then the good wyf sayd that she understod hym wel. Loe, what sholde a man in thyse dayes now wryte, egges or eyren? Certaynly it is harde to playse every man by cause of dyversite and chaunge of langage.

Old English legal codes, confirmed by William the Conqueror, recognised differences between the laws of Wessex (Kent, Surrey, Sussex, Berkshire, Hampshire, Wiltshire, Dorset, Somerset and Devon), Mercia (Oxfordshire, Warwickshire, Gloucestershire, Worcestershire, Herefordshire, Shropshire, Staffordshire and Cheshire) and the Danelaw, which comprised the rest of eastern and midland England and Yorkshire. The difference between Wessex and Mercian law was slight and technical; but the Danelaw had its own quite separate institutions (such as presentment by 12 jurymen, who could take a majority vote – a refinement only recently adopted by English Law), and its own system of land tenure. In many parts of the Danelaw, Domesday Book reveals a much higher proportion of peasant freeholders than elsewhere in England; these were undoubtedly the descendants of Danish soldiers. Danish influence was by no means confined to the Danelaw; it is to be found, for instance, in the brilliant and grotesque carvings which adorn the Norman church at Kilpeck in Herefordshire. The Danes can claim to be the most important element in the composition of the English nation, after the Anglo-Saxons and the Britons.

APPENDIX II

Cromwell and Ireland

IRELAND had been the grave of English military reputations. It did not destroy Cromwell's: his operations there were masterly and highly successful. But it has proved the grave of his moral reputation. There is no doubt that he took the view, shared by Spenser (who knew Ireland well), Bacon and Milton – indeed the overwhelming majority of Englishmen – that the Irish were culturally inferior and their subjection necessary. In 1649 he believed the Irish would be used to overthrow the Revolution:

> If our interest is rooted out there they will in a very short time be able to land forces in England, and put us to trouble here . . . If they shall be able to carry on their work they will make this the most miserable people in the earth for all the world knows their barbarism.

Indeed, he feared the Irish menace more than any other:

> I had rather be overrun with a Cavalierish interest than a Scotch interest; I had rather be overrun with a Scotch interest than an Irish interest; and I think of all this is the most dangerous.

Yet at one point there was a distinct possibility that the native Irish might cooperate with parliamentary forces in opposing the Irish Protestant royalists. The Levellers themselves urged, in the words of Walwyn: 'The cause of the Irish natives in seeking their just freedom was the very same with our cause here in endeavouring our own rescue and freedom from the power of oppression.' The Council of Officers took a vote pledging that the army should not be used either 'to eradicate the natives or to divest them of their estates'. But under pressure from the Catholic clergy, acting on papal instructions, the Irish turned against the New Model.

Cromwell was determined not to get bogged down in a long Irish campaign, like all his predecessors there. He was ill: 'I have been crazy in my health,' he wrote. But there is no evidence that sickness influenced his acts. He believed that the severe measures taken at Drogheda and Wexford would end resistance quickly and 'save more effusion of blood'. Events proved him right. In all other English campaigns in Ireland, a great many people were killed in a casual fashion, because the troops were ill-disciplined and unpaid. Cromwell's strict enforcement of martial law on his men, and still more his success in getting them paid regularly, saved countless Irish lives. He believed that he came as a benefactor of the people:

> We come (by assistance of God) to hold forth and maintain the lustre and glory of English liberty in a nation where we have an undoubted right to do it; where in the people of Ireland . . . may equally participate in all benefits, to use liberty and fortune equally with Englishmen, if they keep out of arms.

He hoped, in particular, to transform the iniquitous administration of justice in Ireland. In December 1649 he begged John Sadler to become Chief Justice of Munster:

> We have a great opportunity to set up, until the Parliament shall otherwise determine, a way of doing justice among these poor people, which, for the uprightness and cheapness of it, may exceedingly gain upon them, who have been accustomed to as much injustice, tyranny and oppression from their landlords, the great men, and those that should have done them right, as, I believe, any people in that which we call Christendom.

But of course he was frustrated by the clergy; indeed it was in Ireland that he discovered what the Reformation had been about. While prepared to tolerate Catholics in England, he found the Irish still enmeshed in a political, pro-Continental religion, dominated by a priestly caste. This he would not accept. As he said, echoing Elizabeth's view: 'We look at ministers as helpers of, not lords over, the faith of God's people.' And he associated the power of the priests with the mass. He told the Governor of Ross:

> For that which you mention concerning liberty of conscience, I meddle not with any man's conscience. But if by liberty of conscience you mean the liberty of exercising the mass, I judge it best to use plain dealing, and to let you know, where the Parliament of England have power, that will not be allowed of.

He told the Irish priesthood that the mass had been forbidden for 80 years before the 1641 rebellion in Ireland, and that he would carry out the law; but of course, before Cromwell, the law had never been effectively enforced. His real motives were amply displayed when he reminded the clergy that 'So anti-Christian and dividing a term as clergy and laity' was unknown to the primitive Church:

> It was your pride that begat this expression, and it is for filthy lucre's sake that you keep it up, that by making the people believe they are not so holy as yourselves, they might for their penny purchase some sanctity from you; and that you might bridle, saddle and ride them at your pleasure.

In short, Cromwell regarded the clergy and the landlords as the real authors of the Irish tragedy, as, indeed, they still are.

Finally, it is a curious fact that in 1651, when General Monck sacked Dundee, he killed as many people as Cromwell in Drogheda, and with far less military justification; yet the episode is rarely mentioned.

Guide to Chronology

1900–1700 BC	First constructions at Stonehenge
c. 250 BC	First British hill-forts
55–54 BC	Julius Caesar's expeditions
AD 43	Claudian invasion
61	Boudicca's revolt and suppression
122	Construction of Hadrian's Wall begun; Antonine Wall 139–42
406	Britain stripped of Roman troops
410	Honorius endorses *de facto* British independence
c. 460	Contact with Rome lost; large-scale Anglo-Saxon settlements in progress
516 (?)	Battle, or Siege, of Badon (Mons Badonicus)
537 (?)	Defeat and death of Arthur in civil war
570–80	Completion of Anglo-Saxon conquest of British lowlands
597	Augustine arrives in Canterbury and restores contact with Rome
663	Synod of Whitby endorses forms of Roman Christianity over Celtic Christianity
c. 690	Laws of Ine of Wessex first codified in writing
c. 700–750	Composition of *Beowulf*
734	Bede completes his *Ecclesiastical History*
757–96	Reign of Offa of Mercia, greatest of Anglo-Saxon Kings before Alfred
825	Wessex replaces Mercia as paramount power, opening the way for the unification of England
866–71	Invasion by the great Danish army
871–99	Reign of Alfred of Wessex; *Anglo-Saxon Chronicle* begun
900–50	Conquest of Danelaw; unitary English state established
978–1016	Reign of Æthelred the Unready; resumption of Danish invasions
1016–36	Reign of Cnut
1042–66	Restoration of Wessex line and reign of Edward the Confessor
1066	Battle of Hastings; William I crowned King on Christmas Day
1070	Lanfranc appointed Archbishop of Canterbury
1086	Domesday Survey
1087–1100	Reign of William II (Rufus)
1093	Anselm appointed Archbishop of Canterbury; later exiled
1100–35	Reign of Henry I; Conquest of Normandy

CHRONOLOGY

1135–54	Reign of Stephen and contest with the Empress Matilda; period known as 'the Anarchy'
1154–89	Reign of Henry II
1170	Murder of Becket
1169–70	Beginning of the Conquest of Ireland
1189–99	Reign of Richard I; his crusade and return, 1190–94
1199–1216	Reign of King John
1204	Loss of Normandy
1206	Struggle with Pope Innocent III over election of Stephen Langton as Archbishop of Canterbury, culminating in Papal Interdict and surrender by John to the Papacy
1215	Magna Carta
1216–72	Reign of Henry III
1258	Provisions of Oxford and beginnings of Barons Revolt
1259	Provisions of Westminster
1264	Simon de Montfort's victory at the Battle of Lewes
1265	The Model Parliament; Battle of Evesham and death of De Montfort
1272–1307	Reign of Edward I
1282–4	Edward I's conquest of Wales; his heir invested as Prince of Wales (1301)
1290	Expulsion of the Jews
1307–27	Reign of Edward II
1314	Battle of Bannockburn
1326–7	Defeat, deposition and murder of Edward II by Queen Isabella and Mortimer
1327–77	Reign of Edward III
1337	Invasion of France and beginning of Hundred Years War
1340	English naval victory of Sluys
1345	Probable birth date of Chaucer; writes *Canterbury Tales* c. 1380; dies 1400
1346	Victory over the French at Crécy, and Scots at Neville's Cross
1348–9	First phase of the Black Death
1351	Statutes of Labourers and Provisors
1353	Statute of Praemunire, limiting Papal authority in England
1356	English victory at Poitiers, followed by Treaty of Bretigny (1361)
1362	English officially replaces French in the courts; Langland's *The Vision of Piers Plowman* written
c. 1375–1400	*Sir Gawain and the Green Knight* written
1377–99	Reign of Richard II
1377–84	Wyclif preaches against papal authority and denounces monks and friars
1381	Peasants' revolt
1399–1413	Reign of Henry IV

1400–15	Welsh revolts of Owen Glendower
1403	Defeat and death of Henry Percy (Hotspur) at Shrewsbury
1413–22	Reign of Henry V
1415	Battle of Agincourt
1420	Treaty of Troyes recognises Henry as heir to the French crown
1422–61	Reign of Henry VI
1430	Franchise limited to 40 shilling freeholders
1431	Burning of Joan of Arc
1453	French conquest of Guienne and end of Hundred Years War
1455	Battle of St Albans and beginning of Wars of the Roses
1461	Battle of Towton and deposition of Henry VI
1461–83	Reign of Edward IV
1469–70	Sir Thomas Malory finishes his *Morte d'Arthur*
1470–1	Temporary restoration of Henry VI; death of the Earl of Warwick at Barnet, and of Prince Edward at Tewkesbury; murder of Henry VI
1476	First printing press set up at Westminster
1483	Reign of Edward V, one of the murdered 'princes in the Tower'
1483–85	Reign of Richard III; his defeat and death at Bosworth (1485)
1485–1509	Reign of Henry Tudor, Henry VII. He marries Elizabeth of York (1486)
1497	The Cabots voyage to North America under royal patronage
1509–47	Reign of Henry VIII
1512–13	War with France and Scotland; Battle of Flodden
1516	Thomas More's *Utopia*
1517	Luther begins Reformation conflict; Henry VIII comes to Pope's defence and is made Fidei Defensor (1521)
1527	Henry begins divorce proceedings
1529	Fall of Wolsey and opening of Reformation Parliament
1531–2	Convocation acknowledges Royal Supremacy; More resigns as Chancellor
1533	Thomas Cranmer made Archbishop of Canterbury and acknowledges Henry's marriage to Anne Boleyn as valid; appeals to Rome abolished by statute
1534	Abolition of papal authority in England; Act of Supremacy; Act of Succession; new Statute of Treasons, followed by executions of More and Bishop Fisher (1535)
1536	Dissolution of smaller monasteries and Pilgrimages of Grace
1538	Great English Bible issued
1539–40	Greater monasteries dissolved
1540	Attainder and execution of Thomas Cromwell
1547–53	Reign of Edward VI

1547–9	Somerset's Protectorate
1553–8	Reign of Mary
1554	Execution of Lady Jane Grey
1554–6	England reunited with Rome; repeal of anti-papal legislation; burning of Latimer and Ridley (1555) and of Cranmer (1556) at Oxford
1558	Loss of Calais
1558–1603	Reign of Elizabeth I
1559	Acts of Supremacy and Uniformity restore Anglican church
1560	Treaty of Edinburgh gives English endorsement to the Scots Reformation and creates Anglo-Scottish alliance
1563	The Thirty-Nine Articles; Statute of Apprentices and Labourers
1564	Birth of Shakespeare (died 1616)
1568	Mary Queen of Scots flees to England; Northern Rising defeated (1569)
1570	Excommunication of Elizabeth by Pope Pius V
1577–80	Drake's first voyage round the world
1581	Jesuit mission to England under Campion and Parsons
1587	Execution of Mary Queen of Scots; Drake's raid on Cadiz
1588	Defeat of Spanish Armada
1590	Spenser's *Faerie Queen* published
1598	Revolt in Ireland, followed by Essex's expedition (1599) and his rebellion against Elizabeth and his execution (1601)
1601–2	*Hamlet* written
1603–25	Reign of James I
1611	Failure of Robert Cecil, Lord Salisbury, to negotiate surrender of ancient Crown prerogatives for accountable financial grants from Parliament: end of political consensus
1620	Departure of the Pilgrims to New England
1625–49	Reign of Charles I
1628	Petition of Right; assassination of Buckingham
1629	Dissolution of parliament and beginning of Charles' 11–year personal rule
1633	Laud made Archbishop of Canterbury
1637	Hampden test-case over ship money
1638	Scots revolt over attempt to impose the Laudian liturgy
1639	Thomas Wentworth, Earl of Strafford, Charles' chief minister, advises recall of Parliament; war with Scotland
1640	'Short' parliament refuses money for war against Scots; England invaded; Long Parliament meets (November)
1641	Irish revolt; prerogative courts abolished; Strafford executed; Grand Remonstrance passed
1642–6	First civil war
1647	Scots hand over Charles I to parliament; the Putney Debates
1648	Purge of parliament

1649	Execution of Charles I; England declared a Free Commonwealth
1651	Navigation Act; publication of Hobbes' *Leviathan*
1653–8	Cromwell Lord Protector; women first appear on English stage, and first English opera performed
1660–85	Reign of Charles II
1660	Foundation of the Royal Society
1661–5	The 'Clarendon Code'; Lord Chancellor Clarendon dismissed and flees into exile (1667)
1663	*Paradise Lost* completed by John Milton (1608–74)
1665	The last Great Plague
1666	Great Fire of London
1675	Purcell composes *Dido and Aeneas*; Wren chosen to design St Paul's (completed 1710)
1679	Habeas Corpus Act passed
1681	Charles II dissolves his last parliament; Dryden publishes *Absalom and Achitophel*
1685–9	Reign of James II, culminating in the Glorious Revolution (1688)
1687	Publication of Newton's *Principia*
1689–1702	Reign of William III and Mary (1689–94)
1689	Bill of Rights passed
1690	Battle of the Boyne; publication of Locke's *Essay Concerning Human Understanding*
1692	Massacre of Glencoe
1694	Foundation of the Bank of England
1701	Act of Settlement determines protestant succession
1702–14	Reign of Anne
1704	Battle of Blenheim
1707	Union of England and Scotland
1713	Treaty of Utrecht
1714	Pope's *Rape of the Lock* published
1714–27	Reign of George I
1720	Collapse of South Sea Bubble
1726	Swift completes *Gulliver's Travels*
1721–42	Walpole First Minister
1729	Wesley founds the Methodist Society
1727–60	Reign of George II
1735	Hogarth paints *The Rake's Progress*
1739	The War of Jenkins' Ear
1740–8	War of the Austrian Succession
1746	Crushing of the Jacobite 'Forty-Five' rebellion at Culloden
1755	Publication of Dr Johnson's *Dictionary*
1756–3	Seven Years' War
1757	Clive wins Battle of Plassey
1759	Wolfe captures Quebec

1760–1820	Reign of George III
1763	Beginning of the Wilkes controversy over freedom of the press
1764	Watt's first commercial steam-engine
1765	Colonial Stamp Duty (repealed 1766)
1768	Cook's first voyage to Australia; Sir Joshua Reynolds first President of the new Academy of Arts
1770–82	The North Ministry
1770 (?)	Gainsborough exhibits *The Blue Boy*
1773	The 'Boston Tea Party'
1775–81	The War of American Independence
1776	Publication of Adam Smith's *Wealth of Nations*; of the first volume of Gibbon's *Decline and Fall of the Roman Empire*; and of Bentham's *Fragment on Government*
1780	Gordon riots in London
1783–1801	Pitt's First Ministry
1785	First power-loom
1788–95	Impeachment of Warren Hastings
1791	Publication of first part of Tom Paine's *Rights of Man*; church and king riots in Birmingham
1793–1802	War with France
1795	Speenhamland Poor Relief system introduced
1798	Publication by Wordsworth and Coleridge of *Lyrical Ballads*
1801	Union of Great Britain and Ireland
1803–15	War with France
1804–6	Pitt's Second Ministry; coalition of 'All the Talents' and death of Fox (1806)
1805	Battle of Trafalgar
1807	Abolition of the Slave Trade
1808–14	Peninsula War
1811	George III finally insane
1812	Publication of Byron's *Childe Harold*; War with the United States (till 1814)
1813	Publication of Jane Austen's *Pride and Prejudice*
1815	Battle of Waterloo
1818	Publication of Keats' *Endymion*
1819	Peterloo Massacre
1820–30	Reign of George IV
1820	Trial of Queen Caroline; Shelley writes *Adonais*
1821	Constable exhibits *The Hay Wain*
1822	Death of Castlereagh; Canning Foreign Secretary
1825	Stockton and Darlington railway opened
1829	Catholic Emancipation Act
1830–37	Reign of William IV

1830	Liverpool and Manchester Railway opened
1831–2	Struggle for the First Reform Bill
1833	First comprehensive Factory Act; birth of the Oxford Movement
1834	Poor Law Reform Act
1835	Municipal Corporations Act
1836	First episodes of Dickens' *Pickwick Papers*
1837–1901	Reign of Queen Victoria
1839	Publication of the Durham Report on Canada; climax of Chartism
1841–6	Peel's Ministry
1844	Turner exhibits *Rain, Steam and Speed*; Disraeli publishes *Coningsby*
1845–6	Great Potato Famine; repeal of the Corn Laws (1846)
1847	Publication of the first episodes of Thackeray's *Vanity Fair* and of Charlotte Brontë's *Jane Eyre*
1848	Publication of the Marx-Engels *Communist Manifesto*; pre-Raphaelite Brotherhood founded; publication of first volumes of Macaulay's *History of England*
1850	Tennyson Poet Laureate
1851	The Great Exhibition; Landseer exhibits *The Monarch of the Glen*
1854–6	Crimean War
1857	Indian Mutiny
1859	Publication of Darwin's *Origins of Species* and of George Eliot's *Adam Bede*
1861	Publication of J.S. Mill's *Representative Government*
1867	Second Reform Act
1868–74	Gladstone's First Ministry
1870	Forster's Education Act; civil service entrance by competitive examination
1871	Religious tests at Universities abolished; Trade Union Act
1872	The Ballot Act
1874–80	Disraeli's great Ministry; Thomas Hardy's *Far From the Madding Crowd* published (1874)
1875	Britain acquires Suez Canal; First Gilbert and Sullivan Opera, *Trial by Jury*
1877	Victoria Empress of India
1878	Treaty of Berlin
1880–85	Gladstone's Second Ministry
1884	Third Reform Act
1885	Death of Gordon at Khartoum
1886	Gladstone's Third Ministry, and defeat of first Home Rule Bill
1886–92	Salisbury's Ministry
1888	Publication of Kipling's *Plain Tales from the Hills*

1892–5	Gladstone's last Ministry; second Home Rule Bill defeated; Rosebery Ministry
1895–1902	Salisbury's last Ministry
1898	Battle of Omdurman
1899–1902	Boer War
1900	Foundation of Labour Party; first performance of Elgar's *Dream of Gerontius*
1902–5	Balfour's Ministry
1904	Entente Cordiale
1905	Publication of H.G. Wells's *Kipps*
1905–15	Campbell-Bannerman and (from 1908) Asquith governments
1908	First Old Age Pensions Act
1909	Lords reject Lloyd George Budget
1910	Publication of E.M. Forster's *Howards End*
1911	Parliament removes House of Lords veto; beginning of unemployment and health insurance
1913	Publication of D.H. Lawrence's *Sons and Lovers*
1914	Home Rule Bill passed; suspended for duration of war
1914–18	First World War
1915–16	Asquith coalition government
1916–22	Lloyd George coalition government
1918	Universal adult male suffrage established; votes for women of 30 and over
1920	Government of Ireland Act establishes partition
1922–4	Bonar Law and first Baldwin governments; James Joyce's *Ulysses* published (in Paris) and T.S. Eliot's *The Waste Land* (in London) in 1922
1924	First Labour government
1924–9	Second Baldwin government
1925	Britain returns to the Gold Standard
1926	General Strike
1928	Vote extended to all women of 21 and over
1929–31	Second Labour government
1931	National government formed; Britain leaves Gold Standard
1932	Exchange control established; Aldous Huxley's *Brave New World* published
1933	Unemployment reaches 2,955,000 (January)
1935	Baldwin Prime Minister
1936	Abdication crisis; publication of Keynes' *General Theory of Employment, Interest and Money*
1937	Chamberlain Prime Minister
1939–45	Second World War
1940–45	Churchill's coalition government
1942	Publication of the Beveridge Plan
1944	Butler Education Act

1945–51	Third Labour government; nationalisation of coal, gas, electricity, railways and Bank of England
1948	National Health Service created; abolition of plural voting; independence given to India, Pakistan, Ceylon and Burma
1949	Devaluation of Sterling
1951	Korean Rearmament Budget and beginning of Bevanite revolt
1951–64	Conservative governments under Churchill, Eden, Macmillan and Home
1956–7	Suez Crisis
1957	Ghana becomes first Black African colony to attain independence
1964–70	Fourth Labour government
1968	Devaluation of Sterling
1970–	Conservative government under Heath

Index